N & N Science Series

Physics

*A Comprehensive Review of Physics
with a Special Section on*

The College Board Achievement Test in Physics [©]

Author:

Nancy Ann Moreau

Physics Teacher
N.Y.S. Physics Mentor
Roy C. Ketcham High School
Wappingers Falls, NY 12590

Cover Design and Artwork:

Eugene B. Fairbanks

Computer Graphics:

Wayne H. Garnsey

The College Board and *The A*   he
registered trademarks of the C   ard.
N & N Science Series — Ph
trademark of the N&N

D1212318

N & N Publishing
18 Montgomery Street
(914) 34

Dedicated to:
Wayne, **Dave**, and **Debbie**

Special Credits

Thanks to:
Kevin Hahn
Mike Pullar
Dean Tomasi
Gloria Tonkinson

and my 1991 & 1992 *AP Physics C Course* students
for their assistance in the preparation of this manuscript

N & N Publishing expresses our appreciation to the College Board and the Educational Testing Service for providing their permission to use materials from The College Board Achievement Tests in Science.

Also, for students who need a more "in depth" explanation of the Achievement Test in Physics, we recommend purchasing the official publication of The College Board:
The College Board Achievement Tests (14 Tests in 13 Subjects)
ISBN: 0-87447-162-1.

It is our hope that the addition of these materials in the **N&N Science Series — Physics** will further enhance the education of our students.

College Board Achievement test questions and related material selected from The College Board Achievement Tests in Science, College Entrance Examination Board, 1986, 1983. Reprinted by permission of the College Board and Educational Testing Service , the copyright owner of the test questions.
Permission to reprint the College Board Test questions and related material does not constitute review or endorsement by Educational Testing Service or the College Board of this publication as a whole or of any other testing information it may contain.

N&N Science Series — Physics was produced on the Macintosh IIci and LaserMax 1000. The Claris *MacWrite II* and *MacDraw II* were used to produce text, graphics, and illustrations. Original line drawings were reproduced on a Microtek MSF-300ZS and edited with *Adobe Photoshop*. Formatting, special designs, graphic incorporation, and page layout was accomplished with *Ready Set Go!* by Manhattan Graphics. Technical assistance was provided by Computer Productions, Newburgh, New York. "To all, thank you for your excellent software, hardware, and technical support."

© Copyright 1988, 1993
N & N Publishing Company, Inc.

SAN # - 216-4221   IBSN # - 0935487 55 7

Revised:
7/1/93   67890 MP 09876543

Table Of Contents

Introduction to Problem Solving 5

UNIT I - Mechanics   *Page*

I.   Matter in Motion (Kinematics)................................... 9
II.   Graphing Motion 16
III.   Forces on Matter.. 25
IV.   Momentum... 37

Optional UNIT I - Motion In A Plane

I.   Two Dimensional Motion and Trajectories 51
II.   Uniform Circular Motion 57

UNIT II - Energy

I.   Work and Energy.. 71
II.   Power.. 75
III.   Energy 78

Optional UNIT II - Internal Energy

I.   Temperature 89
II.   Internal Energy and Heat 92
III.   Kinetic Theory of Gases 98
IV.   The Laws of Thermodynamics 102

UNIT III - Electricity And Magnetism

I.   Static Electricity... 109
II.   The Electric Field 115
III.   Electrons in Motion 121
IV.   Energy and Power in Electrical Problems 131
V.   Magnetism 133

Optional UNIT III - Electromagnetic Applications

I.   Torque on a Current Carrying Loop 153
II.   Electron Beams 157
III.   Other Charged Particle Beams 159
IV.   Electromagnetic Induction 163
V.   Changing Voltage 164

UNIT IV - *Wave Phenomena*

I.	Waves	175
II.	Wave Characteristics	176
III.	Periodic Wave Phenomena	181
IV.	Light	183
V.	Wave Nature of Light	189

Optional UNIT IV - Geometric Optics

I.	Real and Virtual Images	201
II.	Images Formed by Reflection	201
III.	Images Formed by Refraction	207

UNIT V - *Modern Physics*

I.	Dual Nature of Light	219
II.	Models of the Atom	224
III.	Uncertainty Principle	229

Optional UNIT V_1 - Nuclear Energy

I.	Structure of the Nucleus: The Nucleons	235
II.	Isotopes: Definitions	236
III.	Einstein's Mass - Energy Relationship	238
IV.	Natural Radioactivity	240
V.	Nuclear Stability	247

Optional UNIT V_2 - Solid State Physics

I.	Conduction in Solids	257
II.	Semiconductor Devices	267
III.	Transistors	272

Appendices

I.	Math Review	281
II.	Physics Reference Tables	287
III.	*The College Board - Physics Achievement Test*©	295
IV.	Self—Help Question Answers	305
V.	Glossary - Index	307

Sample Exams

Part I Credits for Practice Examinations in Physics 317
Practice Examinations in Physics 319

Introduction To Problem Solving

The accumulation of facts and equations is not knowledge. Research has shown that experienced physicists organize knowledge in conceptual blocks. Each conceptual block is built around a common conceptual theme. The goals of this brief introduction are twofold. First, to make the student aware of the hierarchical structure in physics. As new topics are introduced, see where they fit into the overall organization of physics material. Look for concepts. The charts on the following pages do not include all of the mathematical concepts found in the basic physics course. It summarizes the main ideas of mechanics and energy only. This material was adapted from the work of Dr. Alan Van Heuvelen at New Mexico State University and was made possible in part by a grant from the U.S. Dept. of Education's Fund for the Improvement of Post Secondary Education.

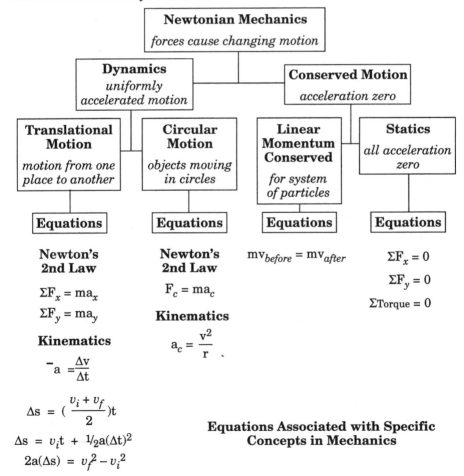

Newtonian Mechanics

forces cause changing motion

Dynamics
uniformly accelerated motion

Conserved Motion
acceleration zero

Translational Motion
motion from one place to another

Circular Motion
objects moving in circles

Linear Momentum Conserved
for system of particles

Statics
all acceleration zero

Equations

Equations

Equations

Equations

Newton's 2nd Law

$$\Sigma F_x = ma_x$$
$$\Sigma F_y = ma_y$$

Kinematics

$$\bar{a} = \frac{\Delta v}{\Delta t}$$

$$\Delta s = \left(\frac{v_i + v_f}{2} \right)t$$

$$\Delta s = v_i t + \tfrac{1}{2}a(\Delta t)^2$$

$$2a(\Delta s) = v_f^2 - v_i^2$$

Newton's 2nd Law

$$F_c = ma_c$$

Kinematics

$$a_c = \frac{v^2}{r}$$

$$mv_{before} = mv_{after}$$

$$\Sigma F_x = 0$$
$$\Sigma F_y = 0$$
$$\Sigma Torque = 0$$

Equations Associated with Specific Concepts in Mechanics

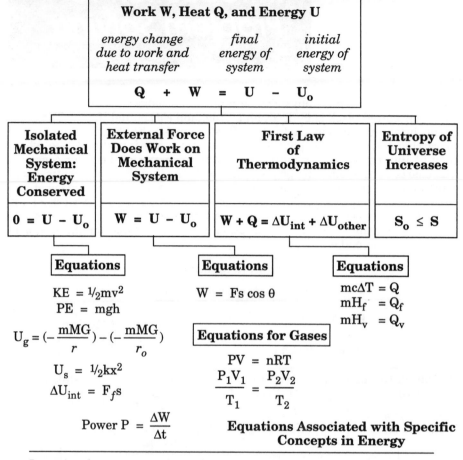

Work W, Heat Q, and Energy U

energy change due to work and heat transfer	final energy of system	initial energy of system

$$Q + W = U - U_o$$

Isolated Mechanical System: Energy Conserved	External Force Does Work on Mechanical System	First Law of Thermodynamics	Entropy of Universe Increases
$0 = U - U_o$	$W = U - U_o$	$W + Q = \Delta U_{int} + \Delta U_{other}$	$S_o \leq S$

Equations

$$KE = \frac{1}{2}mv^2$$
$$PE = mgh$$
$$U_g = (-\frac{mMG}{r}) - (-\frac{mMG}{r_o})$$
$$U_s = \frac{1}{2}kx^2$$
$$\Delta U_{int} = F_f s$$
$$\text{Power } P = \frac{\Delta W}{\Delta t}$$

Equations

$$W = Fs \cos \theta$$

Equations for Gases

$$PV = nRT$$
$$\frac{P_1 V_1}{T_1} = \frac{P_2 V_2}{T_2}$$

Equations

$$mc\Delta T = Q$$
$$mH_f = Q_f$$
$$mH_v = Q_v$$

Equations Associated with Specific Concepts in Energy

Learning how to organize information is the first step in acquiring the skill of problem solving. The following procedure may be applied to new problems as they are encountered.

1. Read the problem carefully. Look for clues to identify the concept that will be applied.
2. Draw a picture of the problem. Include in the sketch the coordinate axes, and symbols for the known and unknown quantities.
3. Draw and label the appropriate diagram of motion, force or free-body to represent the physical situation and any implied quantities.
4. Write down the appropriate mathematical equations to solve for the unknown.
5. Substitute the known values into the equation with appropriate units. Derived units should be expressed as the appropriate fundamental units.
6. Estimate the answer. Try to determine the approximate range of an acceptable answer.
7. Do the actual calculation and simplify. Check the sign, the magnitude and the units of your final answer.

The following is a Problem-Solving Format used to develop problem solving skills. Although the format presented is set up for dynamics problems, the student is encouraged to make the adjustments necessary to solve other types of problems. By using a format similar to this, students will gain a confidence and technique that will last long after the physics facts are forgotten.

Dynamics Problem–Solving Format	
Pictorial Representation ___ *Sketch* ___ *Coordinate axes* ___ *Symbols for known quantities* ___ *Symbol for unknown*	**List Knowns:**
	Identify Unknown:
Physical Representation ___ *Object description* ___ *Motion diagram* ___ *Free–body Diagram and/or Force Diagram*	

Math Representation and Solution

Check

_____ *Sign*

_____ *Magnitude*

_____ *Unit*

Dynamics Problem–Solving Format

Pictorial Representation

___ *Sketch*
___ *Coordinate axes*
___ *Symbols for known quantities*
___ *Symbol for unknown*

List Knowns:

Identify Unknown:

Physical Representation

___ *Object description*
___ *Motion diagram*
___ *Free–body Diagram and / or Force Diagram*

Math Representation and Solution

Check

_____ *Sign*

_____ *Magnitude*

_____ *Unit*

Unit 1

Mechanics

Important Terms To Be Understood

SI units	speed and velocity	friction
fundamental units	instantaneous speed	static friction
derived units	acceleration	kinetic friction
vector	free fall	momentum
scalar	gravity	elastic collision
resultant	force	impulse
equilibrant	newton	gravitational mass
concurrent	static equilibrium	weight
parallelogram rule	dynamic equilibrium	gravitational field
component	Law of Inertia	torques
distance	Newton's Laws	center of mass
displacement	Law of Universal Gravity	coefficient of friction

I. Matter In Motion (Kinematics)

Kinematics deals with the mathematical methods of describing motion without regard to the forces which produce it. These motions are described in terms of their position as a function of time. Motion is relative to a frame of reference. In this review book, the metric (**MKS**) system of units will be used for measurement. The units of the kilogram, meter, second, coulomb, Kelvin, and candela are defined as **fundamental** or SI units. Other units, such as the newton, joule, and volt for example, are considered **derived** units, because they can be written in terms of fundamental units. A newton is therefore a derived unit which can be rewritten as kilogram meter per second squared. In any mathematical calculations, units behave like algebraic quantities. *(The physical units on each side of the equation must be equal.)*

1.1 Vectors And Scalars

A **vector** is a quantity that has magnitude (size) and direction. A **scalar** quantity has magnitude only. In physics, many quantities can be classified as vectors or scalars. Two vector quantities are equal only if they have the same magnitude and direction.

Table 1.1
Examples of
Vectors and Scalars

VECTORS	SCALARS
Displacement	Distance
Velocity	Speed
Acceleration	Energy (work)
Force	Time
Weight	Power
Momentum	Mass
Torque	Charge

Vectors can be represented by a line drawn in a particular direction. The length of the line represents the magnitude of the vector; the direction of the arrow represents the direction of the vector. In print, vector quantities are usually indicated by using bold print (e.g. **F**).

Two vectors are equal only if they have the same magnitude and direction. The resultant of two or more vectors is a single vector which produces the same effect (both in magnitude and direction).

1.2 Distance And Displacement

Distance is a scalar quantity that represents the length of a path from one point to another. **Displacement** is a vector quantity that represents the length and direction of a straight line path from one point to another. Total displacement is a vector sum. A jogger is concerned with distance and a pilot with displacement. The SI unit for distance or displacement is the meter. The height of a student can be approximated at 10^0m or 1m (10^1m represents 10meters).

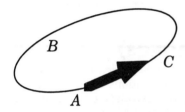

To illustrate the difference between displacement and distance, consider that an object moves from A to C along the path ABC. The magnitude of the displacement is the length of the vector AC. The distance the object actually moves along path ABC is greater than the magnitude of the displacement.

Practical Applications:
- displacement (air distance) verses road distance between two cities
- map reading
- odometer reading on a car

1.3 Some Definitions

Velocity is the time rate of change of displacement.
Speed is the time rate of change of distance.
Acceleration is the time rate of change of velocity.

Note — An object moving along a circular path may have a constant speed. Since its direction is changing, it cannot have a constant velocity. The object is being accelerated because its velocity direction is changing. Any change in either the magnitude or the direction of the velocity vector over a period of time indicates acceleration.

1.4 What About Positive And Negative Signs?

Vectors have a positive or negative sign to indicate the direction of the vector. As an object moves along a straight line, one direction has positive velocity and the other direction has negative velocity. It doesn't matter which direction is selected to be positive in a problem as long as the direction is consistent within the problem. It is true that whenever an object is speeding up, the velocity and the acceleration will have the same sign. If the object is slowing down, the velocity and acceleration will have opposite signs.

1.5 Addition Of Vectors

The resultant of two or more vectors is a single vector which produces the same effect in both magnitude and direction.

The maximum value that any two vectors can have will occur when the angle between the vectors is zero (0) degrees (vectors pointing in the same direction). That maximum value is the algebraic sum of the two vectors. The minimum value that any two vectors can have will occur when the angle between the two vectors is 180 degrees (pointing in the exact opposite directions). That minimum value is the difference between the two vectors. For example, a 30 m/s velocity and a 20 m/s velocity may combine to act as a single 50 m/s velocity if they act in the same direction, or a 10 m/s velocity if they act in opposite directions. **All** magnitudes between 10 m/s and 50 m/s are possible with these two vectors.

The **equilibrant** is a force vector exactly equal in magnitude to the resultant force, but in the opposite direction. When force vectors are in equilibrium, there is no unbalanced force acting on the system. The system may be at rest or in motion with constant velocity.

Example

Consider two forces acting on a point. These forces are said to be **concurrent** vectors. One force acts due east and has a magnitude of 30 N. A second force has a magnitude of 40 N and acts at an angle of 30° North of East. Calculate the resultant force.

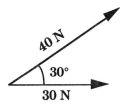

Note — Before beginning any problem, it is helpful to sketch the situation. After the sketch is completed, the problem may be solved either by graphical or mathematical methods.

Graphical Method — Construct a scale vector diagram. In this diagram, 1.0cm represents 10.0 N. Add the vectors "head to tail." The resultant (**R**) starts where the first vector begins and ends where the last vector ends.

Measure the resultant with a ruler and determine the amount of force it represents. Determine the angle with a protractor.

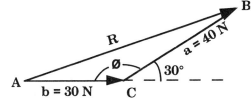

Mathematical Technique — Applying the cosine rule to triangle of the preceding figure:

$$c^2 = a^2 + b^2 - 2ab \cos \varnothing$$

substituting:

$$c^2 = (40\ N)^2 + (30\ N)^2 - 2(40\ N)(30\ N) \cos 150°$$
$$= 1600\ N^2 + 900\ N^2 + 2078\ N^2$$
$$= 4578\ N^2$$

$$c = 68\ N$$

using the sine rule:

$$\frac{a}{\sin A} = \frac{c}{\sin C}$$

$$\sin A = \frac{a(\sin C)}{c} = \frac{40\ N \sin 150°}{68\ N}$$

$$\sin A = .294$$

$$A = 17°$$

1.6 Components Of A Vector

Any vector can be treated as if it is the sum of a pair of vectors. There is an infinite number of these pairs and three are shown in the figure below. The perpendicular pair such as **S** and **T** is most useful.

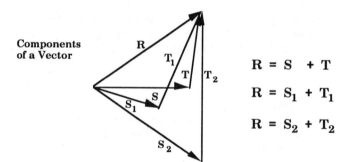

Components of a Vector

$$R = S + T$$
$$R = S_1 + T_1$$
$$R = S_2 + T_2$$

Consider a vector, **F**, resolved into two perpendicular vectors of magnitude F_y and F_x.

Resolving a Vector into two Perpendicular Components

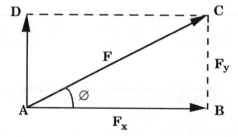

From trigonometry:

$$F_x = F \cos \varnothing \qquad F_y = F \sin \varnothing$$

Example

A plane is traveling at a velocity of 300 m/s in a direction 30° North of East. At what velocity is the plane traveling to the North? To the East?

Graphical Method:

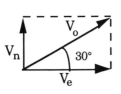

scale: 1 cm = 100 m/s

V_{north} = 1.5 cm

= 150 m/s

V_{east} = 2.6 cm

= 260 m/s

Mathematical Solution:

$$\sin 30° = \frac{V_{north}}{V_o}$$

$$V_n = V_o \sin 30°$$

$$V_n = (300 \text{ m/s}) (0.5) = 150 \text{ m/s}$$

$$\cos 30° = \frac{V_{east}}{V_o}$$

$$V_e = V_o \cos 30°$$

$$V_e = (300 \text{ m/s}) (0.866)$$

$$= 260 \text{ m/s}$$

This problem can be solved by the graphical method or the mathematical method. Graphically, a vector whose length is 3 cm will represent the velocity of 300 m/s. The top of the page is normally considered North. First draw the vector to scale at the proper angle. Next resolve the vector into components by constructing a perpendicular to each axis. Measure the length of the projection along the east and north axis. This length represents the magnitude of the velocity in each direction.

Questions

1   Which vector represents the resultant of **X** and **Y**?

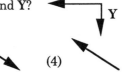

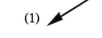 (1)   (2)   (3)   (4)

2   The resultant of two concurrent forces is minimum when the angle between them is
(1) 0 degrees   (3) 90 degrees
(2) 45 degrees   (4) 180 degrees

3   Which pair of terms are vector quantities?
(1) work and velocity   (3) weight and distance
(2) force and momentum   (4) acceleration and mass

4   Two displacement vectors of 9 meters and 4 meters are combined. The maximum resultant is
(1) 5 m   (2) 9 m   (3) 13 m   (4) 36 m

5   A person travels 4 meters North, 6 meters West, and 4 meters South.
    What is the total displacement?
    (1) 14 m East   (3) 4 m South
    (2) 6 m West   (4) 4 m North
6   A boy exerts a force **F** in pulling a wagon by means of a cord making an
    angle A to the ground. $\mathbf{F_x}$ is the horizontal component of **F**, and $\mathbf{F_y}$ is the
    vertical component of **F**. If the angle A is increased while **F** is kept
    constant
    (1) both $\mathbf{F_x}$ and $\mathbf{F_y}$ will increase
    (2) both $\mathbf{F_x}$ and $\mathbf{F_y}$ will decrease
    (3) $\mathbf{F_x}$ will increase and $\mathbf{F_y}$ will decrease
    (4) $\mathbf{F_y}$ will increase and $\mathbf{F_x}$ will decrease
7   Mass is to weight as kilogram is to
    (1) joule   (3) force
    (2) newton   (4) watt
8   A 800 N store sign is supported by two slanting wire cables of equal
    length. The cables make an angle of 90° at the point of the sign to which
    they are attached. The component of force exerted by the sign along each
    cable is about
    (1) 400 N   (3) 720 N
    (2) 560 N   (4) 800 N
9   If the angle between the cables in item 8 is increased, the force acting
    along the cables will
    (1) increase   (2) decrease   (3) remains the same
10 A man walks 40 meters north, then 70 meters east and then 40 meters
    south. What is his displacement from the starting point?
    (1) 150 meters east   (3) 70 meters east
    (2) 150 meters west   (4) 70 meters west
11 A force of 100. newtons is applied to an
    object at an angle of 30° from the hori-
    zontal as shown in the diagram. What
    is the magnitude of the vertical compo-
    nent of this force?
    (1) 0 N   (3) 6.0 N
    (2) 50.0 N   (4) 100 N

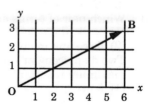

12 As the angle between two concurrent forces increases from 45 to
    90degrees, the magnitude of their resultant
    (1) decreases   (2) increases   (3) remains the same
13 A resultant force of 10. newtons is made up of two component forces act-
    ing at right angles to each other. If the magnitude of one of the compo-
    nents is 6.0newtons, the magnitude of the other component must be
    (1) 16 N   (2) 8.0 N   (3) 6.0 N   (4) 4.0 N

14 What is the magnitude of the horizontal or
    x-component of vector **OB** in the diagram?
    (1) 9
    (2) 6
    (3) 3
    (4) 0

15 A man pulls a wagon by applying a force of 100 N to the handle which is held at an angle of 15° to the ground. The horizontal component of the man's force is
   (1) 26 N   (2) 44 N   (3) 50 N   (4) 97 N
16 Referring to item 15, the vertical component of the man's force is
   (1) 26 N   (3) 50 N
   (2) 44 N   (4) 97 N
17 The maximum number of components that a single vector can be resolved into is
   (1) one   (2) two   (3) three   (4) unlimited
18 A man in a car going northward at 16 m/s throws a ball through the window at a speed of 12 m/s in the eastward direction. The velocity of the ball with respect to the ground is
   (1) 12 m/s east   (3) 20 m/s northeast
   (2) 9 m/s north   (4) 28 m/s northeast
19 Which pair of concurrent forces may have a resultant of 20 N?
   (1) 5.0 N and 10 N   (3) 20 N and 50 N
   (2) 20 N and 20 N   (4) 30 N and 5.0 N
20 Three forces act concurrently on an object in equilibrium. These forces are 10N, 8 N, and 6 N. The resultant of the 6 N force and the 8 N force is
   (1) 0   (3) 10 N
   (2) between 0 and 10 N   (4) greater than 10 N
21 If the force vector shown in the diagram at the right is resolved into two components, these two components could best be represented by which diagram?

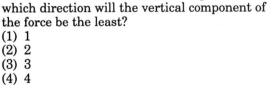

   (1)   (2)   (3)   (4)

22 A ball is fired with a velocity of 12 meters per second from a cannon pointing north, while the cannon is moving eastward at a velocity of 24meters per second. Which vector best represents the resultant velocity of the ball as it leaves the cannon?

   (1)   (2)   (3)   (4)

23 In the diagram, the numbers 1, 2, 3, and 4 represent possible directions in which a force could be applied to a cart. If the force applied in each direction has the same magnitude, in which direction will the vertical component of the force be the least?
   (1) 1
   (2) 2
   (3) 3
   (4) 4

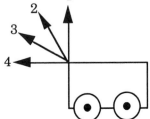

24 Which force could act concurrently with
    Force A to produce Force **B** as a resultant?

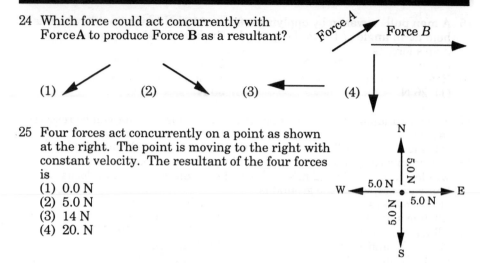

(1)   (2)   (3)   (4)

25 Four forces act concurrently on a point as shown
    at the right. The point is moving to the right with
    constant velocity. The resultant of the four forces
    is
    (1) 0.0 N
    (2) 5.0 N
    (3) 14 N
    (4) 20. N

II. Graphing Motion
2.1 Uniform Motion

The motion of a body may be described in terms of its displacement, velocity and acceleration . Uniform motion indicates that the object is traveling at a constant velocity. Velocity is a vector quantity which represents the time rate of change in displacement for an object. There are no unbalanced forces acting on the body. A car moving at 55 m/s due north is an example of an object traveling with uniform velocity.

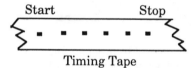

Start   Stop

Timing Tape

One way to analyze the motion of the car is to plot a graph of displacement versus time. The data for such a graph could come from a timing tape such as the one shown. A tape which shows uniform motion will have equally spaced marks. The timing device marks the tape at equal time intervals.

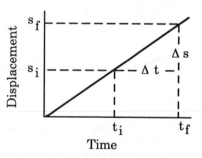

The displacement-time graph is a graph of an object in uniform motion in the positive direction. The slope of the graph represents the time rate of change of position (speed).

$$\text{slope} = \frac{s_f - s_i}{t_f - t_i} = \frac{\Delta s}{\Delta t} = \frac{m}{s}$$

Equation for Uniform Motion:

$$\overline{v} = \frac{\Delta s}{\Delta t}$$

average velocity = displacement/time

Note: Since velocity is a vector quantity, a direction must also be specified. If no direction is specified, the vector $\overline{v}$ is known as the average speed.

A straight line represents constant speed. Time, the independent variable, is placed on the x-axis. Displacement, which is dependent on time, is placed on the y–axis. Below are other displacement-time graphs.

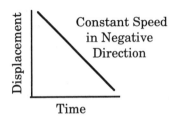

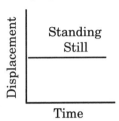

The velocity-time graph indicates motion at a constant velocity. The area under a velocity-time graph represents the distance covered by the moving object.

Area = base x height
= time x velocity
= s x m/s = meters

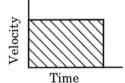

2.2 Accelerated Motion

When the speed of a body changes with time, the object is said to be accelerating. Acceleration is the time-rate of change in velocity: $a = \Delta v/\Delta t$. The units for acceleration in the MKS system is m/s^2. **Free fall** is uniformly accelerated motion. The value for **"g"** is 9.8 m/s^2 or approximately 10 m/s^2.

A timing tape for accelerated motion is illustrated below. The displacement during each time interval is not the same. Increasing displacements indicate that the object is accelerating or increasing speed.

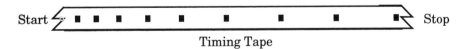

Timing Tape

Decreasing displacements indicate that the object is decelerating or decreasing speed. The displacement-time graph indicates that the displacement increases each second during the accelerated motion. The slope of the graph continues to increase indicating increasing speed. This graph is sometimes known as an **s - t** graph. The **s** represents displacement, not speed.

Below are additional displacement-time graphs for accelerated objects:

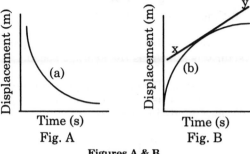

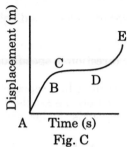

| Fig. A | Fig. B | Fig. C |

Figures A & B
Decreasing Velocity (a) & (b)

Figure C
Constant Velocity - AB,
Decreasing - BC, Stop - CD,
Increasing - DE

If the velocity is changing, the displacement - time graph is curved. The slope of the tangent to the distance - time curve at any point represents the **instantaneous speed** at that point. (see Figure B, line x y) The reading of the speedometer of a car represents instantaneous speed.

The velocity - time graph below illustrates accelerated linear motion. Velocity continues to increase at a constant rate. Equal changes in velocity occur in equal time intervals. The slope of a v-t graph represents acceleration. The area under a velocity - time graph represents displacement.

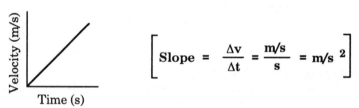

$$\left[\text{Slope} = \frac{\Delta v}{\Delta t} = \frac{m/s}{s} = m/s^2 \right]$$

Additional velocity-time graphs are included below:

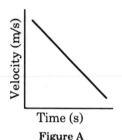

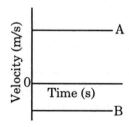

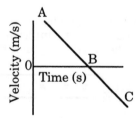

Figure A
Decelerating Uniformly

(Negative Acceleration)

Figure B
Uniform Speed
(No Acceleration)
A is traveling faster
than B, in the
opposite direction.

Figure C
Decelerating - AB, Stopped
at B (Accelerating in opposite
direction BC)
Note - This could be a v-t graph
of a baseball thrown straight up
into the air. The slope would
represent the acceleration
of gravity.

Example

A toy train heads north on a long straight track as plotted on the graph below:

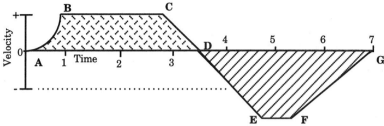

From the data on the graph (above), answer the following questions:

(1) At what time(s) was the train stopped?
(2) During what interval(s) was acceleration constant and not zero?
(3) Identify the interval(s) of non-uniform acceleration.
(4) Identify the interval where the uniform acceleration was the greatest.
(5) Did the train arrive back at the starting point?

Answers:
(1) At t = 0 sec, 3.5 sec and 7 sec when v = 0 the train was stopped.
(2) During CDE and FG a straight line slope shows constant acceleration.
(3) During AB, the slope and therefore, the acceleration is changing and nonuniform.
(4) CDE represents the greatest uniform acceleration, because it has the steepest uniform slope.
(5) No. The area above the horizontal axis indicated the distance traveled in the North direction. The area below the horizontal represents the distance traveled in the South direction. Since they are not equal, the train did not arrive back at the starting point, but arrived South of it.

Equations For Accelerated Motion:

$$\overline{v} = \text{average velocity} = \frac{v_f + v_i}{2}$$

Where:
v_i = initial velocity
v_f = final velocity
acceleration = change in velocity/change in time
s = displacement

$$a = \frac{\Delta v}{\Delta t} = \frac{v_f - v_i}{t_f - t_i}$$

$v_f = v_i + a\Delta t$ if $v_i = 0$,   then $v_f = at$
$v_f^2 = v_i^2 + 2a\Delta s$
$\Delta s = v_i\Delta t + \frac{1}{2}a(\Delta t)^2$
$\Delta s = \overline{v}\Delta t$

These relationships are valid only for cases where there is a constant acceleration.

Example

An object increases its speed from 10. m/s to 24 m/s in 2.0 s.

Find: (a) the average speed of the object,
           (b) the acceleration of the object , and
           (c) the distance the object traveled during 6 seconds.

Solution: **Given**:

$$v_i = 10 \text{ m/s} \qquad v_f = 24 \text{ m/s} \qquad \Delta t = 2.0 \text{ s}$$

Find: $\bar{v}$, **a**, and **distance**

(a) for average speed use:

$$\bar{v} = (v_i + v_f)/2 = (10 \text{ m/s} + 24 \text{ m/s})/2 = 17 \text{ m/s}$$

(b) for acceleration use:

$$a = \Delta v/ \Delta t = (24 \text{ m/s} - 10 \text{ m/s})/2 \text{ s} = 7 \text{m/s}^2$$

(c) for distance use:

$$\Delta s = v_i \Delta t + \tfrac{1}{2}a(\Delta t)^2$$

$$\Delta s = (10 \text{ m/s})(2.0\text{s}) + \tfrac{1}{2}(7 \text{ m/s}^2)(4\text{s})^2 = 34 \text{ m}$$

2.3 Freely Falling Objects

Objects freely falling for short distances without appreciable air resistance may be considered as examples of objects with constant acceleration of gravity. ("Short distances" can be considered to be from sea level to an altitude of 16 kilometers.) With air friction, freely falling objects soon reach a terminal velocity.

Practical Application
- Sky Diver

Free fall is uniformly accelerated motion. The acceleration is given by the symbol **g** which equals 9.8m/s^2 on earth. Students are encouraged to estimate the answers using g as 10 m/s^2. After t seconds, the velocity of an object in free fall is calculated by

$$v_f = v_i + a\Delta t$$

If: $v_i = 0$, $v_f = a\Delta t$ or $v_f = g\Delta t$

The distance the object falls is:

$$\Delta s = v_i \Delta t + \tfrac{1}{2} a(\Delta t)^2 \quad \text{or} \quad s = \bar{v}\Delta t$$

The distance fallen after one second if $v_i = 0$:

$$\Delta s = \frac{1}{2} (10. \text{ m/s}^2) (1.0 \text{ s})^2 = 5.0 \text{ m}$$

Time of fall (s)	0.00	1.00	2.00	3.00	4.00	5.00
Speed (m/s)	0.0	10.	20.	30.	40.	50.
Distance traveled (m)	0.0	5.	20.	45.	80.	125.

Table – Free fall of an Object Starting from Rest

2.4 The Pendulum

When a pendulum is swinging over a small arc, its bob is in simple harmonic motion. The period of the pendulum can be determined mathematically to be:

$$T = 2\pi \sqrt{\frac{1}{g}}$$

Where:
T = period of pendulum
l = length of pendulum
g = acceleration of gravity

The period of a pendulum depends only on its length. It does not depend on the mass of the bob or on the amplitude of the swing, provided that the swing is small.

Example

Find the length of a pendulum that has a period of 1.00 seconds.

Given: $g = 9.80 \text{ m/s}^2$ $T = 1.00 \text{ sec}$

Find: l

Solution:

$$T = 2\pi \sqrt{\frac{1}{g}}$$

$$T^2 = 4\pi^2 \frac{1}{g}$$

$$l = \frac{T^2 g}{4\pi^2} \quad \text{or} \quad \frac{(1.00 \text{ s}^2)(9.80 \text{ m/s})^2}{4(3.14)^2}$$

$$l = .252 \text{ m}$$

How would you adjust a pendulum clock that runs too fast?

The second, s, is the SI unit of time and is a fundamental unit. The size of a second can be approximated by counting "one - thousand - one, etc."

Practical Applications
- Pendulum Clock
- Stop Watches

Questions

1   An object, starting from rest, accelerates at a rate of 3.0 meters per
    second squared for 6.0 seconds. The velocity of the object at the end of
    this time is
    (1) 0.50 m/s   (2) 2.0 m/s   (3) 3.0 m/s   (4) 18 m/s

2   Which graph best represents the relationship between velocity and time
    for an object accelerating at a constant positive rate?

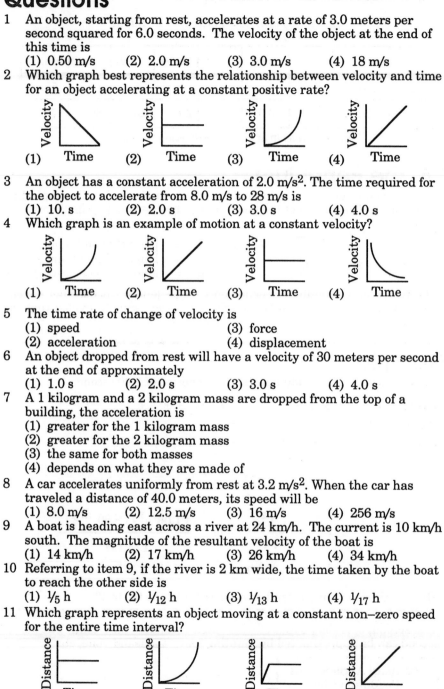

3   An object has a constant acceleration of 2.0 m/s². The time required for
    the object to accelerate from 8.0 m/s to 28 m/s is
    (1) 10. s   (2) 2.0 s   (3) 3.0 s   (4) 4.0 s

4   Which graph is an example of motion at a constant velocity?

5   The time rate of change of velocity is
    (1) speed   (3) force
    (2) acceleration   (4) displacement

6   An object dropped from rest will have a velocity of 30 meters per second
    at the end of approximately
    (1) 1.0 s   (2) 2.0 s   (3) 3.0 s   (4) 4.0 s

7   A 1 kilogram and a 2 kilogram mass are dropped from the top of a
    building, the acceleration is
    (1) greater for the 1 kilogram mass
    (2) greater for the 2 kilogram mass
    (3) the same for both masses
    (4) depends on what they are made of

8   A car accelerates uniformly from rest at 3.2 m/s². When the car has
    traveled a distance of 40.0 meters, its speed will be
    (1) 8.0 m/s   (2) 12.5 m/s   (3) 16 m/s   (4) 256 m/s

9   A boat is heading east across a river at 24 km/h. The current is 10 km/h
    south. The magnitude of the resultant velocity of the boat is
    (1) 14 km/h   (2) 17 km/h   (3) 26 km/h   (4) 34 km/h

10 Referring to item 9, if the river is 2 km wide, the time taken by the boat
    to reach the other side is
    (1) 1/5 h   (2) 1/12 h   (3) 1/13 h   (4) 1/17 h

11 Which graph represents an object moving at a constant non–zero speed
    for the entire time interval?

Base your answers to questions 12 and 13 on the data recorded on the timing tape below.

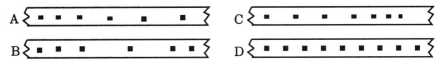

12 Free fall is best represented by tape
    (1) A   (2) B   (3) C   (4) D
13 Uniform motion is best represented by tape
    (1) A   (2) B   (3) C   (4) D

Base your answers to questions 14 through 17 on the graph which represents the motion of cars A and B on a straight track. Car B passes car A at the same instant that car A starts from rest at t = 0 seconds.

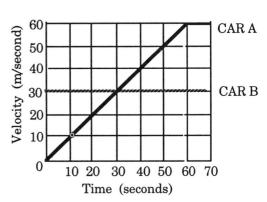

14 What is the acceleration of car A during the interval between t = 0 and   t = 60?
    (1) 1 m/s^2 (2) 10 m/s^2 (3) 20 m/s^2 (4) 30 m/s^2
15 How far did car A travel in the interval between t = 0 and t = 60?
    (1) 30 m   (2) 360 m   (3) 1,800 m   (4) 3,600 m
16 How long after t = 0 did it take car A to catch up to car B?
    (1) 10 s   (2) 20 s   (3) 30 s   (4) 60 s
17 During the time interval given below, which car traveled the greatest distance?
    (1) car A from t = 0 to t = 30   (3) car B from t = 0 to t = 30
    (2) car A from t = 30 to t = 60   (4) car B from t = 30 to t = 60

Questions 18 through 20 are based on the following information.

A toy projectile is fired vertically from the ground with an initial velocity of + 29 m/s. The projectile arrives at its maximum altitude in 3.0 seconds. [Neglect air resistance.]

18 The greatest height the projectile reaches is approximately
    (1) 23 m   (2) 44 m   (3) 87 m   (4) 260 m
19 What is the velocity of the projectile when it hits the ground?
    (1) 0.0 m/s   (2) -9.8 m/s   (3) -29 m/s   (4) +29 m/s
20 What is the displacement of the projectile from the time it left the ground until it returned to the ground?
    (1) 0.0 m   (2) 9.8 m   (3) 44 m   (4) 88 m

Base your answers to questions 21 through 25 on the accompanying graph which represents the motions of four cars on a straight road.

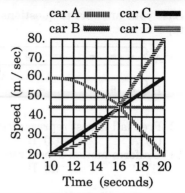

21 The speed of car C at time t = 20 seconds is closest to
(1) 60 m/s   (2) 45 m/s   (3) 3.0 m/s   (4) 600 m/s
22 Which car has a zero acceleration?
(1) A   (2) B   (3) C   (4) D
23 Which car is decelerating?
(1) A   (2) B   (3) C   (4) D
24 Which car moves the greatest distance in the time interval t = 10 seconds to t = 16 seconds?
(1) A   (2) B   (3) C   (4) D
25 Which graph best represents the relationship between distance and time for car C?

(1)   (2)   (3)   (4)

26 The graph at the right represents the motion of a body that is moving with
(1) increasing acceleration   (3) increasing speed
(2) decreasing acceleration   (4) constant speed

27 The graph at the right represents the relationship between distance and time for an object in motion. During which time interval is the speed of the object changing?
(1) AB
(2) BC
(3) CD
(4) DE

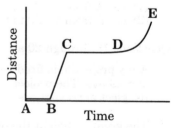

28 Which is constant for a freely falling body?
(1) displacement (2) speed   (3) velocity   (4) acceleration
29 A car moving at a speed of 8.0 m/s enters a highway and accelerates at 3.0 m/s^2. How fast will the car be moving after it has accelerated for 56 meters?
(1) 24 m/s   (2) 20 m/s   (3) 18 m/s   (4) 4.0 m/s

III. Forces On Matter

A **force** is a vector quantity that may be defined as a push or a pull. The MKS unit of force is the **newton (N)**. Forces may act upon an object at a distance without physical contact. The four forces are classified as gravitational (the weakest), the weak interaction, electromagnetic, and the strong interaction. The **newton** is the force which imparts to a mass of one kilogram an acceleration of one meter per second squared. It is a derived unit. A medium size apple weighs approximately 1 N. The English unit for force is the **pound**. The English unit for mass is the **slug**. An average student has a mass of 60 kg and a weight of approximately 600 N.

3.1 Equilibrium

If the vector sum of the concurrent forces acting on an object is zero, the object is in equilibrium. In physics, the word normal means perpendicular.

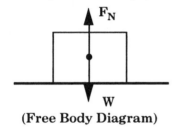

(Free Body Diagram)

Example of Static Equilibrium

F_N = normal force of table pushing up

(the table can push)

W = force of block pushing down

An object in static equilibrium experiences no relative motion. An object in dynamic equilibrium moves at a constant velocity.

Example of Dynamic Equilibrium

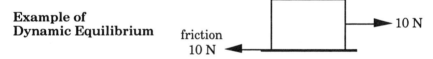

An object is *moving* with a 10N vector force pulling forward to overcome a 10N friction force. Vector sum is zero force. Object keeps moving at constant speed.

Force Table

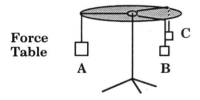

If you are told that a system is in equilibrium, the sum of all vectors acting on that system must equal zero. If any vector force is removed, the resultant of all the remaining vectors is equal to the magnitude of that vector at a direction 180° from the original vector.

In this diagram, the force table is balanced. **B** has a force of 10 Newtons balancing **A** and **C**.

1. What is the resultant of **A** + **C**?
2. If **B** were removed, what would happen to the system?

3.2 Newton's Three Laws Of Motion

First Law — an object remains at rest or in uniform motion unless acted upon by an unbalanced force. An object in uniform motion will continue to move in a straight line unless acted upon by an unbalanced force. This law is sometimes referred to as the **Law of Inertia**. The first law is a special case of the second law, when F = 0. The inertia of an object is proportional to the object's mass.

Practical Applications
- Seat Belt
- Pulling a tablecloth out from under dishes.
- Tossing and catching a ball on a moving train.
- A space probe continues moving after the engines are turned off.

Second Law — an unbalanced force acting on an object causes an acceleration which is directly proportional to the force and in the direction of the force. The law is represented by the equation **F = ma**, where **F** is the net force in newtons, **m** is the mass in kilograms, and **a** is the acceleration in m/s². Graphic and timing tape examples of Newton's Second Law are illustrated below:

Start ▪ ▪ ▪ ▪ ▪ ▪   ▪   ▪   ▶ Stop

Timing Tape

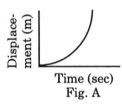

Fig. A

The timing tape above indicates accelerated motion. When the distance between the marks on the tape increases, the object is being accelerated. Figure A is a displacement-time graph of an object acted upon by a constant force. The graph indicates constant acceleration.

Force can be plotted against acceleration according to the formula **F = ma**. The slope of this graph (Fig. B) represents the inertial mass. *Inertial mass is a scalar quantity.*

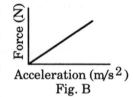

Acceleration (m/s²)
Fig. B

Acceleration can be plotted against mass. The graph (Fig.C) indicates that the acceleration varies inversely with the mass, with the force held constant.

Fig. C

Weight, the measurement of Earth's gravitational attraction for any object, is an example of Newton's Second Law (**W = mg**). The slope of a weight-mass graph (Fig. D) is the acceleration of gravity.

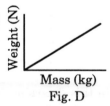

Mass (kg)
Fig. D

Practical Applications
- Elevator
- Spring Scales

The elevator is an interesting application of Newton's Second Law. The force exerted by the cable of an elevator at rest, is equal to the weight of the elevator (**Tension = mg**). When the elevator begins to rise, it is accelerated upward. The cable must now provide an additional force to lift and accelerate the elevator (**Tension = mg + ma**). Soon the elevator reaches a **constant** speed. There is no acceleration and the force on the cable is calculated by the formula **F = mg + m(0)**. As the elevator begins to decelerate, the cable now provides a force calculated by **F = mg + m(-a)**. The process is repeated on the way down as indicated in the diagram:

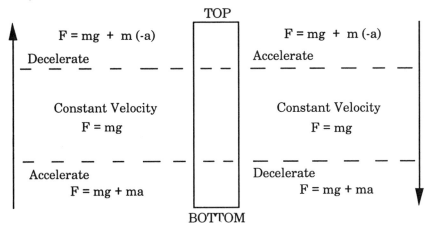

Newton's Third Law — forces always occur in pairs. For any applied force there is an equal and opposing force.

When a book is placed on a table, the book pushes down with a force equal to the book's weight. The table pushes up on the book with an equal and opposite reaction force. Each force acts on a different object. The Law of Conservation of Momentum is the basis for Newton's Third Law.

Practical Application
· Walking

Questions

1 Which of these groups of three concurrent forces cannot be in equilibrium?
   (1) 3 N; 4 N; 2 N   (3) 2 N; 3 N; 2 N
   (2) 5 N; 5 N; 1 N   (4) 4 N; 3 N; 8 N
2 A 100. kilogram mass starts from rest and is accelerated by a force of 200.newtons. The acceleration of the mass is
   (1) 0.25 m/s^2 (2) 2.0 m/s^2 (3) 0.50 m/s^2 (4) 4 m/s^2
3 A 1 kilogram mass will have a weight of about
   (1) 1 N   (2) 10. N   (3) 100. N   (4) 1000. N
4 A student pushes a 200. kilogram mass with a force of 10. newtons, but the mass does not move. What force does the mass exert on the student?
   (1) 0 N   (2) 10. N   (3) 20. N   (4) 200. N

5   A lamp placed on a table has a mass of 2 kilograms. What force does the
    table exert on the lamp?
    (1) 0. N   (2) 2 kg   (3) 20. N   (4) 40. N

6   Two frictionless blocks, having masses of 8.0 kilograms and 2.0 kilo-
    grams, rest on a horizontal surface. If a force applied to an 8.0 kilogram
    block gives it an acceleration of 5.0 m/s², then the same force will give
    the 2.0 kilogram block an acceleration of
    (1) 1.2 m/s²   (2) 2.5 m/s²   (3) 10. m/s²   (4) 20. m/s²

7   A car on which there is no accelerating force
    (1) must be at rest   (3) is speeding up
    (2) may be in motion   (4) is slowing down

8   The graph shows the relationship between the
    acceleration of an object and the unbalanced force
    producing the acceleration. The ratio (ΔF/Δa) of
    the graph represents the object's
    (1) mass
    (2) momentum
    (3) kinetic energy
    (4) displacement

9   Which object had the retarding force acting on it?

    (1)   Time (sec) (2)   Time (sec) (3)   Time (sec) (4)   Time (sec)

10 An 800. newton person is standing in an elevator. If the upward force of
    the elevator on the person is 600. newtons, the person is
    (1) at rest
    (2) accelerating upward
    (3) accelerating downward
    (4) moving downward at constant speed

11 A table exerts a 2.0 newton force on a book lying on the table. The force
    exerted by the book on the table is
    (1) 20. N   (3) 0.20 N
    (2) 2.0 N   (4) 0. N

12 A force F newtons gives an object with mass M, an acceleration of A. The
    same force F will give a second object with mass of 2 M, an acceleration of
    (1) A/2   (2) 2A   (3) A   (4) A/4

13 The diagram represents a constant force F
    acting on a box located on a frictionless
    horizontal surface. As the angle between
    the force and the horizontal increases, the
    acceleration of the box will
    (1) decrease
    (2) increase
    (3) remain the same

14 An elevator containing a man weighing 800 newtons is rising at a constant speed. The force exerted by the man on the floor of the elevator is
(1) less than 80 N   (3) 800 N
(2) between 80 and 800 N   (4) more than 800N

15 When an object is moving with constant velocity, which is true?
(1) An unbalanced force is acting.
(2) The object is being accelerated.
(3) The object is undergoing a change in momentum.
(4) The vector sum of all the forces acting on the object is zero.

16 The graph represents the net force acting on an object as a function of time. During which time interval is the velocity of the object constant?
(1) 0 to 2
(2) 2 to 3
(3) 3 to 4
(4) 4 to 5

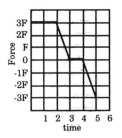

17 An object accelerates at 2.5 meters per second squared, when an unbalanced force of 10 newtons acts on it. What is the mass of the object?
(1) 1.0 kg   (2) 2.0 kg   (3) 3.0 kg   (4) 4.0 kg

18 Which graph best represents the relationship between the acceleration and the unbalanced force applied to an object?

(1)   (2)   (3)   (4)

19 A cart is uniformly accelerating from rest. The net force acting on the cart is
(1) decreasing   (2) zero   (3) constant   (4) increasing

20 An object weighing 4 N rests on a horizontal table. The force of the table-top on the object is
(1) 0.0 N   (3) 4.0 N downward
(2) 4.0 N horizontally   (4) 4.0 N upward

21 Which graph best represents an object in equilibrium?

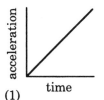

(1)   (2)   (3)   (4)

22 As the vector sum of all the forces on an object increases, the acceleration of the object
(1) decreases   (2) increases   (3) remains the same

23 If the mass of a moving object was doubled, its inertia would be
(1) halved   (2) doubled   (3) unchanged   (4) quadrupled

3.3 Universal Gravitation

Newton's Law of Universal Gravitation states that the force of attraction between any two point masses is directly proportional to the product of their masses, and inversely proportional to the square of the distance between them. It can be illustrated by the following formula:

$$F = Gm_1 m_2/r^2$$

Where:
   F   is the force between the two masses, measured in Newtons
   G   is the Universal Gravitation Constant, 6.67×10^{-11} N - m^2/kg^2
   m_1 and m_2 are masses, measured in kg
   r   is the distance between the two masses, measured center to center in meters

   The law indicates that there is a force of the earth pulling on the moon and an equal and opposite force of the moon pulling on the earth. Every object experiences this force of attraction, but when the masses are relatively small, the forces are also small. This law is limited to point or spherical sources with uniform mass distribution.

> **Practical Applications**
> · Moon orbiting the Earth
> · Tides

Example

   Calculate the force of attraction between a 50. kg mass and a 3000. kg mass two meters apart.

   Given: G = constant = 6.67×10^{-11} N m^2/kg^2
          m_1 = 50. kg
          m_2 = 3000. kg
          r = 2.0 m

   Find: Force

   Solution:
          F = $Gm_1 m_2/r^2$
             = $(6.67 \times 10^{-11}$ N $m^2/kg^2)(50.$ kg$)(3000.$ kg$)/(2.0$ m$)^2$
             = 2.5×10^{-6} N

3.4 Gravitational Fields

   The concept of field was introduced by Michael Faraday to deal with the problems of forces acting at a distance. Every mass may be considered to be surrounded by a gravitational field. The interaction of the fields results in attraction. An object can push or pull you without touching you.

Mass measured by gravitational attraction is called **gravitational mass.** The most sensitive experiments possible to date conclude that gravitational mass is equivalent to inertial mass, and both are expressed in the same units.

The magnitude of the strength of a gravitational field at any point is the force per unit mass at that point in the gravitational field. The relationship:

$$g = F/m$$

Where: **g** is the acceleration of gravity (in m/s^2)
　　　　F is the force (in Newtons)
　　　　m is the mass (in kilograms)

The direction of the gravitational field is the direction of the force on a test mass. The force to mass ratio is the same for all objects for short distances above the Earth's surface.

3.5 Weight

The weight of an object is equal to the net gravitational force acting on it. The weight of an object is equal to the product of its mass and the gravitational force.

$$W = mg$$

Where: **W** is weight in newtons
　　　　m is mass in kilograms
　　　　g is acceleration of gravity in m/s^2

Weight is a vector quantity that varies in magnitude with the location of the object with reference to the Earth. Spring scales may be used in weight measurements, but comparison of masses ("weighing") on a balance should be referred to as measurement of mass. To one significant figure, 100 grams weighs one newton. The average apple has a weight of one newton. Kilograms and grams are commonly misused as if they were weight units, especially on grocery package labeling.

Questions

1   Which is the most likely weight of a high school student?
    (1) 10 N   (3) 600 N
    (2) 50 N   (4) 2,500 N
2   Which property of an object is a measure of the Earth's gravitational attraction for the object?
    (1) kinetic energy   (3) weight
    (2) momentum   (4) volume
3   If the distance between a spaceship and the center of the Earth is increased from one Earth radius to two Earth radii, the gravitational force acting on the spaceship becomes approximately
    (1) ¼ as great   (3) ½ times greater
    (2) 2 times as great   (4) 4 times greater

4 What is the gravitational acceleration on a planet, where a 2 kilogram mass has a weight of 16 newtons on the planet's surface?
   (1) $\frac{1}{8}$ m/s^2 (2) 8. m/s^2 (3) 10. m/s^2 (4) 32. m/s^2

5 An astronaut drops a stone near the surface of the moon. Which graph best represents the motion of the stone, as it falls towards the moon's surface?

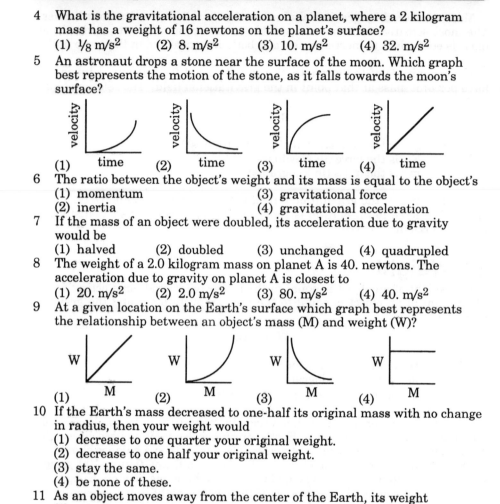

6 The ratio between the object's weight and its mass is equal to the object's
   (1) momentum   (3) gravitational force
   (2) inertia   (4) gravitational acceleration

7 If the mass of an object were doubled, its acceleration due to gravity would be
   (1) halved   (2) doubled   (3) unchanged   (4) quadrupled

8 The weight of a 2.0 kilogram mass on planet A is 40. newtons. The acceleration due to gravity on planet A is closest to
   (1) 20. m/s^2 (2) 2.0 m/s^2 (3) 80. m/s^2 (4) 40. m/s^2

9 At a given location on the Earth's surface which graph best represents the relationship between an object's mass (M) and weight (W)?

10 If the Earth's mass decreased to one-half its original mass with no change in radius, then your weight would
   (1) decrease to one quarter your original weight.
   (2) decrease to one half your original weight.
   (3) stay the same.
   (4) be none of these.

11 As an object moves away from the center of the Earth, its weight
   (1) remains constant.
   (2) varies inversely as its distance from the center of the Earth.
   (3) varies inversely as the square of its distance from the center of Earth.
   (4) increases.

12 A small test mass is placed on the line joining a mass of 8 kg and a mass of 2 kg at such a point that their combined gravitational attractions on it cancel out. The two masses are 12 meters apart. The position of the test mass is
   (1) midway between the two masses.
   (2) closer to the smaller mass.
   (3) closer to the larger mass.
   (4) unable to be determined based on the information given.

13 On the planet Gamma, a 4.0 -kilogram mass experiences a gravitational force of 24 newtons. What is the acceleration due to gravity on planet Gamma?
   (1) 0.17 m/s^2 (2) 6.0 m/s^2 (3) 9.8 m/s^2 (4) 96 m/s^2

14 The fundamental units for a force of one newton are
    (1) meters/second^2 (3) meters/second^2/kilogram
    (2) kilograms   (4) kilogram·meters/second^2
15 Which two quantities are measured in the same units?
    (1) velocity and acceleration   (3) mass and weight
    (2) weight and force   (4) force and momentum
16 The ratio of the gravitational attraction between a mass of 2 kg and a
    mass of 5 kg and between a mass of 3 kg and a mass of 4 kg when the
    pairs of masses are the same distance apart is
    (1) 1 to 1   (2) 12 to 10   (3) 5 to 6   (4) 6 to 20
17 The strength of the Earth's gravitational field at a point in space at which
    a mass **m** is acted on by a force **f** is
    (1) mg   (2) f/m   (3) f/mg   (4) m/f
18 A small test mass is placed on the line joining a mass of 4 kg and a mass
    of 1 kg at such a point that their combined gravitational attractions on it
    cancel out. The two masses are 12 meters apart. The position of the test
    mass is
    (1) midway between the two masses
    (2) 2 meters from the smaller mass
    (3) 3 meters from the smaller mass
    (4) 4 meters from the smaller mass

3.6 Static And Kinetic Friction

Friction is a *force that opposes motion*. It is the result of contact of irregular surfaces. It is calculated by multiplying the coefficient of friction (μ) by
the force perpendicular to the surface or the normal force. The coefficient of
friction is defined as the frictional force divided by the normal force. When
the object is on a level surface, the force perpendicular is equal to the weight
of the object.

```
Practical Applications
  •   traction on dry, wet, and icy roads
  •   waxing skis
  •   amusement park rides
  •   teflon coating
```

There are two types of friction encountered in elementary physics problems: **static** and **kinetic** friction. Static friction is the force that must be
overcome in order to start an object moving. Kinetic friction is frictional force
that must be overcome to keep the object moving. *The static friction is always
greater than the kinetic friction.* Both frictions are calculated using the same
formula with the proper coefficent for static or kinetic friction.

As illustrated on the next page, if a force of 10.0 newtons is applied to a
200.0 newton block and the block does not move, the force of friction is exactly 10.0 newtons. The force of friction will never oppose more than the force
applied. If the force is increased to exactly 20.0 newtons, and the block just
starts moving, there are no unbalanced forces acting on the block. Therefore,
the force of static friction was 20.0 newtons.

The coefficient of static friction can be calculated from the equation:

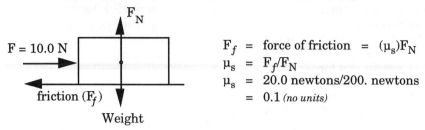

$$F_f = \text{force of friction} = (\mu_s)F_N$$
$$\mu_s = F_f/F_N$$
$$\mu_s = 20.0 \text{ newtons}/200. \text{ newtons}$$
$$= 0.1 \text{ (no units)}$$

Starting friction is the maximum static friction force. Once started a force of 10. newtons keeps the block moving at constant velocity. Therefore, the coefficient of kinetic friction is

$$\mu_k = \frac{F_f}{F_N} = \frac{10 \text{ N}}{200 \text{ N}} = .05$$

The rotor - type amusement park ride is an example of static friction. As the coefficient of friction increases, a larger force is necessary to start or move objects of the same mass.

Dry, sliding kinetic friction is practically independent of the apparent surface area and the relative velocity of the object. Two pieces of very smooth glass experience high sliding friction when the pieces of glass are slid across each other.

Rolling friction is the friction that occurs when objects are rolled over a surface. Rolling friction is generally less than sliding friction. Examples include roller skates, ball bearings, and bicycles.

Fluid friction is the friction that results from an object moving through a fluid such as water or air. Examples include parachutes, terminal velocity, and the streamlining of cars.

When the object is on an incline, the force perpendicular to the surface of the ramp is calculated by:

$$F_1 = W \cos \varnothing$$
$$F_2 = W \sin \varnothing$$
$$F_N = F_1$$

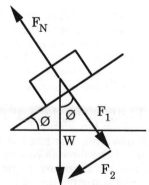

$$\cos \varnothing = F_1/\text{weight} \quad \text{or} \quad F_1 = W \cos \varnothing$$

The force of friction decreases as the angle increases since:

$$\textbf{F}_f = \textbf{ Force of friction } = (\mu)\textbf{F}_1 = (\mu)\textbf{W} \cos \varnothing$$

F_2 is pushing the block down the ramp. It is calculated from:

$$\sin \varnothing = \textbf{F}_2 / \textbf{weight} \quad \text{or} \quad \textbf{F}_2 = \textbf{W} \sin \varnothing$$

As the angle increases, the component of the force pushing the block down the ramp increases. F_2 increases; F_1 decreases. When one tries to push the block up the ramp, friction acts in a downhill direction. The uphill force to move the block must overcome F_2 and friction (F_f).

The block will slide down by itself, if $\textbf{F}_2 > \textbf{F}_f$. The force necessary to push the block up is $\textbf{F}_2 + \textbf{F}_f = \textbf{F}$.

When the force applied is greater than the force of friction, the object will accelerate according to Newton's 2nd Law.

Questions

1   A 500 - newton box rests on a horizontal surface. A force of 50 newtons parallel to the surface is required to start the box moving. What is the maximum coefficient of static friction between the box and the surface?
    (1) 0.1   (2) 10   (3) 0.5   (4) 25,000

2   In order to keep an object weighing 20. newtons moving at constant speed along a horizontal surface, a force of 10. newtons is required. The force of friction between the surface and the object is
    (1) 0 N   (2) 10. N   (3) 20. N   (4) 30. N

3   A constant unbalanced force acts on a 15.0 kilogram mass moving along a horizontal surface at 10.0 meters per second. If the mass is brought to rest in 1.50 seconds, what is the magnitude of the force of friction?
    (1) 10.0 N   (2) 100. N   (3) 147 N   (4) 150. N

4   Block A is pulled with constant velocity up an incline as shown in the diagram. Which arrow best represents the direction of the force of friction acting on block A?

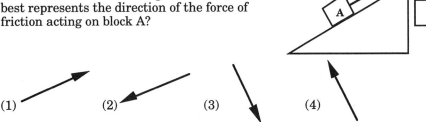

5   An empty wooden crate is slid across a warehouse floor. If the crate were filled, the coefficient of kinetic friction between the crate and the floor would
    (1) decrease   (2) increase   (3) remain the same

6   An empty wooden crate is slid across a warehouse floor. If the crate were filled, the force of kinetic friction between the crate and the floor would
    (1) decrease   (2) increase   (3) remain the same

IV. Momentum

Momentum is a vector quantity. It is the product of mass and velocity. The direction of the momentum is the same as the direction of the velocity. The unit for momentum is kg-m/s or newton-second.

When a body accelerates, its velocity changes; therefore, acceleration always produces a change in momentum. The momentum of an object will increase or decrease when a net external force acts on it.

In a system of objects, the individual components of a system will gain or lose momentum as they interact with one another. These interaction forces are internal forces and thus the momentum of the system does not change. As some components of the system gain momentum, other components of the system lose an equal amount of momentum. The momentum of the **system** is conserved.

In special cases, called **elastic collisions**, there is also conservation of kinetic energy. That is, the sum of the kinetic energy before the collision equals the sum of the kinetic energy after the collisions. Unless otherwise specified, collisions are inelastic, meaning that kinetic energy is not conserved, but momentum is conserved.

Important Equations for Momentum:

$$p = \text{momentum} = \text{mass x velocity}$$

and

$$\text{momentum before} = \text{momentum after}$$

The following examples illustrate applications of the Law of Conservation of Momentum:

Examples

Case 1: Cart A approaches cart B (initially at rest) with an initial velocity of 30 m/s. After the collision, cart A stops and cart B continues with what velocity?

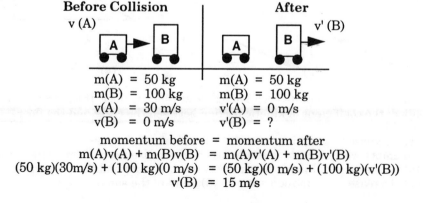

Before Collision   After
v (A)   v' (B)

m(A) = 50 kg   m(A) = 50 kg
m(B) = 100 kg   m(B) = 100 kg
v(A) = 30 m/s   v'(A) = 0 m/s
v(B) = 0 m/s   v'(B) = ?

momentum before = momentum after
m(A)v(A) + m(B)v(B) = m(A)v'(A) + m(B)v'(B)
(50 kg)(30m/s) + (100 kg)(0 m/s) = (50 kg)(0 m/s) + (100 kg)(v'(B))
v'(B) = 15 m/s

Case 2: Cart A approaches cart B (initially at rest) with an initial velocity of 30 m/s. After the collision, cart A locks together with cart B. Both travel with what velocity?

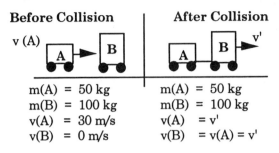

Before Collision	After Collision
m(A) = 50 kg	m(A) = 50 kg
m(B) = 100 kg	m(B) = 100 kg
v(A) = 30 m/s	v(A) = v'
v(B) = 0 m/s	v(B) = v(A) = v'

$$\text{momentum before} = \text{momentum after}$$
$$m(A)v(A) + m(B)v(B) = m(A)v' + m(B)v'$$
$$(50\ kg)(30\ m/s) + (100\ kg)(0\ m/s) = (50\ kg)v' + (100\ kg)\ v'$$
$$v' = 10\ m/s$$

Case 3: Cart A moving with an initial velocity of 30 m/sec approaches cart B moving at an initial velocity of 20 m/s towards cart A. The two carts lock together and move as one. Calculate the magnitude and the direction of the final velocity. (note that velocity to the left is denoted by a (-) negative sign)

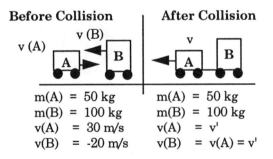

Before Collision	After Collision
m(A) = 50 kg	m(A) = 50 kg
m(B) = 100 kg	m(B) = 100 kg
v(A) = 30 m/s	v(A) = v'
v(B) = -20 m/s	v(B) = v(A) = v'

$$\text{momentum before} = \text{momentum after}$$
$$m(A)v(A) + m(B)v(B) = m(A)v' + m(B)v'$$
$$(50\ kg)(30\ m/s) + (100\ kg)(-20\ m/s) = (50\ kg + 100\ kg)(v')$$
$$v' = (-500\ kg\text{-}m/s)/150\ kg$$
$$v' = -3.33\ m/s$$
$$= 3.33\ m/s\ \text{(to the left)}$$

Case 4: A 1 gm bullet is fired from a 10 kg rifle with a speed of 300 m/s. Calculate the recoil velocity of the gun.

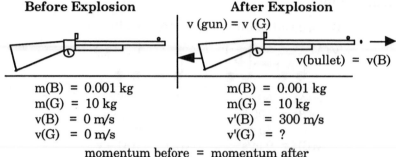

Before Explosion	After Explosion
m(B) = 0.001 kg	m(B) = 0.001 kg
m(G) = 10 kg	m(G) = 10 kg
v(B) = 0 m/s	v'(B) = 300 m/s
v(G) = 0 m/s	v'(G) = ?

momentum before = momentum after
m(B)v(B) + m(G)v(G) = m(B)v'(B) + m(G)v'(G)
(.001 kg)(0 m/s) + (10 kg)(0 m/s) = (.001 kg)(300 m/s) + (10 kg) (v'(G))
0 = .3 kg-m/s + 10kg (v'(G))
v'(G) = -0.03m/s

After an explosion, momentum in one direction is equal to momentum in the other direction.

Questions

1 What is the momentum of a 12 kilogram ball that has a velocity of 3 m/s?
   (1) 36 kg-m/s   (2) 15 kg-m/s   (3) 9 kg-m/s   (4) 4 kg-m/s

2 What is the recoil velocity of the cannon shown in the diagram?
   (1) 1 m/s
   (2) 2 m/s
   (3) 100 m/s
   (4) 200 m/s

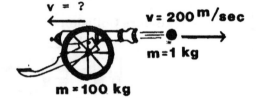

3 Two lab carts are separated by a compressed spring as shown. When the spring is released, the 2 kilogram cart moves with a speed of 2 meters per second to the right. What will be the speed of the 4 kilogram cart to the left?
   (1) 1 m/s   (3) 8 m/s
   (2) 2 m/s   (4) 4 m/s

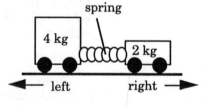

4 Two lab carts, A and B, are held together by a cord that keeps a spring compressed between them as shown in the diagram. The cord is cut. Compared to the momentum of cart B, what will be the momentum of cart A?
   (1) less   (2) greater   (3) the same

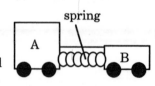

5 The direction of an object's momentum is always the same as the direction of the object's
   (1) inertia   (3) velocity
   (2) potential energy   (4) weight

6 The diagram represents two identical carts, attached by a cord, moving to the right at speed V. If the cord is cut, what would be the speed of cart A?
   (1) 0   (3) V
   (2) 2V   (4) V/2

7 Momentum may be expressed in
   (1) joule   (2) watts   (3) kg-m/s^2 (4) N-s

8 As an object falls freely toward the Earth, its momentum
   (1) decreases   (2) increases   (3) remains the same

9 If a 3.0 kilogram object moves 10. meters in 2.0 seconds, its average momentum is
   (1) 60 kg-m/s   (2) 30 kg-m/s   (3) 15 kg-m/s   (4) 10 kg-m/s

10 A 1.0 kilogram mass changes speed from 2.0 meters per second to 5.0 meters per second. The change in the object's momentum is
   (1) 9.0 kg-m/s   (2) 21 kg-m/s   (3) 3.0 kg-m/s   (4) 29 kg-m/s

11 A 4.0 kilogram mass is moving at 3.0 meters per second toward the right and a 6.0 kilogram mass is moving 2.0 meters per second toward the left on a horizontal frictionless table. If the two masses collide and remain together after the collision, their final momentum is
   (1) 1.0 kg-m/s   (2) 24 kg-m/s   (3) 12 kg-m/s   (4) 0 kg-m/s

12 A 2 kilogram object traveling 10 meters per second north has a perfectly elastic collision with a 5 kilogram object traveling 4 meters per second south. What is the total momentum after collision?
   (1) 0 kg-m/s   (3) 20 kg-m/s south
   (2) 20 kg-m/s north   (4) 40 kg-m/s east

13 Two unequal masses, originally at rest, are being drawn together by a stretched spring. The masses have the same
   (1) momentum   (3) velocity
   (2) acceleration   (4) kinetic energy

14 An 80. -kilogram skater and a 60. -kilogram skater stand at rest in the center of a skating rink. The two skaters push each other apart. The 60. -kg skater moves with a velocity of 10. m/s east. What is the velocity of the 80. -kilogram skater? (Neglect any frictional effects.)
   (1) 0.13 m/s west   (3) 10. m/s east
   (2) 7.5 m/s west   (4) 13. m/s east

4.1 Impulse

When an unbalanced force acts on an object for a period of time, a change in momentum is produced. **Impulse** is defined as the product of force and time. It is the cause of a change in momentum, and it is equal to the change in momentum.

$$\text{Impulse} = F\Delta t = m\Delta v = \Delta p = J$$

The units for impulse are the same as the units for momentum, newton-second or kg-m/s.

Practical Applications
- hitting a ball with a bat
- the "follow through" in a golf swing or baseball throw
- pushing a swing

Examples

1. An impulse of 50 newton-seconds is applied to a 10 kg mass in the direction of motion. If the mass had a speed of 20 m/s before the impulse, its speed after the impulse could be

Given: Impulse = **50 N-s**
v_i = 20 m/s
m = 10 kg
$m\Delta v$ = impulse

Find: new speed (v) = $v_i + \Delta v$

Δv = impulse/m = 50 N-s/10 kg = 5 m/s

Therefore: v = $v_i + \Delta v$
So: **v** = **25 m/s**

2. A 30 kg mass moving at a speed of 3 m/s is stopped by a constant force of 15 newtons. How many seconds must the force act on the mass to stop i?

Given: m = 30 kg
v_i = 3 m/s
F = -15 N
v_f = 0 m/s

Find: time

Solution: Impulse = $F \Delta t = m \Delta v$
$F \Delta t$ = $m \Delta v$
$N \cdot s$ = $kg \cdot m/s$
Δt = $m \Delta v/F$
Δt = 30 kg (0 m/s − 3 m/s) / -15 N
Δt = 6 s

Questions

1 A net force of 12 newtons acting north on an object for 4.0 seconds will produce an impulse of
(1) 48 kg-m/s north   (3) 3.0 kg-m/s north
(2) 48 kg-m/s south   (4) 3.0 kg-m/s south

2 A 5.0 kilogram cart moving with a velocity of 4.0 meters per second is brought to a stop in 2.0 seconds. The magnitude of the average force used to stop the car is
(1) 20. N   (2) 2.0 N   (3) 10 N   (4) 4.0 N

3 An impulse of 30.0 newton-seconds is applied to a 5.00 kilogram mass. If the mass had a speed of 100. meters per second before the impulse, its speed after the impulse could be
(1) 250 m/s   (2) 106 m/s   (3) 6.0 m/s   (4) 0 m/s

4 An object is brought to rest by a constant force. Which factor other than the mass and velocity of the object must be known in order to determine the magnitude of the force required to stop the object?
(1) the time that the force acts on the object
(2) the gravitational potential energy of the object
(3) the density of the object
(4) the weight of the object

5 A bat applies an average force of 500 newtons on a baseball for 0.20second. What was the average force applied by the ball on the bat?
(1) 100 N   (2) 200 N   (3) 500 N   (4) 1,000 N

6 What is the magnitude of the change in momentum produced when a force of 5.0 newton acts on a 10 kilogram object for 3.0 seconds?
(1) 1.5 kg-m/s   (2) 5.0 kg-m/s   (3) 10 kg-m/s   (4) 15 kg-m/s

7 A 5.0 newton force imparts an impulse of 15 newton-seconds to an object. The force acts on the object for a period of
(1) 0.33 s   (2) 20 s   (3) 3.0 s   (4) 75 s

8 A 20 kilogram mass moving at a speed of 3.0 meters per second is stopped by a constant force of 15 newtons. How many seconds must the force act on the mass to stop it?
(1) 0.20   (2) 1.3   (3) 5.0   (4) 4.0

9 A force of 10 newtons acts on an object for 0.010 second. What force, acting on the object for 0.050 seconds, would produce the same impulse?
(1) 1.0 N   (2) 2.0 N   (3) 5.0 N   (4) 10 N

Base your answers on question 10 through 14 on the diagram which represents carts **A** and **B** being pushed apart by a spring which exerts an average force of 50. newtons for a period of 0.20 seconds. (Assume frictionless conditions.)

10 What is the magnitude of the impulse applied by the spring on cart **A**?
(1) 5.0 N-s   (2) 10 N-s   (3) 50 N-s   (4) 100 N-s

11 Compared to the magnitude of the impulse acting on cart **A**, the magnitude of the impulse acting on cart **B** is
(1) one-half as great   (3) the same
(2) twice as great   (4) four times as great

12 Compared to the velocity of cart B at the end of the 0.20 sec interaction, the velocity of cart A is
(1) one-half as great   (3) the same
(2) twice as great   (4) four times as great

13 What is the average acceleration of cart **B** during the 0.20 second interaction?
(1) 0 m/s^2 (2) 10. m/s^2 (3) 25 m/s^2 (4) 50 m/s^2

14 Compared to the total momentum of the carts before the spring is released, the total momentum of the carts after the spring is released, is
(1) one-half as great   (3) the same
(2) twice as great   (4) four times as great

15 An impulse I is applied to an object. The change in the momentum of the object is
    (1) I   (2) 2I   (3) I/2   (4) 4I

16 An experiment consists of throwing balls straight up with varying initial velocities. Which quantity will have the same value in all trials?
    (1) initial momentum   (3) time of travel
    (2) maximum height   (4) acceleration

Thinking Physics

Can you explain these apparent discrepancies?

1   Two cars moving at 88 km/hr in opposite directions have different velocities.
2   On Earth, a coin will hit the ground before a feather when dropped from the same height at the same time. On the Moon, both objects hit the surface at the same time.
3   A 50 kg girl can out-pull two 90 kg football linemen.
4   An object can accelerate while traveling at constant speed, but not at constant velocity.
5   Electric power lines sometimes break in the winter when a small amount of ice forms on the wires.
6   A vector whose magnitude is 3 N and when added to a vector whose magnitude is 4 N can equal a vector whose magnitude is 5 N.
7   A slow steady pull unwinds a roll of toilet paper, but a sudden jerk tears a single piece.
8   A car that is stopped at a traffic light is struck from the rear. The occupants of the struck car usually suffer neck injuries.
9   To tighten the head of a loose hammer, its better to bang the handle against the top of the work table instead of the head of the hammer.
10 When you throw a heavy object from your hands while standing on a skateboard, you roll backwards.
11 A space probe continues moving after the engines are turned off.
12 When you sit on a chair, the chair pushes you up.
13 An object can push or pull you without touching you.
14 The weight of a person in an elevator can change as the elevator goes up and down.
15 Force of friction is no greater for a wide tire than for a narrow tire.
16 Raisins dipped in flour remain distributed through a dough while undipped raisins sink to the bottom of the dough.
17 As a rocket is launched from its pad it increases both its velocity and its acceleration.
18 An automobile moving at 30 m/s has more momentum than a bullet moving at 500 m/s.
19 Modern uni-body cars are designed to collapse in an accident, and it is considered a "safety" feature.
20 You can throw a raw egg into a sagging sheet without breaking the egg.

Free Response Questions

1   An object is traveling in a straight-line path. The data were collected for the object's distance from a reference point on this path and time of travel.

Distance (meters)	4.0	8.0	20.	40.	68.
Time (seconds)	0	1.0	2.0	3.0	4.0

Using the information in the data table above, construct a line graph following the instructions below.

a) Label the axes with the variables to be plotted and mark an appropriate scale on each axis.

b) Plot the data points on the grid provided and sketch the curve.

c) The motion of the object is best described as _____.

d) The slope of the curve at t = z seconds represents
    1) distance traveled
    2) acceleration
    3) instantaneous velocity
    4) average speed for two seconds of travel

2   A student moves along a path and records the directions and distances she moves on the table shown. To her surprise, she arrives back at the original starting point.

a) Draw a vector diagram of the student's route.

b) What is the total distance traveled?

Number	Direction	Number of Paces
1	N	2
2	W	1
3	N	3
4	E	4
5	N	1
6	W	2
7	S	8
8	W	2
9	N	2
10	E	1

3   A student walks from her house toward the bus stop, located 50. meters to the east. After walking 20. meters, she remembers that she has left her lunch at the door. She runs home, picks up her lunch, walks again, and arrives at the bus stop.

a) Sketch a displacement versus time graph for the student's motion.

b) Label your graph with appropriate values for time and displacement.

4 In a laboratory exercise, a student collected the following data as the
   unbalanced force applied to a body of mass M was changed.

   a) Construct a graph and label the axis of
       the graph with the appropriate values for
       force and acceleration.

   b) Plot an acceleration versus force graph
       for the laboratory data provided.

   c) Using the data of your graph, determine
       the mass, M, of the body. (Show all
       calculations.)

Data Table

Force (newtons)	Acceleration (m/s²)
4.0	2.1
8.0	4.0
12.0	6.0
16.0	7.9
20.0	10.0

5 An aluminum block weighing 20 newtons,
   sliding from *left to right* in a straight line on a
   horizontal steel surface, is acted on by a
   2.4-newton friction force. The block will be
   brought to rest by the friction force in a
   distance of 10 meters.

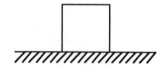

   a) On the diagram of the block, draw an arrow to identify the direction
       of *each* force acting on the block while it is still moving, but is being
       slowed by the friction force. Identify *each* force by appropriately
       labeling the arrow that represents its line of direction.

   b) Determine the magnitude of the acceleration of the block as it is
       brought to rest by the friction force. (Show all work.)

6 A pizza sign is to be hung from a wall. Assume
   that the sign weighs 100N, and the angle between
   the boom and the wires is 30°. (Assume the boom
   and wire are weightless.)

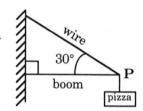

   a) Draw a force diagram for all forces acting on
       point P. Determine the tension in the wire.

   b) If the sign is hung as indicated at the right, compute the
       new tension in the wire.

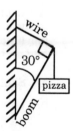

7   A student wants to hang a 100 N picture in his room as shown at the right.

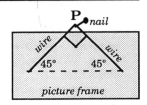

   a)   Make a vector diagram of the forces acting on point **P** (the nail).

   b)   Determine the tension in the wire.

   c)   Describe in full sentences what would happen to the tension in the wire if the angle was reduced to 30° on each side.

Self–Help Questions

1   The appropriate height of a high school physics student is
   (1) 10^1 m   (3) 10^0 m
   (2) 10^2 m   (4) 10^{-2} m

2   What is the approximate thickness of this piece of paper?
   (1) 10^1 m   (3) 10^{-2} m
   (2) 10^0 m   (4) 10^{-4} m

3   Which is the most likely mass of a high school student?
   (1) 1 kg   (3) 60 kg
   (2) 5 kg   (4) 250 kg

4   Distance is to displacement as
   (1) force is to weight   (3) velocity is to acceleration
   (2) speed is to velocity   (4) impulse is to momentum

5   Which terms represent vector quantities?
   (1) power and force   (3) time and energy
   (2) work and distance   (4) displacement and velocity

6   Forces of 6.0 N North and 8.0 N West act concurrently. What is the magnitude and direction of the equilibrant force?
   (1) 10 N Northwest   (3) 14 N Northeast
   (2) 10 N Southeast   (4) 14 N Southwest

7   Which diagram represents the vector with the largest horizontal component? (Assume each vector has the same magnitude.)

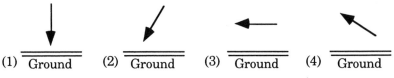

   (1) Ground   (2) Ground   (3) Ground   (4) Ground

8   Which vector best represents the resultant of forces F_1 and F_2 acting concurrently on point **P** as shown in the diagram?

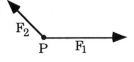

   (1)   (2)   (3)   (4)

9   Two forces (*OA* and *OB*) act simultaneously at point O as shown on the diagram at the right. The magnitude of the resultant force is closest to

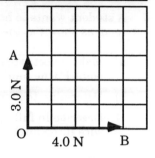

   (1) 5.0 N
   (2) 7.0 N
   (3) 3.0 N
   (4) 4.0 N

10   A resultant force of 20 newtons is made up of two component forces acting at right angles to each other. If the magnitude of one of the components is 12.0 newtons, the magnitude of the other component must be
   (1) 32 N   (2) 16 N   (3) 12 N   (4) 8 N

11   A student walks 3 blocks North, 4 blocks East, and 3 blocks South. What is the displacement of the student?
   (1) 10 blocks East   (3) 4 blocks East
   (2) 10 blocks West   (4) 4 blocks West

12   What is the distance traveled by an object that moves with an average speed of 6.0 meters per second for 4.0 seconds?
   (1) 0.67 m   (2) 1.3 m   (3) 24 m   (4) 48 m

13   The graph at the right represents the relationship between distance and time for an object in motion. During which interval is the speed of the object constant but not zero?

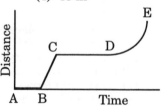

   (1) *AB*
   (2) *BC*
   (3) *CD*
   (4) *DE*

14   An object initially traveling in a straight line with a speed of 5.0 meters per second is accelerated at 2.0 meters per second squared for 4.0 seconds. The total distance traveled by the object in the 4.0 seconds is
   (1) 36 m   (2) 40 m   (3) 16 m   (4) 4.0 m

15   A blinking light of constant period is situated on a lab cart. Which diagram best represents a photograph of the light as the cart moves with constant velocity?

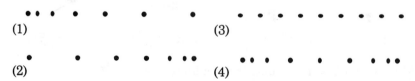

   (1)   (3)

   (2)   (4)

16   An object with an initial velocity of 3.0 meters per second accelerates at a rate of 3.0 meters per second squared for 6.0 seconds. The velocity of the object at the end of this time is
   (1) 0.50 m/s   (2) 18 m/s   (3) 21 m/s   (4) 54 m/s

17   A ball dropped from a bridge takes 2.0 seconds to reach the water below. How far is the bridge above the water?
   (1) 15 m   (2) 20 m   (3) 44 m   (4) 88 m

18 An object originally moving at a speed of 20 meters per second accelerates uniformly for 5.0 seconds to a final speed of 50 meters per second. What is the acceleration of the object?
    (1) 14 m/s^2 (2) 10 m/s^2 (3) 6 m/s^2 (4) 4.0 m/s^2

19 The diagram at the right shows a graph of velocity as a function of time for an object in straight-line motion. According to the graph, the object most likely has
    (1) a constant momentum
    (2) an increasing acceleration
    (3) a decreasing mass
    (4) an increasing speed

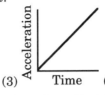

20 An object is acted upon by a constant unbalanced force. Which graph best represents the motion of this object?

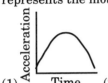

(1) Time   (2) Time   (3) Time   (4) Time

21 Which graph best represents the motion of an object initially at rest and accelerating uniformly?

(1) Time   (2) Time   (3) Time   (4) Time

22 An object near the surface of planet Y falls freely from rest and reaches a speed of 10.0 meters per second after it has fallen 20.0 meters. What is the acceleration due to gravity on planet X?
    (1) 2.50 m/s^2 (2) 5.00 m/s^2 (3) 9.80 m/s^2 (4) 10.0 m/s^2

Base your answers to questions 23 through 25 on the information and diagram at the right. The diagram represents a block sliding along a frictionless surface between points A and G.

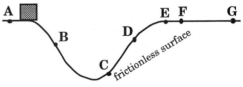

23 As the block moves from point A to point B, the speed of the block will be
    (1) decreasing   (3) constant, but not zero
    (2) increasing   (4) zero

24 Which expression represents the magnitude of the block's acceleration as it moves from point C to point D?
    (1) $\dfrac{m}{F}$ (2) $\dfrac{\Delta v}{\Delta t}$ (3) $m\Delta v$ (4) $\dfrac{2\Delta s}{\Delta t}$

25 Which formula represents the velocity of the block as it moves along the horizontal surface from point E to point F?
    (1) $\overline{v} = \dfrac{\Delta s}{\Delta t}$ (2) $\overline{v} = \dfrac{\Delta v}{2}$ (3) $v_f^2 = 2a\Delta s$ (4) $\Delta v = \dfrac{1}{2}a(\Delta t)^2$

26 A flashing light of constant 0.20 second period is situated on a lab cart.
    The diagram below represents a photograph of the light as the cart moves
    across a tabletop. How much time elapsed as the cart moved from
    position *A* to position *B*?

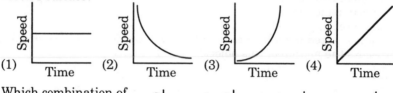

(1) 1.0 s   (2) 5.0 s   (3) 0.80 s   (4) 4.0 s

27 Which graph best represents the motion of a freely falling body near the
    Moon's surface?

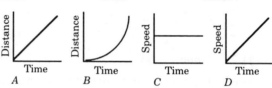

28 Which combination of
    graphs best describes
    free-fall motion?
    (Neglect air resistance.)

(1) *A* and *C*   (2) *B* and *D*   (3) *A* and *D*   (4) *B* and *C*

29 A copper coin resting on a piece of cardboard
    is placed on a beaker as shown in the
    diagram at the right. When the cardboard
    is rapidly removed, the coin drops into the
    beaker. The two properties of the coin
    which best explain its fall are its weight
    and its
    (1) temperature   (3) volume
    (2) electrical resistance   (4) inertia

30 A force of 50. newtons causes an object to accelerate at 10. meters per
    second squared. What is the mass of the object?
    (1) 500 kg   (2) 60. kg   (3) 5.0 kg   (4) 0.20 kg

31 A 75.0 kilogram object in outer space is attracted to a nearby planet with
    a net force of 300. newtons. What is the magnitude of the object's
    acceleration?
    (1) 4.00 m/s^2 (2) 9.81 m/s^2 (3) 157 m/s^2 (4) 2250 m/s^2

32 An object with a mass of 0.5 kilograms starts from rest and achieves a
    maximum speed of 20 meters per second in 0.01 second. What average
    unbalanced force accelerates this object?
    (1) 1,000 N   (2) 10 N   (3) 0.1 N   (4) 0.001 N

33 A cart rolls down an inclined plane with
    constant speed as shown in the diagram at
    the right. Which arrow represents the
    direction of the frictional force?
    (1) *A*
    (2) *B*
    (3) *C*
    (4) *D*

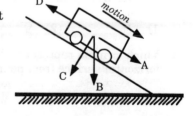

34 In the diagram at the right, surface A of
the wooden block has twice the area of
surface B. If it takes F newtons to keep
the block moving at a constant speed
across the table when it slides on
surface A, what force is needed to keep
the block moving at constant speed
when it slides on surface B?

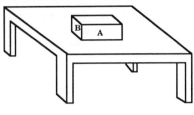

(1) F (2) $2F$ (3) $\frac{1}{2}F$ (4) $4F$

35 The table at the right lists the coefficients of kinetic
friction for four materials sliding over steel. A
10.-kilogram block of each of the materials in the table
is pulled horizontally across a steel floor at constant
velocity. Which block would require the *smallest*
applied force to keep it moving at constant velocity?

Material	μ_k
brass	0.44
copper	0.36
steel	0.57
wood	0.32

(1) brass   (3) steel
(2) copper   (4) wood

36 A 100-newton box rests on a horizontal surface. A force of 10-newtons
parallel to the surface is required to start the box moving. What is the
maximum coefficient of static friction between the box and the surface?
(1) 0.1   (2) 10   (3) 0.5   (4) 1000

37 A 20-kilogram cart traveling east with a speed of 6 meters per second
collides with a 30-kilogram cart traveling west. If both carts come to rest
immediately after collision, what was the speed of the westbound cart
before the collision?
(1) 6 m/s   (2) 2 m/s   (3) 3 m/s   (4) 4 m/s

38 Compared to the inertia of a 1-kilogram mass, the inertia of a 4-kilogram
mass is
(1) $\frac{1}{4}$ as great   (3) 16 times as great
(2) $\frac{1}{16}$ as great   (4) 4 times as great

39 The diagram at the right shows spheres A and B
with masses of M and $2M$, respectively. If the
gravitational force of attraction of sphere A on
sphere B is 3 newtons, then the gravitational force of
attraction of sphere B on sphere A is

Mass   Mass
M $2M$

(1) 9 N   (3) 3 N
(2) 2 N   (4) 4 N

40 Gravitational force of attraction F exists between two point masses A and
B when they are separated by a fixed distance. After mass A is tripled
and mass B is halved, the gravitational attraction between the two
masses is
(1) $\frac{1}{6}F$ (2) $\frac{2}{3}F$ (3) $\frac{3}{2}F$ (4) $6F$

41 Two point masses are located a distance, D, apart. The gravitational
force of attraction between them can be quartered by changing the
distance to
(1) $\frac{1}{2}D$ (2) $2D$ (3) $\frac{1}{4}D$ (4) $4D$

42 A 2.0-kilogram rifle initially at rest fires a 0.002-kilogram bullet. As the
bullet leaves the rifle with a velocity of 500 meters per second, what is
the momentum of the rifle-bullet system?
(1) 2.5 kg·m/s   (2) 2.0 kg·m/s   (3) 0.5 kg·m/s   (4) 0 kg·m/s

43 If the distance between a spaceship and the center of the Earth is increased from one Earth radius to four Earth radii, the gravitational force acting on the spaceship becomes approximately
(1) $\frac{1}{16}$ as great   (3) 16 times as great
(2) $\frac{1}{4}$ as great   (4) 4 times greater

44 Which graph best represents the relationship between the mass of an object and its distance from the center of the Earth?

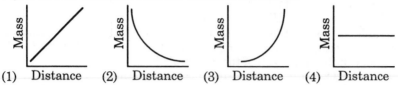

(1)   Distance   (2)   Distance   (3)   Distance   (4)   Distance

45 Four forces are acting on an object as shown in the diagram at the right. If the object is moving with a constant velocity, the magnitude of force F must be
(1) 0 N
(2) 20 N
(3) 100 N
(4) 40 N

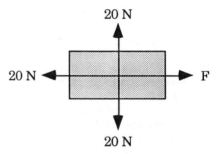

46 A spring is compressed between two stationary blocks as shown in the diagram at the right. Block A has a mass of 6.0 kilograms.
After the spring is released, block A moves west at 8.0 meters per second and block B moves east at 12 meters per second. What is the mass of block B? (Assume no frictional effects.)
(1) 16 kg   (2) 12 kg   (3) 4.0 kg   (4) 6.0 kg

47 A constant braking force of 20 newtons applied for 5 seconds is used to stop a 4.0-kilogram cart traveling at 25 meters per second. The magnitude of the impulse applied to stop the cart is
(1) 20 N·s   (2) 60 N·s   (3) 100 N·s   (4) 200 N·s

48 A 25-kilogram mass travels east with a constant velocity of 40. meters per second. The momentum of this mass is
(1) 1.0×10^3 kg·ms east   (3) 1.0×10^3 kg·ms west
(2) 9.8×10^3 kg·ms east   (4) 9.8×10^3 kg·ms west

49 An unbalanced 6.0-newton force acts eastward on an object for 3.0 seconds. The impulse produced by the force is
(1) 18 N·s east   (3) 18 N·s west
(2) 2.0 N·s east   (4) 2.0 N·s west

50 A rocket with a mass of 1,000 kilograms is moving at a speed of 20 meters per second. The magnitude of the momentum is
(1) 50 kg·m/s   (3) 20,000 kg·m/s
(2) 200 kg·m/s   (4) 4,000,000 kg·m/s

1 Optional Unit

Motion In A Plane

Important Terms To Be Understood

horizontal component	centripetal acceleration	satellite
vertical component	centripetal force	geosynchronous orbit
projectile	Kepler's Laws	escape velocity

In the previous Mechanics Unit, the straight up and down motion of an object was discussed. The pull of gravity on the object caused it to accelerate downward with free-fall acceleration, **g**. We are now going to consider what happens if an object is shot at an angle.

Once the object is free, it is subject to only one force, the force of gravity. Air friction will be ignored in the following discussion.

The motion of an object traveling in a plane (two dimensions) may be described by separating the motion of the object into horizontal (**x**) and vertical (**y**) components of the vector quantities displacement, velocity and acceleration.

I. Two Dimensional Motion And Trajectories
1.1 Horizontal Projectile

When the initial velocity (speed and direction) of a projectile in the gravitational field of the Earth is given, the subsequent motion of the projectile may be described.

Example Of A Fired Bullet

A projectile fired horizontally has an initial vertical velocity of zero.

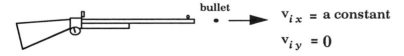

bullet

v_{ix} = a constant

$v_{iy} = 0$

Assuming no friction from the air, there is nothing in the horizontal direction to accelerate the bullet. As soon as the bullet leaves the gun, gravity accelerates the bullet in the vertical direction. The same principle is used in supply drops from an airplane.

Example Of A Falling Package

Suppose that a package is dropped from an airplane which is moving parallel to the Earth. Its horizontal velocity is **v**, the same as the airplane. Once it is dropped, no horizontal force acts on the package; therefore, the horizontal velocity will remain unchanged. Its **x** component (horizontal) remains v until the package hits the ground, as indicated in the illustration below.

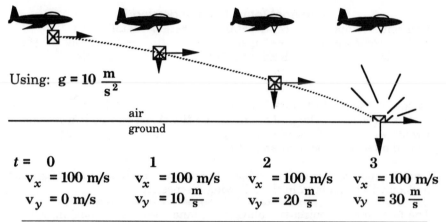

Using: $g = 10 \dfrac{m}{s^2}$

air
ground

$t =$	0	1	2	3
	$v_x = 100$ m/s	$v_x = 100$ m/s	$v_x = 100$ m/s	$v_x = 100$ m/s
	$v_y = 0$ m/s	$v_y = 10 \dfrac{m}{s}$	$v_y = 20 \dfrac{m}{s}$	$v_y = 30 \dfrac{m}{s}$

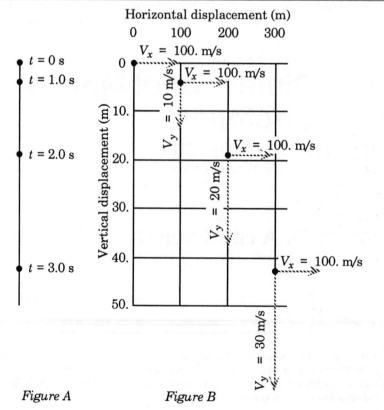

Figure A   *Figure B*

In the *falling package* example (opposite top of page), a plane traveling at 100 meters per second releases a package. When the package is first released, it has no vertical velocity. The vertical pull of gravity will accelerate it downward at a rate of 9.8 m/s². The vertical (**y** component) velocity will increase by 9.8 m/s (or approximately 10 m/s) each second. As shown in the example, the package falls downward with an ever increasing velocity. All of this happens as the package moves horizontally with unchanging velocity v_x.

In the *drop object* example (opposite bottom of page), the graph (A) represents an object at rest dropped from a height of 45 meters. Graph (B) represents an object moving at 100 m/s in the horizontal direction which is then released from a height of 45 meters.

A projectile in flight is doing two things at the same time:

1) It is moving horizontally with constant speed.
2) It is moving up or down with acceleration **g**.

Once you recognize this, the solution of projectile problems is easy. Simply split each problem into two problem parts. One part involves a horizontal motion at constant velocity. For that motion:

$$v_{ix} = v_{fx} = \overline{v}_x$$

The motion equation that is important for the horizontal problem is:

$$\Delta s = \overline{v}_x \, \Delta t$$

The vertical part of the motion is exactly the same as the free-fall motion discussed in the Mechanics Unit.

The following mechanics formulas are applied to trajectory motion:

$$\Delta s_y = v_{iy}\Delta t + \tfrac{1}{2}\,a_y(\Delta t)^2 \quad \text{and} \quad \Delta s_x = v_{ix}\Delta t + \tfrac{1}{2}\,a_x(\Delta t)^2$$

Where:

Δs_y = vertical displacement
a = gravity = -9.8 m/s²
Δt = time in the air
v_{iy} = 0 (zero), when fired horizontally
Δs_x = horizontal displacement
a = zero in horizontal direction

Example

An airplane traveling at 100. m/s drops a package from a height of 3000. meters. Calculate the time it takes to reach the ground. How far in front of the target must the package be dropped? (Example continued on the next page.)

Given: v_{iy} = 0.0 m/s
                 v_{ix} = 100. m/s
                 s_y = -3000. m
                 a_y = g = -9.8m/s^2

Take the downward direction as positive to eliminate negative sign.

Find: (1) time to reach ground
           (2) distance from drop point to target

Solution: *Divide the problem into two portions, vertical and horizontal:*

(1) Vertical: At the beginning, the plane was moving horizontally; therefore: $v_{iy} = 0$

$$\Delta s_y = v_{iy}\Delta t + \tfrac{1}{2} a_y(\Delta t)^2$$

$$3000. \text{ m} = (0)(\Delta t) + \tfrac{1}{2}(9.8 \text{ m/s}^2)(\Delta t^2)$$
$$25 \text{ sec} = t$$

(2) Horizontal: The plane and the package were traveling at a speed of 100m/s horizontally. Since we can ignore friction, $\bar{v}_{ix}$ = 100m/s, there is no force acting on it in the **x** direction; therefore: $a_x = 0 \text{ m/s}^2$

$$\Delta s_x = v_{ix}\Delta t + \tfrac{1}{2} a_x(\Delta t)^2$$
$$\Delta s_x = (100 \text{ m/s})(25s) + 0$$
$$\Delta s_x = 2500 \text{ m}$$

The package should be released 2500 m before the target.

1.2 Projectile Fired At An Angle

For a projectile fired at an angle with the surface of the Earth, the initial vertical and horizontal components of the velocity may be determined and the motion treated as two separate linear motion problems.

Projectile Motion

When an object is fired into the air at an angle, the vertical component of the velocity is accelerated and the horizontal component remains in uniform motion. To solve a projectile problem, first resolve the vector into components. The vertical component, v_{iy}, is calculated using the sine function.

$$\sin \varnothing = v_{iy}/v_i \quad \text{or} \quad v_{iy} = v_i \sin \varnothing$$

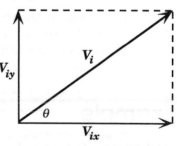

An initial velocity vector separated into horizontal and vertical components.

The horizontal component, v_{ix}, is calculated using the cosine function.

$$\cos \emptyset \;=\; v_{ix}/\,v_i \qquad \text{or} \qquad v_{ix} \;=\; v_i \cos \emptyset$$

Next, determine the time to reach the maximum height at which time velocity final, v_f, equals zero.

Since gravity is acting to slow down the projectile use -9.8 m/s^2 for g.

$$v_{fy} \;=\; v_{iy} + a_y \Delta t \qquad \text{where:} \quad \mathbf{g} = \mathbf{-9.8\ m/s^2}$$

The maximum height can be calculated from the formula:

$$\Delta s_y \;=\; v_{iy}\Delta t + \tfrac{1}{2}\,a(\Delta t)^2$$

Since the time to fall is the same as the time to reach the highest point, the total time in the air is 2t. During the total time in the air, the horizontal velocity remains constant. The distance traveled horizontally can be calculated from the equation:

$$s_x \;=\; v_{ix}2t$$

Example

A projectile is shot into the air at a 30° angle with a velocity of 100 meters per second. Calculate the maximum height, the time in the air and the horizontal range of the projectile.

Given:

$$v_i = 100\ \text{m/s} \qquad \emptyset = 30°$$

Find: time in air, maximum height, horizontal range

Solution:

a)
$$
\begin{aligned}
v_{iy} &= v_i \sin \emptyset = 100\ \text{m/s}(.5)\\
&= 50\ \text{m/s}\\
v_f &= v_{iv} + a\Delta t\\
0 &= 50\ \text{m/s} + (-9.8\ \text{m/s}^2)(\Delta t)\\
t_{up} &= 5.1\ \text{sec time to reach maximum height}\\
\text{total time} &= 2t_{up}\\
&= 10.2\ \text{sec} \approx 10\ \text{sec}
\end{aligned}
$$

from the top down

b)
$$
\begin{aligned}
\Delta s_y &= v_{iy}\Delta t + \tfrac{1}{2}\,a(\Delta t)^2\\
&= 0 + (\tfrac{1}{2})(-9.8\text{m/s}^2)(5.1\text{s})^2\\
&= -127\ \text{m} \quad \text{or} \quad 130\ \text{m down}
\end{aligned}
$$

c)
$$
\begin{aligned}
\text{horizontal } s_x &= v_{ix}\ (\text{total time})\\
&= v_i \cos \emptyset(2t_{up})\\
&= (86.6\text{m/s})10\ \text{s}\\
&= 866\ \text{m} \approx 870\ \text{m}
\end{aligned}
$$

Axis **x** and axis **y** must be perpendicular to each other so that the **x** and **y** motions are independent. The horizontal motion has no force acting on it horizontally so there is uniform motion. The vertical motion has a constant gravitational force and undergoes uniformly accelerated motion. The connection between these independent motions is that they take place simultaneously on the same time scale.

It can be shown that for a given velocity the maximum range of the projectile will occur when Ø = 45°.

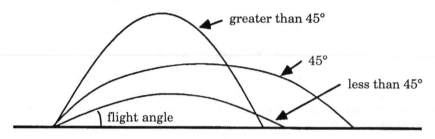

A projectile's maximum distance is obtained at a 45° projection angle (if friction is ignored).

Questions

Base your answers to questions 1 through 3 on the information below.

A rocket is launched at an angle of 60° to the horizontal. The initial velocity of the rocket is 500. meters per second (neglect friction).

1 The vertical component of the initial velocity is
   (1) 250. m/s   (2) 433 m/s   (3) 500. m/s   (4) 1,000 m/s
2 Compared to the horizontal component of the rocket's initial velocity, the horizontal component after 10 seconds would be
   (1) less   (2) greater   (3) the same
3 The horizontal component of the rocket's initial velocity is
   (1) 250 m/s   (2) 300 m/s   (3) 433 m/s   (4) 500 m/s

Use the following description to answer questions 4 through 7.

An object is projected upward at an angle of 45° with an initial velocity of 50 m/s.

4 How high will the object rise?
   (1) 22 m   (2) 42 m   (3) 63 m   (4) 104 m
5 How long will the object be in the air?
   (1) 3.6 s   (2) 6.1 s   (3) 7.2 s   (4) 12.2 s
6 How far will the object travel in the horizontal direction?
   (1) 35.8 m   (2) 132 m   (3) 255 m   (4) 330 m
7 If the projection angle were changed to 30°, the horizontal distance covered would
   (1) decrease   (2) increase   (3) remain the same

Base your answers to questions 8 through 10 on the information below.

A cannon fires a projectile at an angle with the horizontal. The horizontal component of the projectile's initial velocity is 866 meters per second and its initial vertical component is 500. meters per second. (Neglect air resistance.)

8   What is the shape of the path that the projectile will follow?
(1) circular   (2) straight   (3) hyperbolic   (4) parabolic
9   After 5.00 seconds, approximately what is the vertical component of the projectile's velocity?
(1) 450 m/s   (2) 500. m/s   (3) 49 m/s   (4) 0 m/s
10   The maximum height to which the projectile rises is approximately
(1) 2.50×10^3 m   (3) 1.54×10^4 m
(2) 1.25×10^4 m   (4) 4.42×10^4 m

II. Uniform Circular Motion

Uniform circular motion is the motion of an object at constant speed along a circular path.

a) **Centripetal acceleration** is a vector quantity directed toward the center of curvature. Its magnitude is calculated by the formula:

$$a_c = v^2/r$$

Where: a_c represents centripetal acceleration measured in m/s^2
              v represents velocity in m/s
              r is radius measured in meters

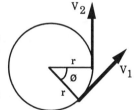

When an object moves in a circle, its velocity is constantly changing. This *change in direction* indicates that the object is being accelerated.

Since velocity is a vector quantity, any change in magnitude (speed) or direction indicates acceleration.

Centripetal acceleration occurs when magnitude and/or direction changes.

b) **Centripetal force** is a vector quantity directed towards the center of the circle. Its magnitude is calculated by using Newton's Second Law of Motion.

$$\begin{aligned} F_c &= ma_c \\ &= mv^2/r \end{aligned}$$

Where: F_c represents the force measured in newtons
              m is the mass in kilograms
              a_c is centripetal acceleration

Centripetal force is the *net force* acting on a body. A body maintains its constant speed since there are no tangential forces. It is *not* in equilibrium.

Example

An object weighing 49. newtons moves in a circular path of radius 0.50meters at a speed of 10. meters per second. Calculate the (a) mass, (b) centripetal acceleration, and (c) the centripetal force.

    Given: $W = 49.\ N$ $r = .50\ m$ $v = 10.\ m/s$

    Find: mass, acceleration, and force

    Solution:

a)
$$m = \frac{W}{g}$$
$$= 49.\ N/9.8\ m/s^2$$
$$= 5.0\ kg$$

b)
$$a_c = \frac{v^2}{r}$$
$$= (10.\ m/s)^2/(.50\ m)$$
$$= 200\ m/s^2$$

c)
$$F_c = \frac{mv^2}{r}$$
$$= (5.0\ kg)(10.\ m/s)^2/(.50\ m)$$
$$= 1000.\ N$$

2.1 Vertical Circles

In the diagram, a plane is going around a vertical loop. Its speed at the two points indicated are v_1 and v_2. We can calculate the force exerted on the pilot of mass m as he passes by each of the points in the loop.

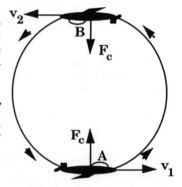

Two forces act on the pilot: (1) the force of gravity (mg) and (2) the push of the seat on him (P). Together, these forces must supply the required centripetal force:

$$F_c = mv^2/r$$

At the bottom of the loop, the seat pushes up with a force P_A and gravity pulls down. Together, these forces must supply F_c which is directed upward.

Therefore: $P_A - mg = F_c$

Because: $F_c = \dfrac{mv_1^2}{r}$

We have: $P_A = mg + m\left(\dfrac{v_1^2}{r}\right)$

The push of the seat must do two things at position **A**. It must support the pilot's weight (**mg**), and it must furnish the required centripetal force ($mv_1{}^2/r$).

The centripetal acceleration adds to the gravitational acceleration. Because of this, people talk about the pilot experiencing "**g-forces.**" If ($mv_1{}^2/r$) was **3 g**, then the pilot would be said to experience an acceleration of **4 g's** as seen when the above equation is rewritten as:

$$P_A = m \left(g + \frac{v_1{}^2}{r} \right)$$

At the top of the loop, the pull of gravity and the push of the seat are in the same direction. Together, they supply the necessary centripetal force.

Therefore: $P_B + mg = F_c$

Substituting for F_c and solving for P_B gives: $P_B = m \left(\dfrac{v_2{}^2}{r} \right) - mg$

The seat supplies less than the required centripetal force. The pull of gravity is helping the seat hold the pilot in a circular path.

P can become zero. When this occurs, the plane is moving slow enough so that the centripetal acceleration v^2/r just equals **g** (the gravitational acceleration). At that instant the pilot is falling freely. What would happen if the plane was going slower than this?

Questions

1   Which diagram represents a tangential velocity vector of an object in orbit?

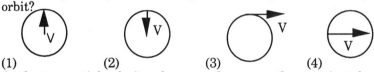

   (1)   (2)   (3)   (4)

2   As the tangential velocity of a mass decreases, the centripetal acceleration of the mass
   (1) decreases
   (2) increases
   (3) remains the same

3   The magnitude of centripetal force is directly proportional to
   (1) mass, only   (3) velocity, only
   (2) acceleration, only   (4) mass and acceleration

4   An object is traveling in a circular path at 3 meters per second. If the radius is 6 meters, the centripetal acceleration of the object will be
   (1) 1 m/s^2 (2) 0 m/s^2 (3) 1.5 m/s^2 (4) 13 m/s^2

5   A ball travels in a circle at a speed of 6 meters per second on the end of a cord which is 3meters long. The centripetal acceleration of the ball will be
   (1) 18 m/s^2 (2) 12 m/s^2 (3) 3 m/s^2 (4) 54 m/s^2

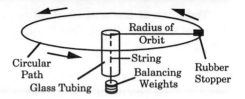

Base your answers to questions 6 through 8 on the diagram which shows a rubber stopper being whirled in a circular path.

6 If the number of balancing weights remains the same while the speed of rotation increases, the radius of the orbit
(1) decreases   (2) increases   (3) remains the same

7 As the speed of rotation increases, the number of weights required to balance the rubber stopper at the same orbital radius will
(1) decrease   (2) increase   (3) remain the same

8 The radius of the circular path is increased with constant speed of rotation. The number of weights required to balance the rubber stopper at the new orbital radius will
(1) decrease   (2) increase   (3) remain the same

9 As a car rounds a curve, the centripetal force holding it on the road will be directed
(1) inward towards the center of curvature
(2) outward away from the center of curvature
(3) forward in the direction of the car's motion
(4) backward away from the direction of the car's motion

Base your answers to questions 10 and 11 on the diagram which represents a 4.0×10^2 - kilogram satellite, S, in a circular orbit at an altitude of 5.6×10^6 meters. The orbital speed of the satellite is 5.7×10^3 meters per second and the radius of the Earth, R, is 6.4×10^6 meters.

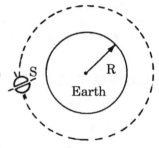

10 The centripetal acceleration of the satellite is approximately
(1) 9.8 m/s^2 (3) 2.7 m/s^2
(2) 4.9 m/s^2 (4) 1.4 m/s^2

11 If the altitude of the satellite decreased, its centripetal acceleration would
(1) decrease   (2) increase   (3) remain the same

Base your answers to questions 12 and 13 on the information below:

A 4 kilogram mass is traveling in uniform circular motion at 2 meters per second on a 1-meter cord.

12 The centripetal acceleration of the mass is
(1) 1 m/s^2 (2) 2 m/s^2 (3) 16 m/s^2 (4) 4 m/s^2

13 The centripetal force is
(1) 1 newton   (3) 16 newtons
(2) 2 newtons   (4) 4 newtons

14 An object travels in a circular orbit. If the speed of the object is doubled, its centripetal acceleration will be
(1) halved   (2) doubled   (3) quartered   (4) quadrupled

15 In the diagram, satellite S moves in a clockwise
    circular orbit around the Earth (E). The direction of
    the acceleration of the satellite is toward point
    (1) A
    (2) B
    (3) C
    (4) D

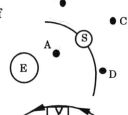

Base your answers to questions 16 through 20 on the
diagram and information below:

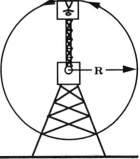

    At an amusement park, a passenger whose mass
is 50. kilograms rides in a cage. The cage has a
constant speed of 10. meters per second in a
vertical circular path of radius R, equal to
10.meters.

16 What is the magnitude of the centripetal acceleration of the passenger?
    (1) 1.0 m/s^2 (3) 5.0 x 10^2 m/s^2
    (2) 2.0 x 10^3 m/s^2 (4) 10. m/s^2
17 What is the direction of the centripetal acceleration of the passenger?
    (1) to the left   (3) up
    (2) to the right   (4) down
18 What is the centripetal acceleration of the passenger at the instant the
    cage reaches the highest point in the circle?
    (1) 1.0 m/s^2 (3) 5.0 x 10^2 m/s^2
    (2) 0 m/s^2 (4) 10. m/s^2
19 What does the 50.-kilogram passenger weigh at rest?
    (1) 1600 N   (2) 490 N   (3) 50. N   (4) 0 N
20 What is the magnitude of the centripetal force acting on the passenger?
    (1) 0 N   (2) 50. N   (3) 4.9 x 10^2 N   (4) 5.0 x 10^2 N

2.2 Kepler's Laws

    Throughout the last few decades of the sixteenth century, **Tycho Brahe**
made precise measurements of the positions of the planets and various other
bodies in the Solar System. **Johannes Kepler** made detailed analysis of the
measurements and by 1619 had announced three laws which describe planetary motion.

    1)   The path of each planet is an ellipse which has the Sun at one of its
        foci. A circle is a special case of an ellipse when the two foci coincide
        at the center of the circle. Neptune will be further from the Sun than
        Pluto until 1996.

    2)   Each planet moves in such a way that the (imaginary) line joining it
        to the Sun sweeps out equal areas in equal times. The sun is closest
        to the Earth in January and furthest away in July.

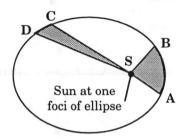

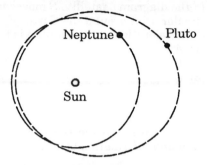

Kepler's Second Law
The average speed of a planet between points A and B is greater than its average speed between points C and D.

Area ABS = Area CDS

Orbits of Neptune & Pluto
Since Pluto's orbit is more elliptical than Neptune's, Pluto is closer to the Sun while traveling over part of its orbit than Neptune is at any point in its orbit.

The upper left figure illustrates **Kepler's Second Law**, but it gives an exaggerated idea of the eccentricity of most planetary orbits. With the exception of Mercury and Pluto, most planets follow very nearly circular paths.

3) The ratio of the mean radius of the orbit cubed to the orbital period of its motion squared is the same for all planets. The value for this ratio, R^3/T^2, for all planets orbiting our Sun is 3.35×10^{18} m³/s².

$$\frac{R^3}{T^2} = K$$

Where **K** is a constant for this ratio. **K** is not a universal constant. It is a constant for satellites of a particular body being orbited, such as the Sun, a star, a planet, etc.

It can be shown that **Kepler's Third Law** is consistent with **Newton's Law of Universal Gravitation**.

$$F_B = F_c \quad \text{or} \quad \frac{Gm_p m_s}{R^2} = m_p \frac{v^2}{R}$$

Where: m_p is the mass of the planet F_B is the force at position B
         m_s is the mass of the Sun F_C is the centripetal force

Substituting: $v = 2\dfrac{R}{T}$

Then: $\dfrac{Gm_s}{R^2} = \dfrac{(2R/T)^2}{R} = 4^2\dfrac{R}{T^2}$ or $\dfrac{R^3}{T^2} = \dfrac{Gm_s}{4^2}$

Note that the constant depends on the mass of the body being orbited, but not on the mass of the planet.

2.3 Earth As A Planet

The Earth's orbit is an ellipse. On January 3rd, perihelion, the Earth is 147 million kilometers from the Sun. On July 4th, aphelion, the Earth is 152 million kilometers from the Sun. The Earth's axis is tilted at 23.5° to a perpendicular to the plane of its orbit. This angle is maintained throughout the Earth's revolution about the Sun. In July in the Northern hemisphere, the Earth is tilted towards the Sun. In January in the Northern hemisphere, the Earth is tilted away from the Sun. The angle at which the Sun's rays strike the Earth, rather than the distance from the Sun, determines the Earth's seasonal changes in temperatures.

A beam of sunlight striking the surface at 90° delivers twice as much energy per square meter as an angle of 30°. Therefore, in the Northern Hemisphere we have summer when the Sun is further away from the Earth than it is in the winter.

2.4 Satellite Motion

A **satellite** is defined as a smaller body which revolves around a larger one. Satellites of the Sun include the planets. Satellites of the Earth include the Moon and human made objects placed in orbits about the Earth. Satellite orbits around the Earth are frequently circular.

You may remember from the study of projectiles fired horizontally, that the greater the initial speed, the farther the body will travel before hitting the ground. If a projectile was fired with sufficient speed, assuming no air friction, it could circle the Earth without ever landing. The reason is that the Earth's surface is curved. There exists a launching speed in which the Earth will curve away just as rapidly as the object falls towards it. The projectile will remain in a circular orbit always accelerating toward the Earth but never getting any closer.

At a speed of 7,900 m/s, a satellite will remain in a circular orbit around the Earth. As speeds drop below 7,900 m/s, the acceleration of gravity becomes greater than the centripetal acceleration v^2/r, and the satellite cannot remain in a circular orbit. At speeds greater than 7,900 m/s, the satellite will be moving too fast to remain in a circular orbit and rise higher above the Earth. Additional height *increases* the potential energy of the satellite and *decreases* the kinetic energy. It will then orbit in an elliptical rather than in a circular orbit.

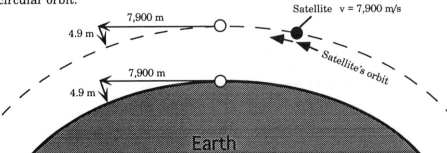

The above illustration represents the displacement of an Earth satellite in a low circular orbit compared with the curvature of the Earth's surface. The Earth's surface curves away from a tangent to it at the rate that the satellite falls toward the Earth. (not drawn to scale)

At sufficiently high speeds, the satellite moves farther and farther away from the Earth until the gravitational field is so weak that the object will continue to travel into space and never return. This speed is called the **escape velocity**.

A **geosynchronous orbit** is one in which the period of the satellite is the same as the period of the Earth's rotation around its own axis. In a geosynchronous orbit, the satellite always remains the same position over a point on the Earth's equator.

The radius of orbit of a human-made satellite of the Earth may be determined by equating R^3/T^2 for the Moon and R^3/T^2 for the satellite.

Using Kepler's Third Law:
$$\frac{R_m^3}{T_m^2} = \frac{R_s^3}{T_s^2}$$

The Moon's orbit = 2.35×10^6 seconds; R_{Moon} = 3.84×10^8 m

To determine the radius for a geosynchronous orbit, substitute

$$T_s = 8.64 \times 10^4 s = 24 \text{ hours}$$

The value obtained will be from the center of the Earth. To determine the distance of the satellite from the Earth's surface, subtract the radius of the Earth, 6.38×10^6m. It is also possible to determine the period or the speed of the satellite, if the radius of the orbit is known.

2.5 Escape Velocity

As an object is projected upward from the surface of the Earth, the height which the object ultimately obtains depends upon the speed at which it is projected. If an object were required to escape from the Earth, it would have to be projected with a velocity which is at least great enough for the object to reach infinity before coming to rest. The *minimum* velocity that achieves this is known as the **escape velocity**.

The work done in moving an object from the surface of the Earth to infinity is given by:

$$W = G \frac{mm_e}{r_e}$$

The escape velocity can be calculated by:

$$\tfrac{1}{2}mv^2 = G \frac{mm_e}{r_e} \qquad \text{i.e.} \qquad v = \sqrt{\frac{2\,Gm_e}{r_e}}$$

This calculation applies only to bodies which are not being driven, for example projectiles. The escape velocity does not depend on the direction of the projectile because the kinetic energy a body loses in reaching any particular height depends only on the height concerned and not on the path taken. The mass of the object is not important. As long as the speed is 11 km/s, the object will escape from the Earth.

Questions

1 As the distance between two objects decreases, the gravitational force of attraction between them will
   (1) decrease   (2) increase   (3) remain the same
2 If the mass of one of two objects is increased, the force of attraction between them will
   (1) decrease   (2) increase   (3) remain the same
3 The planet Mars is orbiting the Sun in an elliptical path. As the distance between Mars and the Sun decreases, the speed of the orbiting planet
   (1) decreases   (2) increases   (3) remains the same
4 An Earth satellite makes 3 complete revolutions about the Earth in 6hours. The period of revolution for satellite is
   (1) ½ hour   (2) 2 hours   (3) 9 hours   (4) 18 hours
5 What is the shape of a natural orbit?
   (1) circular   (2) parabolic   (3) elliptical   (4) hyperbolic
6 The centripetal force that holds the Moon in its orbit is directed towards
   (1) Earth   (2) the Moon   (3) the Sun   (4) Mars
7 To achieve weightlessness, an object and its surroundings must have the same
   (1) mass   (2) speed   (3) inertia   (4) acceleration
8 As the radius of a satellite's orbit is decreased, its speed will
   (1) decrease   (2) increase   (3) remain the same
9 As a satellite moves from orbit A to orbit B as shown in the diagram, its period
   (1) decreases
   (2) increases
   (3) remains the same

10 In order for a rocket to go beyond the pull of the Earth's gravitational force, it must reach
    (1) orbital velocity   (3) orbital speed
    (2) escape velocity   (4) circular speed
11 As a rocket rises above the surface of the Earth, the gravitational force on the rocket
    (1) decreases   (2) increases   (3) remains the same
12 The gravitational force between two objects is inversely proportional to
    (1) mass squared   (3) mass
    (2) distance squared   (4) distance

13 The diagram at the right shows positions of a satellite as it orbits the Earth. At which position will the satellite achieve its highest velocity?
    (1) A
    (2) B
    (3) C
    (4) D

14 The orbital period for a satellite in geosynchronous orbit around the Earth is
    (1) one hour   (2) one day   (3) one month   (4) one year

15 Two 1 kilogram masses are moving apart at a constant speed. Which graph best represents the force of gravitational attraction between the two masses during the time they are moving?

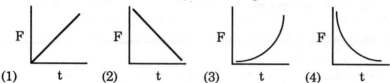

F   (1)   t   F   (2)   t   F   (3)   t   F   (4)   t

16 Two identical planets are orbiting the Sun. The mean radius of the orbit of the second planet is four times the mean radius of the orbit of the first planet. Compared to the orbital period of the first planet, the period of the second planet will be
   (1) one-fourth as great
   (2) one-half as great
   (3) eight times greater
   (4) four times greater

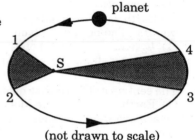

(not drawn to scale)

17 The diagram above represents the motion of a planet around the Sun, *S*. The time it takes the planet to go from point *1* to point *2* is identical to the time it takes the planet to go from point *3* to point *4*. Which statement must be true?
   (1) The two shaded regions of the diagram have equal areas.
   (2) The centripetal acceleration of the planet is constant.
   (3) The planet moves at a constant speed.
   (4) The planet moves faster when it is farthest from the Sun.

Thinking Physics

1 How can you explain that a bullet fired horizontally and a bullet dropped from the same height hit the ground at the same time?

2 How can the planet Neptune be further away from the Sun than Pluto?

3 Why is it warmer in the summer when the Sun is further away from the Earth than in the winter?

4 If you jump off a diving board, while holding a 5 kg weight, you would observe that the weight seems to become "weightless." Can you explain this?

5 Is an astronaut ever "weightless" in space?

Self–Help Questions

1   A projectile is launched at an angle of 60° above the horizontal.
    Compared to the vertical component of the initial velocity of the
    projectile, the vertical component of the projectile's velocity when it has
    reached its maximum height is
    (1) less   (2) greater   (3) the same

2   A ball is fired vertically upward at 5.0 meters per second from a cart
    moving horizontally to the right at 2.0 meters per second. Which best
    represents the resultant velocity of the ball when fired?

    (1)   (2)   (3)   (4)

3   A 1-kilogram object is thrown horizontally and a 2-kilogram object is
    dropped vertically at the same instant and from the same point above the
    ground. If friction is neglected, at any given instant both objects will
    have the same
    (1) kinetic energy   (3) total velocity
    (2) momentum   (4) height

Base your answers to questions 4
through 6 on the diagram at the right
which represents a ball being kicked by
a foot and rising at an angle of 30.° from
the horizontal. The ball has an initial
velocity of 5.0 meters per second.
(Neglect friction.)

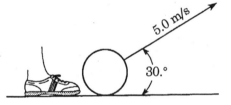

4   What is the magnitude of the horizontal component of the ball's initial
    velocity?
    (1) 2.5 m/s   (2) 4.3 m/s   (3) 5.0 m/s   (4) 8.7 m/s

5   As the ball rises, the vertical component of its velocity
    (1) decreases   (2) increases   (3) remains the same

6   If the angle between the horizontal and the direction of the 5.0-m/s veloci-
    ty decreases from 30.° to 20.°, the horizontal distance the ball travels will
    (1) decrease   (2) increase   (3) remains the same

7   The diagram at the right shows an object traveling
    clockwise in a horizontal, circular path at constant
    speed. Which arrow best shows the direction of the
    centripetal acceleration of the object at the instant
    shown?

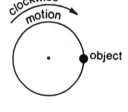

    (1)   (2)   (3)   (4)

8 A motorcycle of mass 100 kilograms travels around a flat, circular track of radius 10 meters with a constant speed of 20 meters per second. What force is required to keep the motorcycle moving in a circular path at this speed?
(1) 200 N   (2) 400 N   (3) 2000 N   (4) 4000 N

9 Two masses, *A* and *B*, move in circular paths as shown in the diagram at the right. The centripetal acceleration of mass *A*, compared to that of mass *B*, is
(1) the same
(2) twice as great
(3) one-half as great
(4) four times as great

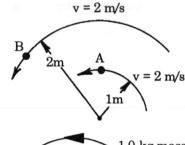

Base your answers to questions 10 through 12 on the diagram at the right which shows an object with a mass of 1.0 kilogram attached to a string 0.50meter long. The object is moving at a constant speed of 5.0meters per second in a horizontal circular path with center at point *O*.

10 What is the magnitude of the centripetal force acting on the object?
(1) 2.5 N   (2) 10. N   (3) 25 N   (4) 50. N

11 While the object is undergoing uniform circular motion, its acceleration
(1) has a magnitude of zero
(2) increases in magnitude
(3) is directed toward the center of the circle
(4) is directed away from the center of the circle

12 If the string is cut when the object is at the position shown, the path the object will travel from this position will be
(1) toward the center of the circle
(2) a curve away from the circle
(3) a straight line tangent to the circle

13 The path of a satellite orbiting the Earth is best described as
(1) linear   (2) hyperbolic   (3) parabolic   (4) elliptical

14 A satellite orbits the Earth in a circular orbit. Which statement best explains why the satellite does not move closer to the center of the Earth?
(1) The gravitational field of the Earth does not reach the satellite's orbit.
(2) The Earth's gravity keeps the satellite moving with constant velocity.
(3) The satellite is always moving perpendicularly to the force due to gravity.
(4) The satellite does not have any weight.

15 The Earth is closest to the Sun during January and farthest from the Sun during July. During which month is the gravitational potential energy of the Earth with respect to the Sun the greatest?
(1) January   (2) March   (3) July   (4) September

16 In the diagrams below, P represents a planet and S represents the Sun. Which best represents the path of planet P as it orbits the Sun? (The diagrams are not drawn to scale.)

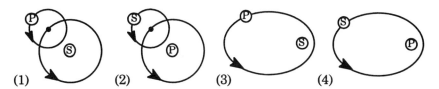

(1)   (2)   (3)   (4)

17 As the distance of a satellite from the Earth's surface increases, the time the satellite takes to make one revolution around the Earth
(1) decreases   (2) increases   (3) remains the same

18 Which condition is required for a satellite to be in the geosynchronous orbit about the Earth?
(1) The period of revolution of the satellite must be the same as the rotational period of the Earth.
(2) The altitude of the satellite must be equal to the radius of the Earth.
(3) The orbital speed of the satellite around the Earth must be the same as the orbital speed of the Earth around the Sun.
(4) The daily distance traveled by the satellite must be equal to the circumference of the Earth.

Base your answers to questions 19 through 23 on the diagram at the right which represents a satellite in an elliptical orbit about the Earth. The highest point, A, is four Earth radii ($4R$) from the center of the Earth. The lowest point, B, is two Earth radii ($2R$) from the center of the Earth. The mass of the satellite is 3.0×10^6 kilograms.

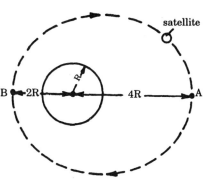

19 Which vector represents the direction of the satellite's velocity at A?

(1)   (2)   (3)   (4)

20 Which vector represents the direction of the centripetal force on the satellite at B?

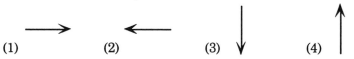

(1)   (2)   (3)   (4)

21 As the satellite moves from point A toward point B, the velocity of the satellite
(1) decreases   (2) increases   (3) remains the same

22 Neglecting atmospheric friction, as the satellite moves from A to B, its
    total mechanical energy
    (1) decreases   (2) increases   (3) remains the same
23 Compared to the magnitude of the force of the satellite on the Earth, the
    magnitude of the force of the Earth on the satellite is
    (1) less   (2) greater   (3) the same
24 For planets orbiting the Sun, the ratio of the mean radius of the orbit
    cubed to the orbital period of motion squared is
    (1) greatest for the most massive planet
    (2) greatest for the least massive planet
    (3) constantly changing as the planet rotates
    (4) the same for all planets
25 The diagram at the right shows the
    movement of a planet around the Sun.
    Area 1 equals area 2. Compared to the
    time the planet takes to move from C to
    D, the time it takes to move from A to B
    is
    (1) less
    (2) greater
    (3) the same

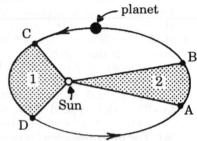

Unit 2 — Energy

Important Terms To Be Understood

work	watt	spring constant
energy	potential energy	elastic potential energy
joule	kinetic energy	Conservation of Energy
power	Hooke's Law	conservative force
horsepower	non-conservative force	work–energy relationship

I. Work And Energy

Our modern society is dependent upon energy for its existence. We need energy to do the many kinds of work that our civilization requires. Today, energy is in short supply.

This unit deals primarily with the concept of *mechanical energy*. There are other forms of energy such as *thermal, chemical, nuclear, and electromagnetic*. Energy can be transformed from one form to another.

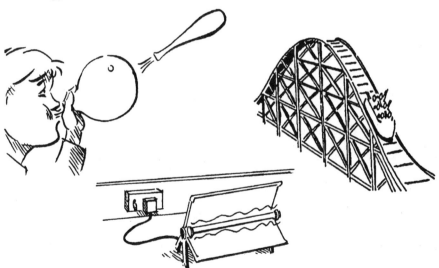

1.1 Work: Technical Meaning

In general usage, the word covers a great variety of occupations and jobs. In scientific terms, the term "work" has a definite meaning. Scientists say that work is done whenever a force causes movement. Work is done on an object when a force displaces an object. Work is a scalar quantity. The SI unit for work is the **joule (J)**. In fundamental units, one joule is a kilogram meter per second squared. Energy is needed to do work.

Pushing a wall produces:	Pulling a box produces:
No Movement, No Work	***Movement and Work***

1.2 Work: How To Measure

The quantity of work done depends on both the size of the force and how far the object moves. Work is defined as:

Work = force · displacement (moved in direction of the force) **= F∆s**

To move a force of 1 N through 1 m requires 1 newton-meter of work. A newton-meter is given the name joule. 1 joule (J) of work is done when a force of 1 N moves through 1 meter.

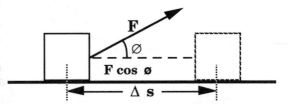

In the above illustration, the force causes the block to slide a distance (∆s) along the floor. The component of the force in the direction of the displacement is **F cos ø**. The more general way to write the work equation is

$$W = F∆s \cos ø$$

The work equation can be stated in words as follows:

The work done by a force (**F**) during a displacement (**∆s**) is equal to the product of the three quantities:

1) the force magnitude,
2) the magnitude of the displacement, and
3) the cosine of the angle between them.

```
Practical Applications
•   an elevator
•   block and tackle
•   jack lifting a car
```

Examples

1. How much work is done in slowly lifting a 5.0 kg box from the floor to a height of 1.2 m above the floor?

Given: $\mathbf{m}$ = 5.0 kg
$\Delta\mathbf{s}$ = $\mathbf{h}$ = 1.2 m

Find: Work against gravity

$$W = F\Delta s$$

Solution: F = mg = (5.0 kg)(9.8 m/s^2) = 49 N
W = $F\Delta s$ = (49 N)(1.2 m) = 58.8 J = 59 J
W = 59 J

2. A 600 N force pulls a 50 kg block a distance of 2.0 m along the floor. The force acts at a 37° angle to the horizontal. Calculate the work done.

Given: $\mathbf{F_x}$ = $\mathbf{F}$ cos ø = 600 N (cos 37°) = 480 N
$\Delta\mathbf{s}$ = 2.0 m

Find: Work

Solution: W = FΔs = (480 N)(2.0 m) = 960 J

1.3 Force vs Displacement Graph

The area under a force versus displacement graph is the work done by the force. Consider a block that is pulled at constant speed along a table with a force of 10. N over a distance of 1.0 meter. The force is constant in this example.

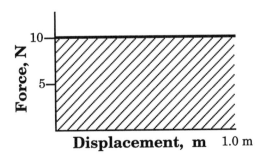

Area = **base x height**
= 1.0 m x 10 N
= 10. N·m
= 10. J

In this case, the 10. joules of work was done against friction and was changed to thermal energy or heat.

In the graph below, the force is proportional to the elongation. The area under the curve represents the work done by the force.

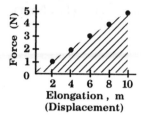

Area = ½base x height

= ½(10 m) 5 N

= 25. N·m

= 25. J

Questions

1   The amount of work done by a force of 20 newtons acting through a distance of 2 meters is
(1) 0.1 joules   (2) 10 joules   (3) 22 joules   (4) 40 joules

2   An object weighing 12 newtons is raised vertically 3 meters. The work done on the object is
(1) 12 J   (2) 36 J   (3) 3 J   (4) 4 J

3   A force of 80. newtons pushes a 50.-kilogram object across a level floor for 8.0meters. The work done is
(1) 10 joules   (2) 400 joules   (3) 640 joules   (4) 3,920 joules

4   A constant force of 20. newtons applied to a box causes it to move at a constant speed of 4.0 meters per second. How much work is done on the box in 6.0seconds?
(1) 480 joules   (2) 240 joules   (3) 120 joules   (4) 80. joules

5   An object has a mass of 8.0 kilograms. A 2.0-newton force displaces the object a distance of 3.0 meters to the east, and then 4.0 meters to the north. What is the total work done on the object?
(1) 10. joules   (2) 14 joules   (3) 28 joules   (4) 56 joules

6   A force 100. newtons is used to push a trunk to the top of an incline 3.0 meters long. Then a force of 50. newtons is used to push the trunk for 10. meters along a horizontal platform. What is the total work done on the trunk?
(1) 8.0×10^2 joules   (3) 3.0×10^2 joules
(2) 5.0×10^2 joules   (4) 9.0×10^2 joules

7   In the diagram, 55 joules of work is needed to raise a 10.newton weight 5.0 meters at a constant speed. How much work is done to overcome friction as the weight is raised?
(1) 5 J
(2) 5.5 J
(3) 11 J
(4) 50. J

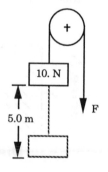

8   One joule is equivalent to one

(1) $\dfrac{\text{newton}}{\text{meter}^3}$

(2) kilogram · meter³

(3) watt² · newton

(4) $\dfrac{\text{kilogram} \cdot \text{meter}^2}{\text{second}^2}$

9   A 2.2 -kilogram mass is pulled by a
    30.newton force through a distance of
    5.0meters as shown in the diagram.
    What amount of work is done?

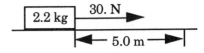

(1) 11 J   (3) 150 J
(2) 66 J   (4) 330 J

II. Power

Power is defined in physics as the time rate of doing work.

$$\text{Power} = \frac{\text{Work done}}{\text{time required}} = \frac{W}{t} = \frac{\text{joules}}{\text{sec}} = \text{watts}$$

Suppose that there were two cranes available for loading a ship. No matter which is chosen to do the job of loading, the work to be done remains the same. But if one crane could load the ship twice as fast as the other one that crane would have twice the power.

The rate of working at 1 joule per second is called a watt. Power is a scalar quantity. It is the slope of a work vs time graph. One kilowatt is equal to 1000watts. These units of power and energy do not only apply to machines and mechanical energy, but are used for the measurement of all forms of energy and power. For example, electrical energy is measured in joules and electrical power in watts.

2.1 Early Types Of Engines

When James Watt invented the steam engine people naturally wanted to know how these new engines compared with the horses previously used to do the same sort of work. How many horses would one of his engines replace? Watt therefore decided to express the power of his engines in terms of the rate at which a horse could do work. To do this he got a horse and measured the rate at which it could pull a known weight up a mine shaft.

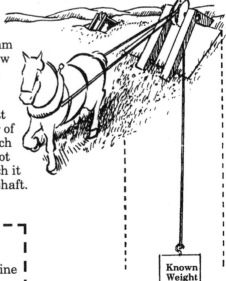

┌─────────────────────────────────────┐
│ **Practical Applications**   │
│ ·   running up stairs   │
│ ·   horsepower rating of a car engine │
└─────────────────────────────────────┘

Examples

1. If the horse could walk at 1.49 m/s when raising a load of 500 N, how much work could it do every second? The power developed could be calculated from the formula:

$$P = \frac{W}{t} = \frac{F\Delta s}{t} = F\left[\frac{\Delta s}{t}\right] = F\overline{v}$$

The expression, $F\Delta s/t$, also represents average power generated when the speed of the object to which the force is applied varies with time.

Given: $\overline{F} = 500$ N
$\overline{v} = 1.49$ m/s

Find: Power

Solution: $P = F \times \overline{v} = (500 \text{ N}) (1.49 \text{ m/s}) = 745 \text{ N-m / s}$
$P = 745 \text{ J/s} = 745 \text{ watts}$

2. Suppose a ¼ horsepower motor (shown right) lifts its load at a speed of 0.20 m/s. What is the maximum load it could lift at this speed? (1 hp = 746 watts)

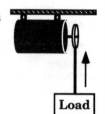

Given: $v = 0.20$ m/s

Ideally, ¼ hp can do work at the rate of:

$$¼ (746)\text{J/s} = 187 \text{ J/s}$$

Find: F

Solution:
$$P = F\overline{v}$$

$$F = \frac{P}{\overline{v}} = \frac{187 \text{ J/s}}{.20 \text{ m/s}} = 935 \frac{J}{m}$$

$$F = 935 \text{ N}$$

Note: Because of friction and other energy loses, the actual maximum load would be less than this.

The power developed by any system, whether mechanical, electrical, or otherwise is measured in watts. Power times time gives energy. When we light a 60 watt light bulb for 60 seconds, we are consuming electrical energy according to the equation:

Energy (W)	=	Power (P) x time (t)
	=	60 watts x 60 sec
	=	60 joules / sec x 60 sec
	=	3600 joules

We purchase **energy** from the electric company. The unit watt-second or kilowatt-hour is a unit of electrical energy.

Questions

1 A machine performs 10 joules of work in 2 seconds. The power developed by this machine is
   (1) 5 watts   (2) 8 watts   (3) 12 watts   (4) 20 watts

2 The unit for power is the
   (1) kilogram   (2) joule   (3) watt   (4) newton

3 The rate at which work is being done is called
   (1) energy   (3) momentum
   (2) force   (4) power

4 Five watts of power is developed by a motor that does 20 joules of work in what amount of time?
   (1) 100 s   (2) 0.5 s   (3) 25 s   (4) 4 s

5 A motor rated at 100. watts accelerates an object along a horizontal frictionless surface with an average speed of 4.0 meters per second. What force is supplied by the motor in the direction of motion? (Assume 100% efficiency.)
   (1) 0.04 N   (2) 25 N   (3) 100 N   (4) 400 N

6 One elevator lifts a mass a given height in 10 seconds and a second elevator does the same work in 5 seconds. Compared to the power developed by the first elevator, the power developed by the second elevator is
   (1) half as great   (3) the same
   (2) twice as great   (4) four times as great

7 A crane raises a 200-newton weight to a height of 50 meters in 5 seconds. The crane does work at the rate of
   (1) 8×10^{-1} watt   (3) 2×10^3 watts
   (2) 2×10^{-1} watts   (4) 5×10^4 watts

8 If 700 watts of power is needed to keep a boat moving through the water at a constant speed of 10 meters per second, what is the magnitude of the force exerted by the water on the boat?
   (1) 0.01 N   (2) 70 N   (3) 700 N   (4) 7,000 N

9 Car *A* and car *B* are of equal mass and travel up a hill. Car *A* moves up the hill at a constant speed that is twice the constant speed of car *B*. Compared to the power developed by car *B*, the power developed by car *A* is
   (1) the same   (3) half as much
   (2) twice as much   (4) four times as much

10 A weightlifter lifts a 2,000 -newton weight a vertical distance of 0.5meter in 0.1second. What is the power output?
   (1) 1×10^{-4} W   (3) 1×10^4 W
   (2) 4×10^{-4} W   (4) 4×10^4 W

11 A machine raises a 160 N weight 4.00 m when an effort of 40.0 N acts over a distance of 20.0 m. The work lost in friction is
   (1) 120 N · m   (2) 160 N · m   (3) 200 N · m   (4) 640 N · m

12 A 10,000 watt motor operates an elevator weighing 5,000 N. Assuming no frictional losses, how high is the elevator raised in 10 seconds?
   (1) 2 m   (2) 20 m   (3) 50 m   (4) 100 m

13 A 2.2 -kilogram mass is pulled horizontally by a 30. newton force through a distance of 5.0 meters. What amount of work is done?
   (1) 11 J   (2) 66 J   (3) 150 J   (4) 330 J

III. Energy

Energy is the ability to do work. The SI unit for energy is the joule (J). A unit of energy, the joule, is the same as a unit of work. Energy can take many forms, such as electrical, nuclear, heat, chemical, light, and sound. Energy is a scalar quantity.

3.1 Kinetic Energy

Kinetic energy (**KE**) is the energy an object has because of its motion. The kinetic energy is calculated from the formula:

$$KE = \tfrac{1}{2}mv^2$$

Where: **KE** = kinetic energy measured in joules
m = mass of moving object in kilograms
v = velocity of object in m/s

3.2 Potential Energy

Potential energy (**PE**) is the energy that results from the position or state of a body. Wound-up clock springs and stretched elastic bands possess potential energy resulting from change of shape or size. In each case, the stored potential energy may later be released.

Under ideal conditions, the change in potential energy is equal to the work required to bring the object to that position or condition. Potential energy is a scalar quantity.

3.3 Gravitational Potential Energy

If work is done **on** an object against gravitational force, there is an increase in the gravitational potential energy of the system.

Any object at a position above some reference level is said to have potential energy relative to the reference level. If the reference level is the floor, a ball on the table has potential energy relative to the floor. If the reference level is the table, the ball which is resting on the table has no potential energy relative to the table.

If work is done **by** gravitational force on an object, there is a decrease in the gravitational potential energy of the system. The change in gravitational potential energy of an object is equal to the product of its weight and the vertical change in height.

$$\Delta PE = mg\Delta h$$

Where: Δ**PE** - potential energy in joules
mg = weight (**w**) or **gravitational force**
Δ**h** - height (which is small compared to the radius of the Earth)

Note: Gravitational potential energy depends on a reference level or point. It really makes no difference which reference point is used in a problem, but once the level is chosen, it must not be changed during the discussion. Usually, the lowest point is used as the reference level.

3.4 Elastic Potential Energy

When work is done on a spring in compressing or stretching it, potential energy is stored in the spring. The data was collected in a Hooke's Law demonstration. Then, it was plotted on a force - elongation graph below.

Data Table	
Force (N)	Elongation (m)
1.0	2.0
2.0	4.0
3.0	6.0
4.0	8.0
	10.

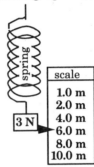

As weights are added to a spring, a set of data similar to that above is generated. The spring behaves according to Hooke's Law. Hooke's Law states that the force is proportional to the elongation (displacement) or in formula form:

$$F = kx$$

Where: **F** is the force in newtons
k is the spring constant in **N/m**
x is elongation or displacement in meters.

$$
\begin{aligned}
\text{Area} \quad &= \quad \tfrac{1}{2}bh \\
&= \quad \tfrac{1}{2}(10.\ \text{m})5.0\ \text{N} \\
&= \quad 25.\ \text{N·m} \\
&= \quad 25.\ \text{J}
\end{aligned}
$$

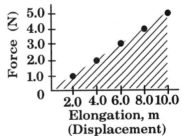

When the data is graphed (above), the slope of the line represents the spring constant (**k**) and the area under the curve represents the stored potential energy of the spring.

$$PE_s = \tfrac{1}{2}kx^2$$

Practical Applications
- toys
- mattress (bed) springs
- spring scale
- shock absorber

Questions

1   A mass of 10 kilograms has a speed of 5 meters per second. The kinetic energy of the mass is
    (1) 25 J   (2) 50 J   (3) 125 J   (4) 250 J

2   As an object moves upward at a constant speed, its kinetic energy
    (1) decreases   (2) increases   (3) remains the same

3   A 10. kilogram object and a 5.0 kilogram object are released simultaneously from a height of 50. meters above the ground. After falling freely for 2.0 seconds, the objects will have different
    (1) accelerations   (3) kinetic energies
    (2) speeds   (4) displacements

4   Which cart shown below has the greatest kinetic energy?

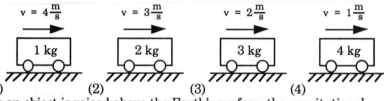

(1)   (2)   (3)   (4)

5   As an object is raised above the Earth's surface, the gravitational potential energy of the object - Earth system
    (1) decreases   (2) increases   (3) remains the same

6   As the velocity of an object falling toward the Earth increases, the gravitational potential energy of the object with respect to the Earth
    (1) decreases   (2) increases   (3) remains the same

7   As an object falls towards the Earth, its kinetic energy
    (1) decreases   (2) increases   (3) remains the same

8   The diagram represents a block suspended from a spring. The spring is stretched 0.200 meter. If the spring constant is 200.newtons per meter, what is the weight of the block?
    (1) 40.0 N
    (2) 20.0 N
    (3) 8.00 N
    (4) 4.00 N

9   Which mass has the greatest potential energy with respect to the floor?
    (1) 50-kg mass resting on the floor
    (2) 2-kg mass 10 meters above the floor
    (3) 10-kg mass 2 meters above the floor
    (4) 6-kg mass 5 meters above the floor

10 In the accompanying diagram, a 1.0kilogram sphere at point *A* has a potential energy of 5.0 joules. What is the potential energy of the sphere at point *B*, halfway down the incline?
    (1) 0.0 joule   (3) 3.0 joules
    (2) 2.5 joules   (4) 6.0 joules

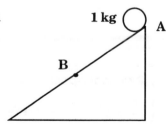

11 Hanging a weight of 5.0 N on a coiled spring stretches it 6.0 cm. A 10 N
    weight will stretch this spring
    (1) 6.0 cm   (3) 30 cm
    (2) 12 cm   (4) 60 cm

12 A 0.10 -meter spring is stretched from
    equilibrium to position A and then to position B
    as shown in the diagram below. Compared to the
    spring's potential energy at A, what is its
    potential energy at B ?
    (1) the same
    (2) twice as great
    (3) half as great
    (4) four times as great

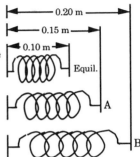

13 A ball is thrown upward from the Earth's surface. While the ball is
    rising, its gravitational potential energy will
    (1) decrease   (2) increase   (3) remain the same

14 The graph at the right shows the force
    exerted on a block as a function of the
    block's displacement in the direction of the
    force. How much work did the force do in
    displacing the block 4.0 meters?
    (1) 1.0 J
    (2) 0 J
    (3) 16 J
    (4) 4.0 J

15 When the speed of an object is halved, its kinetic energy is
    (1) quartered   (2) halved   (3) the same   (4) doubled

16 A mass resting on a shelf 10.0 meters above the floor has a gravitational
    potential energy of 980. joules with respect to the floor. The mass is
    moved to a shelf 8.00 meters above the floor. What is the new
    gravitational potential energy of the mass?
    (1) 960. J   (2) 784 J   (3) 490. J   (4) 196 J

3.5 Conservation Of Energy

    Conservation of energy is one of the fundamental principles in physics. Energy cannot be created or destroyed, but it can be changed from one form into another. The pendulum (page 82) is a good example of this principle.

WORK on   PE of   KE of
bow & arrow   bow & arrow   bow & arrow

The Pendulum Swing (assuming reference level at B)

A.   At the top of the swing (**A**), the potential energy is greatest, and the kinetic energy is zero. When the bob momentarily stops to change direction, the kinetic energy at this point is zero. All the energy is potential energy.

B.   At the bottom of the swing (**B**), the potential energy is zero, and the bob is moving with the greatest velocity. Kinetic energy is now at its greatest. The maximum kinetic energy at the bottom equals the maximum potential energy at the top of the swing.

C.   At any position between the highest and lowest point (**C**), the pendulum has both kinetic and potential energy. The sum of the kinetic and potential energies is always a constant value. When kinetic energy increases, potential energy must decrease to keep the total energy of the system constant.

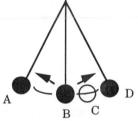

D.   At the top of the swing (**D**), the bob stops. All of the energy is potential energy again as it was at point **A**.

Summary of Pendulum Relationship
(assuming no friction)

Position	General Formula	Special Considerations	Final Equation
A	E = KE + PE	KE = 0	E = PE (max)
B	E = KE + PE	PE = 0	E = KE (max)
C	E = KE + PE	any point	E = KE + PE
D	E = KE + PE	KE = 0	E = PE (max)

Conservation of energy can also be illustrated by a falling object. When the object is released, it initially has only potential energy. At the halfway point in the fall, half of the energy is potential and half is kinetic. The instant before hitting the ground all of the original potential energy has been converted into kinetic energy. Yet, the sum total of **KE** and **PE** always equals a constant value as illustrated in the table below.

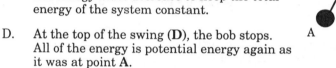

Height	K.E.	P.E.	Total (K.E. + P.E.)
20.0 m	0 J	2000 J	2000 J
10.0 m	1000 J	1000 J	2000 J
5.0 m	1500 J	500 J	2000 J
0.0 m	2000 J	0 J	2000 J

ground level   (using g = 10 m/s^2)

3.6 Work – Energy Relationship

In a system with no friction (or negligible friction), the total mechanical energy (**KE + PE**) remains constant. In a system with friction:

$$\text{Work} = \Delta\text{KE} + \Delta\text{PE} + W_f$$

Where: W_f is the work against friction.

A **conservative force** is one for which the work done on an object is independent of the path taken. Potential energy can be defined only for a conservative force.

```
┌ ─ ─ ─ ─ ─ ─ ─ ─ ─ ─ ─ ─ ─ ┐
 Practical Applications
 •   amusement park roller coaster
 •   "hot wheels" cars
 •   ski jump
└ ─ ─ ─ ─ ─ ─ ─ ─ ─ ─ ─ ─ ─ ┘
```

A **non-conservative force** is one for which the work done on an object depends upon the path taken. Friction is a non-conservative force. In any transfer of energy among objects in a closed system, the total energy of the system remains constant. A closed system is one where no external forces act and no external work can be done on or by the system. The total mechanical energy of a mechanical system is the sum of the kinetic and potential energies. Work done against friction in a mechanical system is converted to heat or internal energy.

Questions

Base your answers to questions 1 through 3 on the information below.

A 6 newton force pushes a 12 kilogram mass a distance of 10 meters on a horizontal surface with a uniform velocity.

1   The work done on the mass is
   (1) 0 J   (2) 60 J   (3) 72 J   (4) 120 J
2   The potential energy gained by the object is
   (1) 0 J   (2) 60 J   (3) 72 J   (4) 120 J
3   The weight of the mass is
   (1) 0 N   (2) 60 N   (3) 72 N   (4) 120 N
4   A pendulum swings in the direction shown in the diagram. At what position is the kinetic energy of the pendulum bob the greatest?
   (1) A
   (2) B
   (3) C
   (4) D
5   A baseball bat strikes a ball with an average force of 2.0×10^4 newtons. If the bat stays in contact with the ball for a distance of 5.0×10^{-3} meter, what kinetic energy will the ball acquire from the bat?
   (1) 1.0×10^2 joules   (3) 2.5×10^1 joules
   (2) 2.0×10^2 joules   (4) 4.0×10^2 joules

6   The diagram shows a ball that starts from rest
    and falls freely in a vacuum. At what point will
    the potential energy be equal to the kinetic
    energy?
    (1) A
    (2) B
    (3) C
    (4) D

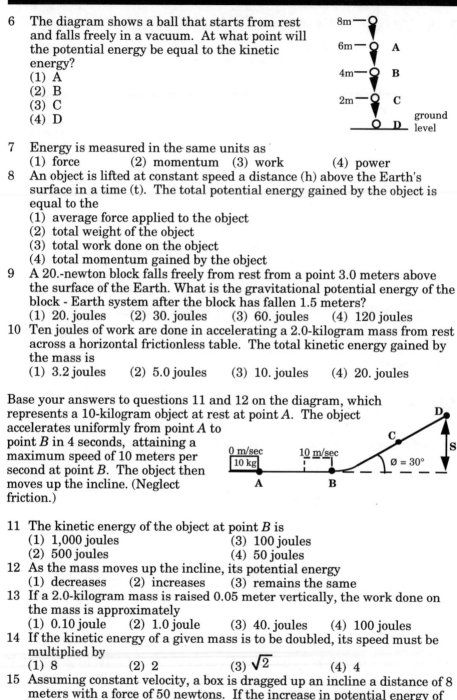

7   Energy is measured in the same units as
    (1) force   (2) momentum   (3) work   (4) power
8   An object is lifted at constant speed a distance (h) above the Earth's
    surface in a time (t). The total potential energy gained by the object is
    equal to the
    (1) average force applied to the object
    (2) total weight of the object
    (3) total work done on the object
    (4) total momentum gained by the object
9   A 20.-newton block falls freely from rest from a point 3.0 meters above
    the surface of the Earth. What is the gravitational potential energy of the
    block - Earth system after the block has fallen 1.5 meters?
    (1) 20. joules   (2) 30. joules   (3) 60. joules   (4) 120 joules
10 Ten joules of work are done in accelerating a 2.0-kilogram mass from rest
    across a horizontal frictionless table. The total kinetic energy gained by
    the mass is
    (1) 3.2 joules   (2) 5.0 joules   (3) 10. joules   (4) 20. joules

Base your answers to questions 11 and 12 on the diagram, which
represents a 10-kilogram object at rest at point A. The object
accelerates uniformly from point A to
point B in 4 seconds, attaining a
maximum speed of 10 meters per
second at point B. The object then
moves up the incline. (Neglect
friction.)

11 The kinetic energy of the object at point B is
    (1) 1,000 joules   (3) 100 joules
    (2) 500 joules   (4) 50 joules
12 As the mass moves up the incline, its potential energy
    (1) decreases   (2) increases   (3) remains the same
13 If a 2.0-kilogram mass is raised 0.05 meter vertically, the work done on
    the mass is approximately
    (1) 0.10 joule   (2) 1.0 joule   (3) 40. joules   (4) 100 joules
14 If the kinetic energy of a given mass is to be doubled, its speed must be
    multiplied by
    (1) 8   (2) 2   (3) $\sqrt{2}$ (4) 4
15 Assuming constant velocity, a box is dragged up an incline a distance of 8
    meters with a force of 50 newtons. If the increase in potential energy of
    the box is 300 joules, the work done against friction is
    (1) 100 joules   (2) 200 joules   (3) 300 joules   (4) 400 joules

Thinking Physics

Can you explain these apparent discrepancies?

1   It is less work for a short person to lift 1400 N overhead than for a tall person.
2   A power company (electric utility) doesn't sell power.
3   If you carry a box up to the attic, then change your mind and bring it back down, physically speaking - you did no work.

4   In the "toy" illustration at the right, what's happening?

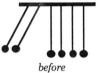 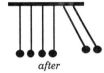

Why does this work as shown?

before   *after*

5   In a class demonstration, you may have seen a pair of "happy" and "sad" spheres. The "happy" sphere bounces. The "sad" sphere doesn't. Is there conservation of energy?

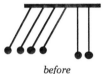 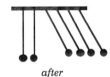

before   *after*

Free Response Questions

1. A block slides down an inclined plane onto a table. It then slides across the table and strikes a spring, compressing the spring.

    a) Draw a labeled diagram illustrating the situation.

    b) The initial height of the block above the table will affect how much the spring is compressed. List three other variables that might affect the amount by which the spring is compressed when the block strikes it.

    c) Consider the height variable. Design a simple experiment to find out how the initial height of the block is related to the compression of the spring.

2. The length of a spring is recorded as different weights are placed on the attached pan. The data is shown in the chart at the right:

    a) Sketch the graph with properly labeled axes.

Weight (N)	Length (cm)
3	5.5
5	6.4
6	7.1
11	9.6

    b) Determine the length of the unstretched spring.

    c) Sketch a graph you might expect if the experiment were repeated with a stiffer spring.

3. Base your answers to parts *a* through *c* on the diagram below which represents a 4.0–kilogram block sliding down a frictionless 30° incline. The top of the incline is 2.5 meters above the ground and the incline is 5.0 meters long. The uniform acceleration of the block down the incline is 4.9 meters per second squared.

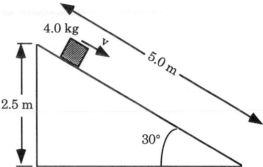

a) On the diagram at the right, draw a vector on the block, representing the weight of the block.

b) What was the potential energy of the block at the top of the incline? (Show all calculations.)

c) The block was initially at rest at the top of the incline. Determine the speed of the block at the bottom of the incline. (Show all calculations.)

Self–Help Questions

1   In the diagram at the right, 65 joules of work is needed to raise a 10.-newton weight 5.0meters. How much work is done to overcome friction as the weight is raised?
    (1) 5 J
    (2) 13 J
    (3) 15 J
    (4) 50. J

2   One newton is equivalent to one

(1) $\dfrac{\text{watt}}{\text{meter}^3}$

(2) kilogram . meter³

(3) watt² . second

(4) $\dfrac{\text{joule}}{\text{meter}}$

3   A 20-newton block is at rest at the bottom of a frictionless incline as shown in the diagram at the right. How much work must be done against gravity to move the block to the top of the incline?
    (1) 10 J   (3) 80 J
    (2) 60 J   (4) 100 J

4   Work is being done when a force
    (1) acts vertically on a cart that can only move horizontally
    (2) is exerted by one team in a tug of war when there is no movement
    (3) is exerted while pulling a wagon up a hill
    (4) of gravitational attraction acts on a person standing on the surface of the Earth

5   Which term is a unit of power?
     (1) joule   (2) newton   (3) watt   (4) hertz
6   What is the maximum distance that a 60.-watt motor may vertically lift a
     90.-newton weight in 7.5 seconds?
     (1) 2.3 m   (2) 5.0 m   (3) 140 m   (4) 1100 m
7   A motor has an output of 500 watts. When the motor is working at full
     capacity, how much time will it require to lift a 50-newton weight
     100 meters?
     (1) 5 s   (2) 10 s   (3) 50 s   (4) 100 s
8   An object gains 10. joules of potential energy as it is lifted vertically
     2.0 meters. If a second object with one-half the mass is lifted vertically
     2.0 meters, the potential energy gained by the second object will be
     (1) 10. J   (2) 20. J   (3) 5.0 J   (4) 2.5 J
9   When a 10-kilogram mass is lifted from the ground to a height of
     10 meters, the gravitational potential energy of the mass is increased by
     approximately
     (1) 1 J   (2) 4 J   (3) 100 J   (4) 1000 J
10   If the speed of an object is halved, its kinetic energy will be
     (1) halved   (2) doubled   (3) quartered   (4) quadrupled
11   A basketball player who weighs 600 newtons jumps 0.5 meter vertically
     off the floor. What is her kinetic energy just before hitting the floor?
     (1) 30 J   (2) 60 J   (3) 300 J   (4) 600 J
12   Which graph best represents the relationship between the kinetic energy
     (KE) of a moving object as a function of its velocity (v)?

     (1)   v   (2)   v   (3)   v   (4)   v
13   The kinetic energy of a 10.0-kilogram mass moving at a speed of
     5.00 meters per second is
     (1) 50.0 J   (2) 2.00 J   (3) 125 J   (4) 250. J

14   As the pendulum swings from position A to position
     B as shown in the diagram at the right, what is the
     relationship of kinetic energy to potential energy?
     (Neglect friction.)

     (1) The kinetic energy decrease is more than the
         potential energy increase.
     (2) The kinetic energy increase is more than the
         potential energy decrease.
     (3) The kinetic energy decrease is equal to the potential energy increase.
     (4) The kinetic energy increase is equal to the potential energy decrease.
15   An object x meters above the ground has Z joules of potential energy. If
     the object falls freely, how many joules of kinetic energy will it have
     gained when it is x/2 meters above the ground?
     (1) Z   (2) 2Z   (3) Z/2   (4) 0
16   A spring of negligible mass with a spring constant of 200 newtons per
     meter is stretched 0.2 meter. How much potential energy is stored in the
     spring?
     (1) 40 J   (2) 20 J   (3) 8 J   (4) 4 J

17 What is the spring constant of a spring of negligible mass which gained
   8joules of potential energy as a result of being compressed 0.4 meter?
   (1) 100 N/m   (2) 50 N/m   (3) 0.3 N/m   (4) 40 N/m

18 The graph at the right represents the relationship
   between the force applied to a spring and the
   elongation of the spring. What is the spring
   constant?
   (1) 20. N/m
   (2) 9.8 N/kg
   (3) 0.80 N·m
   (4) 0.050 m/N

Base your answers to questions 19 through 21 on
the diagram at the right, which shows a
1.0-kilogram aluminum sphere and a 3.0-kilogram
brass sphere. Each of the spheres has the same
diameter, and each is 19.6 meters above the
ground. Both spheres are allowed to fall freely.
(Neglect the effects of air resistance.)

19 Both spheres are released at the same instant.
   They will reach the ground at
   (1) the same time, but with different speeds
   (2) the same time and with the same speeds
   (3) different times, but with the same speeds
   (4) different times and with different speeds

20 If the spheres are 19.6 meters above the ground, the time required for the
   aluminum sphere to reach the ground is
   (1) 1.0 s   (2) 2.0 s   (3) 8.0 s   (4) 4.0 s

21 Which graph shows the relationship between the potential energy and
   height above the ground for each sphere?

   (1) PE vs Height   (2) PE vs Height   (3) PE vs Height   (4) PE vs Height

Base your answers to questions 22 through 25 on
the following information.
   A horizontal force of 10. newtons accelerates a
2.0-kilogram block from rest along a level table
as shown, at a rate of 4.0meters per second squared.

$F = 10.$ N → 2.0 kg

22 The work done in moving the block 8.0meters is
   (1) 8.0 J   (2) 20. J   (3) 80. J   (4) 800 J

23 When the speed of the block is 8.0meters per second, its kinetic energy is
   (1) 8.0 J   (2) 16 J   (3) 64 J   (4) 80. J

24 What is the frictional force retarding the forward motion of the block?
   (1) 8.0 N   (2) 2.0 N   (3) 10. N   (4) 12 N

25 If there were no friction between the block and the table, the acceleration
   of the block would be
   (1) 20. m/s^2 (2) 9.8 m/s^2 (3) 5.0 m/s^2 (4) 4.0 m/s^2

Optional Unit 2

Internal Energy

Important Terms To Be Understood

heat
temperature
Kelvin Scale
Celsius Scale
absolute zero
internal energy
kilocalorie
Mechanical
  Equivalent of Heat
specific heat

diffusion
latent
melting
freezing
fusion
vaporization
evaporation
sublimation
condensation

supercooled
Kinetic Theory
Boyle's Law
Charles' Law
General Gas Law
First Law of
  Thermodynamics
Second Law of
  Thermodynamics
entropy

I. Temperature

Temperature is that property of matter which determines the direction of the exchange of internal energy. Temperature is a scalar quantity. An object at a lower temperature will gain internal energy. Heat, which is a measurement of the total internal energy, does not depend on temperature alone. It also depends on the object's mass, nature, and phase.

Energy always goes from the hot object to the cold object. Temperature determines the direction of heat transfer. In measuring temperatures, the **Celsius scale** is commonly used. This scale has two fixed points, 0°, the freezing point of water and 100°, the boiling point of water at standard pressure. There are 100degrees between the two fixed points.

The **Kelvin scale** also has a fixed point. That point is the triple point of water. This temperature is, by definition, 273.16K. The triple point is that temperature, where all three phases of water can exist simultaneously, in equilibrium. To convert Celsius readings to Kelvin readings use the equation:

$$T \text{ (Kelvin)} = T \text{ (Celsius)} + 273$$

A Kelvin is equal to a Celsius degree. A rise of 10Kelvin is equal to a rise of 10 degrees Celsius. The Kelvin is the SI unit of temperature. It is a fundamental unit.

An object is at **absolute zero** when its internal energy is a minimum and no internal energy can be transferred to another object. On the Kelvin scale, absolute zero is known as the zero point energy. The molecules still have a finite amount of energy. The absolute or Kelvin temperature is directly proportional to the average kinetic energy of the molecules of an ideal gas. An ideal gas is one which consists of perfectly elastic particles of negligible size which exert no forces on each other except during collision. Doubling the absolute temperature doubles the average kinetic energy of the molecules.

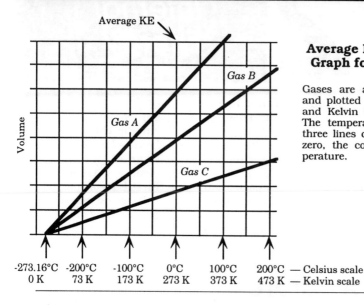

Average Kinetic Energy Graph for Three Gases

Gases are at constant volume and plotted against the Celsius and Kelvin temperature scales. The temperature at which the three lines converge is absolute zero, the coldest possible temperature.

| -273.16°C | -200°C | -100°C | 0°C | 100°C | 200°C — Celsius scale |
| 0 K | 73 K | 173 K | 273 K | 373 K | 473 K — Kelvin scale |

1.1 Temperature Scales

To measure and record temperature we must have a unit and a scale. Body sensation does not provide reliable measurements of temperature. Previous body temperature determines later conditions. Objective temperature scales are reliable and can be reproduced. A scale is obtained by taking two temperatures as fixed points and dividing the distance between them into a number of equal divisions called degrees.

On the Celsius scale the lower fixed point is the temperature at which pure ice melts under normal atmospheric pressure and is taken as 0°C. The upper fixed point is the temperature at which pure water boils under normal atmospheric pressure and is taken as 100°C. There are 100 equal divisions between those points on the thermometer, and, when the mercury or alcohol moves through one division, this corresponds to a change in temperature of 1°C.

On the Fahrenheit scale the freezing point of water is 32 degrees, and the normal boiling point is 212 degrees.

The Kelvin scale of temperature has its zero at absolute zero. Absolute zero is the temperature at which the average kinetic energy of the particles of a substance is at a minimum. This is the lowest possible temperature. Absolute zero is -273°C.

The size of the divisions on the Celsius scale and the Kelvin scale are the same; therefore, intervals of temperature on the two scales are identical. A rise in temperature on the Celsius scale from 12°C to 22°C is 10 C°, and on the Kelvin scale from 285 K to 295 K is 10 K.

$$\text{T (Kelvin)} = \text{T (Celsius)} + 273$$

Complete the following table:

°C	- 180		- 10		50		666
K		50		180		473	

Questions

1 As the temperature of an object approaches absolute zero, its
(1) kinetic energy approaches a maximum
(2) internal energy approaches a minimum
(3) temperature approaches -273 K
(4) temperature approaches 273° C

2 The boiling pint of water is
(1) 273 K   (2) 373 K   (3) 273° C   (4) 373° C

3 On an uncalibrated thermometer, a student correctly marked the positions of the boiling and freezing points of water. For a Celsius scale, how many degrees should he mark between the two points?
(1) 50   (2) 100   (3) 180   (4) 273

4 What is the approximate number of Kelvin degrees between the condensation point and the solidification point of copper?
(1) 1083   (2) 1484   (3) 2567   (4) 3650

5 If only the temperature of two objects is known, it is always possible to determine the
(1) total internal kinetic energy
(2) total internal potential energy
(3) direction of heat flow between them
(4) phase of the objects

6 The temperature of a substance changes from -10°C to 25°C. What is the temperature change expressed on the Kelvin scale?
(1) 15 K   (2) 35 K   (3) 298 K   (4) 318 K

7 As the temperature of an ideal gas increases, the ratio of its average molecular kinetic energy to its absolute temperature
(1) decreases   (2) increases   (3) remains the same

8 The ratio of a Kelvin degree to a Celsius degree is
(1) 1:1   (2) 5:9   (3) 1:273   (4) 273:1

9 A change in temperature of 100° Celsius is equal to a change in Kelvin temperature of
(1) 373 K   (2) 200 K   (3) 100 K   (4) 50 K

10 As heat is added to a solid at a temperature below its melting point, its average molecular kinetic energy
(1) decreases   (2) increases   (3) remains the same

11 The internal energy of a substance is at a minimum when its temperature is
(1) 0° C   (2) 0 K   (3) 273° C   (4) 273 K

12 Which graph best represents the relationship between the absolute temperature and the average molecular kinetic energy of an ideal gas?

(1) Temperature   (2) Temperature   (3) Temperature   (4) Temperature

II. Internal Energy And Heat

Internal energy is the total kinetic and potential energy associated with the motions and relative positions of the molecules of an object, apart from any kinetic or potential energy of the object as a whole. In any given substance, solids have the lowest potential energy; then, liquids and gases have the highest potential energy. An increase in the internal energy of an object either increases the kinetic energy of the random motion of its molecules, which results in a rise in temperature or increases their potential energy of position, which results in a change of phase, or raises the energy levels of the atoms. Energy used in doing work against friction is converted into internal energy.

Heat is energy that flows from a warm body to a cold body because of a temperature difference between them. Heat is a scalar unit. Heat and mechanical energy are both forms of energy. Therefore, they can be measured in the same units. The joule is the SI unit of heat.

2.1 Specific Heat

The specific heat of a substance is the ratio of the quantity of heat required to raise the temperature of a unit mass of the substance one Celsius degree to the quantity of heat required to raise the temperature of a unit mass of water one Celsius degree. Specific heat is represented by the letter c. The units of specific heat are given as joules/kg °C. The specific heat of water is 4.19×10^3 J/kg °C.

The energy absorbed or liberated in a temperature change produces a change in the average kinetic energy of the molecules and the internal kinetic and potential energies associated with the vibrations of the molecules of the substance.

Practical Applications
- crust of pizza cools faster than the topping
- water's modifying effect on climate

2.2 Conservation Of Energy

When there is an exchange of internal energy and no conversion to other forms of energy the total internal energy of the system remains constant. The conservation of heat energy is the **First Law of Thermodynamics**.

When objects of different temperatures come together, the heat lost by the hotter object is equal to the heat gained by the colder object under ideal conditions.

$$Q = mc\Delta T \qquad \text{and} \qquad Q_{lost} = Q_{gained}$$

Where: Q = heat in joules
c = specific heat
m = mass in kilograms
ΔT = change in temperature

Examples

1. How much heat is required to raise the temperature of 10. kg of aluminum from 50.°C to 70.°C?

Given:
m = 10. kg
c = (from table for aluminum) 0.90 kJ/kg °C
ΔT = t(final) - t(initial) = 70.°C - 50.°C = 20.°C

Find: Q

Solution:
Q = $mc\Delta T$
= (10. kg) (0.90 kJ/kg °C) (20. °C)
= 1.8×10^2 kJ

2. A mass of 50. kg of water cools from 90.°C to 20.°C. What is the total amount of heat given up by the water?

Given:
m = 50. kg
c = 4.19 kJ/kg °C
ΔT = t(final) - t(initial) = -70.°C

Find: **Q**
Solution:
Q = $mc\Delta T$
= (50. kg) (4.19 kJ/kg °C) (-70.°C)
= -1.47×10^4 kJ = -1.5×10^4 kJ

The negative sign indicates that the heat was removed or lost. If the question asks for the heat lost, the negative sign is not necessary.

3. If a 200. gm sample of aluminum at an initial temperature of 100.°C is put into a calorimeter containing 500. gm of water at 20.°C. Calculate the final temperature.

Given:
m of aluminum = 200. gm = .200 kg
m of water = 500. gm = .500 kg
c of aluminum = 0.90 kJ/kg °C
c of water = 4.19 kJ/kg °C
$T_{initial}$ of aluminum = 100.°C
$T_{initial}$ of water = 20.°C
Find: T_{final}

Solution:

heat lost (by aluminum) = **heat gained** (by water)

$$mc\Delta T = mc\Delta T$$

$$(.200\text{kg})(0.90 \text{ kJ/kg C°})(100.°C - T_f) = (.500\text{kg})(4.19 \text{ kJ/kg °C})(T_f - 20°C)$$

$$T_f = 26.3°C = 26.°C$$

2.3 Change Of Phase

Three phases of matter are **solid, liquid,** and **vapor.** In the case of water these three phases are called *ice, water,* and *steam.*

To understand what is happening during a change of phase, we must think about the kinetic molecular theory. The kinetic molecular theory is based on the belief that matter consists of tiny particles called molecules. These molecules are in a continual state of motion. **Diffusion,** which is the tendency of a substance to spread out of its own accord, confirms that the molecules are in rapid motion.

According to the theory, in a solid, each molecule vibrates about a mean position. When the molecules receive energy in the form of heat, they convert it into their own vibrational energy, this is seen as an expansion of the solid. If they are provided with sufficient energy, they can overcome the forces holding them together. They are then free to move past each other for short distances. This is the liquid phase.

When given even more energy, the molecules move faster and some have enough energy to escape from the liquid by overcoming the forces of molecular attraction in the liquid. They spread out in the air and move in all directions. In this condition, the vapor phase, the forces of attraction are small and molecules can move at great speeds over large distances. Thus, when an object is changing temperature, the kinetic energy of its molecules is changing. When an object is changing phase, the potential energy of the molecules is changing. Because the heat energy seems to disappear without a trace, it is called **latent** (i.e. hidden) **heat.** The latent heat is stored in the form of potential energy.

Below is a phase diagram for water. During the phase changes, there is no increase in average kinetic energy (temperature). The heat that is absorbed during melting and boiling increases the molecular **potential energy** associated with bonds between molecules. Therefore, there is an increase in the total internal energy. **Total internal energy** is the sum of the kinetic and potential energies of the molecules.

Phase Change Diagram for Water

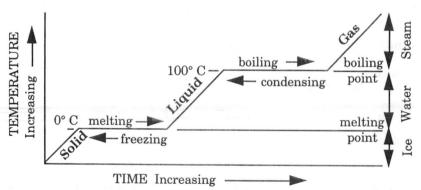

2.4 Melting And Freezing Points

The temperature of a crystalline material remains constant during melting and freezing. The melting point of any crystalline substance occurs at the same temperature as the freezing point. The temperature of a non-crystalline material does not remain constant during "melting and freezing." A non-crystalline solid has no definite melting point.

2.5 Heat Of Fusion

The heat of fusion is the amount of heat that must be added to change one kilogram of substance at its melting point from the solid to the liquid phase with no change in temperature. The formula for heat of fusion is:

$$Q_f = mH_f$$

Where: Q_f = heat in joules
m = mass of material in kilograms
H_f = heat of fusion of material (from the Reference Table)

The heat of fusion also equals the amount of heat that must be removed to change one kilogram of a substance at its freezing point from the liquid to the solid phase.

Example

Calculate the amount of heat necessary to melt 100 grams of copper at its melting point.

Given: m = 100. g = 0.100 kg
H_f = 2.05 x 10^2 kJ/kg

Find: Q_f

Solution: Q_f = mH_f
= (0.100 kg) (2.05 x 10^2 kJ/kg)

Therefore: = 2.05 x 10^1 kg

2.6 Heat Of Vaporization

The heat of vaporization is the amount of heat that must be added to change one kilogram of a substance at its boiling point from the liquid to the gaseous phase without a temperature change. The heat of vaporization also equals the amount of heat that must be removed to change one kilogram of a substance at its condensation point from the gaseous to the liquid phase.

$$Q_v = mH_v$$

Where: Q_v is energy
m is mass
H_v is heat of vaporizatic

Example

Calculate the amount of heat energy necessary to vaporize 50. kg of water at 100.°C.

Given: m = 50. kg
              H_v = 2.26 x 10³ kJ/kg

Find: Q_v

Solution: Q_v = mH_v
              = (50. kg)(2.26 x 10³ kJ/kg)
              = 1.13 x 10⁵ kJ

Therefore: = 1.1 x 10⁵ kJ

Factors affecting boiling and condensing point:
1. **Solute.** The **boiling** point of a substance is raised by dissolving some substance in the liquid. A dissolved salt raises the boiling point of water.
2. **Decreased pressure** lowers the boiling point.
3. **Increased pressure** increases the boiling point.

Factors affecting melting and freezing point:
1. **Solute.** The **freezing** and **melting** point of a substance is lowered by dissolving some substance in the liquid. A dissolved salt lowers the freezing point of water. For example, the application of salt on icy roads.
2. **Increased pressure** lowers the melting point of ice.

The increased pressure under the wire causes the ice to liquefy. The water from the melt refreezes above the wire. This is called **regelation**. A more complete explanation of water molecules is located in section 3.2.

Other important terms:
1. **Evaporation** — a cooling process. The changing of a liquid to a gas. The rate of evaporation is proportional to the rate of cooling.
2. **Sublimation** — the substance passes directly from the solid to the gas phase. Most crystals and dry ice can be sublimed.
3. **Condensation** — a change in phase from gas to liquid.
4. **Supercooled** — the temperature of the liquid is lower than its freezing point.

Questions

1   How much heat must be added to 10.0 kg of water at 20°C to change its
    temperature to 30°C?
    (1) 1 x 10^2J   (2) 5 x 10^3 J   (3) 4.19 x 10^2 J   (4) 4.19 x 10^5 J

2   Which of the following 10.0 kilogram solids will require the most heat to
    change its temperature 1°C?
    (1) iron   (2) lead   (3) copper   (4) aluminum

Base your answers to questions 3
through 5 on the graph which represents the relationship between the
heat energy added and the temperature of substance X. Substance X has
a mass of 1.0 kilogram and is being
heated from 20.°C to 750.°C.

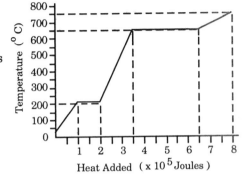

Heat Added (x 10^5 Joules)

3   In which phase is substance X
    when its temperature is 250°C?
    (1) solid
    (2) liquid
    (3) gas
    (4) plasma

4   What is the heat of fusion of substance X?
    (1) 1.0 x 10^5 J/kg   (3) 3.0 x 10^5 J/kg
    (2) 2.0 x 10^5 J/kg   (4) 1.5 x 10^5 J/kg

5   What is the approximate specific heat of the solid phase of substance X?
    (1) 1.0 x 10^5 J/kg.C°   (3) 3.3 x 10^2 J/kg.C°
    (2) 5.0 x 10^2 J/kg.C°   (4) 1.0 x 10^3 J/kg.C°

6   What is the final temperature of the resultant mixture when
    50.kilograms of ice at 0.0° Celsius is added to 50. kilograms of water at
    20.°Celsius?
    (1) 0.0°C   (2) 5.0°C   (3) 10.°C   (4) 20.°C

7   As water changes to ice without sublimation at zero degrees Celsius, its
    mass
    (1) decreases   (2) increases   (3) remains the same

8   What are the two changes of phase a substance can undergo with the
    addition of heat?
    (1) melting and vaporizing   (3) freezing and vaporizing
    (2) melting and freezing   (4) freezing and condensing

9   When a block of ice at zero degrees Celsius melts, the ice absorbs energy
    from its environment. As the ice is melting, the temperature of the
    remaining portion of the block
    (1) decreases   (2) increases   (3) remains the same

10 The internal energy of an object depends on its
    (1) temperature, only   (3) phase, only
    (2) mass, only   (4) temperature, mass and phase

11 Under standard conditions (a pressure of 1.0 atmosphere), water boils at
    a temperature of
    (1) 0 Kelvin   (3) 273 Kelvin
    (2) 100 Kelvin   (4) 373 Kelvin

12 A copper kettle and an aluminum kettle of equal mass contain equal
    amounts of water. Compared to the total heat needed to raise the
    temperature of the water and the aluminum kettle from 20°C to 100°C,
    the amount of heat needed to raise the temperature of the water and the
    copper kettle from 20°C to 100°C would be
    (1) less   (2) greater   (3) the same

13 While the water is boiling, what happens to the average kinetic energy of
    the water molecules? (Assume that the pressure remains constant.)
    (1) It decreases (2) It increases   (3) It remains the same

14 Compared to the heat needed to raise the temperature of 20 kilograms of
    water 10°C, the heat needed to raise the temperature of 20 kilograms of
    steam 10°C is
    (1) less   (2) more   (3) the same

15 Work energy is completely converted to heat energy when all of the work
    done on an object is used to overcome
    (1) momentum   (2) gravity   (3) inertia   (4) friction

16 Which is true if some salt is dissolved in a beaker of water?
    (1) Only the freezing point of the water is affected.
    (2) Only the boiling point of the water is affected.
    (3) Both the freezing and boiling points of the water are affected.
    (4) Neither the freezing point nor boiling point of the water is affected.

17 If pressure is applied to ice, its melting point will
    (1) decrease   (2) increase   (3) remain the same

Base your answers to questions 18
through 20 on the graph at the right
which represents the temperature of
2.0 kilograms of ammonia as heat is
liberated.

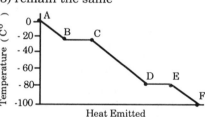

Heat Emitted

18 What is the phase of the substance at -15°C?
    (1) solid   (2) liquid   (3) gas   (4) plasma

19 How much heat will be liberated between points C and D?
    (1) 2.6×10^5 J   (2) 260 J   (3) 5.3×10^5 J   (4) 530 J

20 Between points D and E, the substance is changing phase as it
    (1) becomes a liquid   (3) becomes a gas
    (2) condenses   (4) freezes

III. Kinetic Theory Of Gases

The **Kinetic Theory** accounts for all phases of matter by assuming that
matter is made up of molecules which are in continual motion. This motion is
called **Brownian movement** and exists at all temperatures above absolute
zero. In gases of low density, the average distance of separation of molecules
is large in comparison with their diameters, and the total actual volume of
the gas molecules is negligible in comparison with the volume occupied by the
gas. The forces between molecules in a low density gas are negligible except
when the molecules collide.

Collisions between gas particles will usually result in a transfer of energy between particles, but the total energy of the system remains the same. A gas which would conform strictly to this model would be an **ideal gas**. This model does not exactly represent any gas under all conditions of temperature and pressure. A general rule is that gas well above its boiling point can be considered to be ideal.

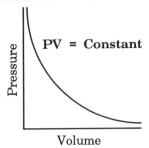

(a) Boyle's Law
Constant Temperature

Pressure exerted by a gas is due to collisions of gas molecules with the walls of the container. Standard Temperature and Pressure (**STP**) of a gas are defined as 0°C (273 K) and 760 mm of mercury (760 torr) or 1 atmosphere of pressure. One atmosphere is a pressure of 1.01×10^5 N/m^2 or 101 kilopascals (101 kPa). The pascal is the SI unit of pressure. (1Pa = 1N/m^2)

In the 17th and 18th centuries, experimental relationships between pressures, volumes, and temperatures resulted in the following three gas laws:

a) **Boyle's Law.** For a fixed mass of gas at constant temperature, the product of pressure and volume is constant.

b) **Charles' Law.** For a fixed mass of gas at constant pressure, the volume is directly proportional to the temperature measured in kelvins.

c) **The Pressure Law.** For a fixed mass of gas at constant volume, the pressure is directly proportional to the temperature measured in kelvins.

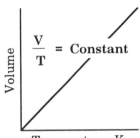

(b) Charles' Law
Constant Pressure

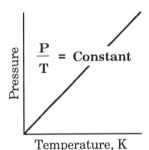

(c) Pressure Law
Constant Volume

The behavior of real gases can be described by the formula, **PV = nRT**, if they are at low pressures and are at temperatures well above those at which they liquefy.

The **Ideal Gas Law** states that the product of pressure and volume of a gas is directly proportional to the number of molecules and the absolute temperature. It is written as:

$$PV = nRT$$

Where:
P	-	pressure in mm of Hg, atm., etc.
V	-	volume in liters, m^3, etc.
n	-	number of moles
R	-	gas constant - in proper units
T	-	absolute temperature in K.

This equation can be used to derive the simple proportion:

$$\frac{P_1 V_1}{T_1} = \frac{P_2 V_2}{T_2}$$

Where: P_1, V_1, T_1, represent initial conditions

P_2, V_2, T_2, represent final conditions

Remember that the temperature must be in Kelvin.

3.1 Expansion
How Does Heat Affect Objects?

Matter generally expands when heated and contracts when cooled. This is due to an increase in the amplitude of vibration of the particles of the substance as a result of an increase in temperature.

The increase in size of a substance when it is heated is called **expansion**. The decrease in size when it is cooled (heat is removed) is called **contraction**. When a substance is heated (or cooled) the expansion (or contraction) takes place in all directions. Bridges and roads are built with expansion joints to allow for changes due to temperature.

Most solids expand when heated and contract when cooled. The rate of expansion of solids is small compared to that of liquids or gases.

Have you ever been presented with the problem of removing a very tight metal cap from a bottle? Someone may have suggested that you place the top in very hot water. You may have been surprised at how easily the cap came off after that. You were heating two different materials, metal and glass. Both materials expanded at different rates when heated. The metal expanded more than the glass and the top was easily removed.

Most liquids expand when heated and contract when cooled. A thermometer is an instrument that uses expansion to measure temperature. When selecting a liquid for a thermometer, it is necessary to choose a liquid that is readily visible and a good conductor of heat. It must expand by a suitable amount over a wide range of temperatures and move freely within a glass container.

Two liquids, mercury and colored alcohol fulfill all these requirements, each compensating for one deficiency in the other: mercury boils at 367°C but freezes at -39°C; alcohol boils at 78°C and freezes at -115°C. If you were going to the South Pole, you would take an alcohol thermometer.

Gases expand when heated and contract when cooled. The rate of expansion of gases is much greater than that of liquids. The expansion rate of all gases is approximately the same. The hot air balloonist uses the principle of gas expansion to lift the craft.

3.2 The Peculiar Expansion Of Water

Most liquids when they solidify do so from the bottom upwards. Water is an exception. As the temperature of water rises from 0°C to 4°C, the water contracts. Above 4°C, the water expands. In fact, water is most dense at 4°C. The dense water sinks to the bottom of the pond.

Water changes density with temperature.

At 4°C, the density of water is at its maximum.

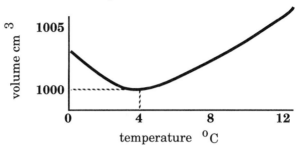

temperature °C

As the outdoor temperature continues to drop to 0°C, the less dense water remains at the surface and freezes. Since ice is less dense than water, it remains at the surface. Many days of very low temperatures are required before the layer becomes thick.

Questions

1. Which graph best represents the relationship between the absolute temperature (T_k) of an ideal gas and the average velocity ($\overline{v}$) of its molecules?

 (1)   (2)   (3)   (4)

2. The oxygen molecule has a mass sixteen times greater than the hydrogen molecule. Compare the average velocity of the hydrogen molecule v_1 to the average velocity of the oxygen molecule v_2 when the two gases are at the same temperature.
   (1) $v_1 = v_2/4$ (3) $v_1 = v_2/16$
   (2) $v_1 = 16 v_2$ (4) $v_1 = 4 v_2$

3. A closed sealed can of air is placed on a hot stove burner. The container will undergo an increase in
   (1) internal energy   (3) pressure
   (2) temperature   (4) all of these

4. Which mathematical expression best represents the relationship between absolute temperature (T) and average molecular kinetic energy (KE)? [k is a constant of proportionality.]

   (1) $T = \dfrac{k}{KE^2}$ (3) $T = k(KE)^2$

   (2) $T = \dfrac{k}{KE}$ (4) $T = k(KE)$

5   An ideal gas occupies 50.0 cubic meters at a temperature of 600. K. If the temperature is lowered to 300. K at constant pressure, the new volume occupied by the gas will be
    (1) $25.0 \, \text{m}^3$ (2) $100. \, \text{m}^3$ (3) $200. \, \text{m}^3$ (4) $400. \, \text{m}^3$

6   According to the kinetic theory of gases, an ideal gas of low density has relatively large
    (1) molecules
    (2) energy loss in molecular collisions
    (3) forces between molecules
    (4) distances between molecules

7   As the pressure of a fixed mass of gas is increased at constant temperature, the density of that gas
    (1) decreases   (2) increases   (3) remains the same

8   A given mass of gas is enclosed in a rigid container. If the velocity of the gas molecules colliding with the sides of the container increases, the
    (1) density of the gas will increase
    (2) pressure of the gas will increase
    (3) density of the gas will decrease
    (4) pressure of the gas will decrease

IV. The Laws Of Thermodynamics

Thermodynamics is the study of the relationship between heat and other forms of energy. When the principle of conservation of energy is stated with reference to heat and work, it is known as the **First Law of Thermodynamics**.

The First Law of Thermodynamics

Key

ΔQ = heat supplied

ΔW = work done

ΔT = change in temperature

ΔKE = change in kinetic energy

ΔPE = change in potential energy

ΔU = $\Delta KE + \Delta PE$

ΔU = internal energy

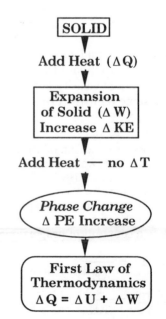

4.1 First Law

The heat energy ($\Delta\mathbf{Q}$) supplied to a system is equal to the increase in the internal energy ($\Delta\mathbf{U}$) of the system plus the work done ($\Delta\mathbf{W}$) by the system on its surroundings, i.e.

$$\Delta\mathbf{Q} = \Delta\mathbf{U} + \Delta\mathbf{W}$$

The internal energy of the system may be increased by:

1) increasing the kinetic energy of the molecules, i.e. increasing the temperature of the system

2) increasing its potential energy (phase change)

For an ideal gas there can be no increase in the internal **potential** energy (since there are no intermolecular forces) and therefore $\Delta\mathbf{U}$ is equal to the increase in the molecular kinetic energies of the gas.

> **Practical Applications**
> * bicycle pump
> * cylinders in an automobile engine

4.2 Second Law

Natural processes go in the direction that increases the total entropy. **Entropy** is a measure of the randomness or disorder of a body. There is a tendency in nature to proceed towards a state of greater disorder. The entropy of the universe is steadily increasing.

Another expression of the second law is that heat will not flow from a cold object to a warmer object. A heat pump is a device that transfers heat from a low temperature reservoir. Heat pumps can be reversed in the summer to be used as air conditioners.

> **Practical Applications**
> * heat pump
> * refrigerator

Some physical changes involving the second law includes: an object breaking, ice changing into water, a house cooling off in the winter and the sun emitting energy.

4.3 Third Law

It is impossible by any set of finite operations to reduce the temperature of a system to absolute zero, although temperatures less than 1×10^{-6} Kelvin have been achieved under laboratory conditions. At absolute zero, the particles would have no kinetic energy (**KE** $= \frac{1}{2}\mathbf{mv}^2$). Since each particle has mass, the particles would have to have zero velocity.

A particle with zero velocity has no momentum (**p = mv**). The de Broglie wavelength of this particle would be given by the formula:

$$\lambda = \frac{h}{p}$$

It would have to be infinite. Scientists believe that it would be impossible to reach a temperature of absolute zero.

Questions

1 As water is vaporized by boiling, its temperature
  (1) decreases   (2) increases   (3) remains the same
2 When a cube of hot metal is placed in a beaker of cold water, the temperature of the water
  (1) decreases   (2) increases   (3) remains the same
3 The melting point of lead is 327°C. At what temperature will liquid lead solidify?
  (1) 6°C   (2) 207°C   (3) 327°C   (4) 1,620°C

4 The graph shows the temperature of a substance as heat is being added at a constant rate. What is the melting point of the substance?
  (1) 10° C
  (2) 20°C
  (3) 30° C
  (4) 40° C

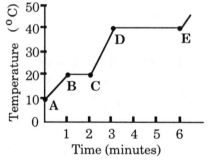

5 The First Law of Thermodynamics is a restatement of
  (1) the principle of entropy   (3) the conservation of energy
  (2) the law of addition of heats   (4) none of these
6 Systems that are left alone tend to move towards a state of
  (1) no entropy
  (2) more entropy
  (3) less entropy
7 When mechanical work is done on a system, there can be an increase in its
  (1) thermal energy   (3) temperature
  (2) internal energy   (4) all of these
8 Two hundred joules of heat is added to a system that performs 120 joules of work. The internal energy of the system is
  (1) 0 J   (2) 80 J   (3) 200 J   (4) none of these
9 Suppose that you rapidly stirred a batch of cake mix. The temperature of the cake mix would
  (1) increase   (2) decrease   (3) remain the same
10 When work is done by a system and no heat is added to it, the temperature of the system will
  (1) increase   (2) decrease   (3) remain the same

11 Opening a refrigerator door in a closed room will cause room temperature to
(1) increase   (2) decrease   (3) remain the same

12 One hundred joules of heat are added to a system whose internal energy changes 80. joules. Calculate the work done by the system
(1) 0.0 J   (2) 20. J   (3) 100. J   (4) 180. J

Thinking Physics

1 In cold regions, a pan of hot water placed outside will freeze faster than a pan of cold water. This phenomenon can be tested in your freezer. Can you explain why?

2 Thermostats in car engines are typically made with bimetallic strips. How do they work?

3 There may be more heat in a large iceberg than in a cup of boiling water. Can you explain why?

4 Why does a slice of pizza burn your mouth when you take a bite, but not burn your hands as you pick it up?

5 What factors cause a pressure cooker to cook food faster?

6 A student claims that she can "boil water until it freezes." How can that be possible?

7 Heat pumps can be used to heat your home when the temperature outside approaches zero. How can it be possible?

Self–Help Questions

1 The minimum average kinetic energy of the molecules in a substance occurs at a temperature of
(1) -273 K   (3) 0°C
(2) 273°C   (4) 0 K

2 As water is vaporized by boiling, its temperature
(1) decreases
(2) increases
(3) remains the same

3 A Celsius temperature reading may be converted to the corresponding Kelvin temperature reading by
(1) subtracting 273   (3) subtracting 180
(2) adding 273   (4) adding 180

4 Heat will always flow from object A to object B if object B has a lower
(1) mass   (3) specific heat
(2) total energy   (4) temperature

5 A temperature change of 51 Celsius degrees would be equivalent to a temperature change of
(1) 51 K   (2) 324 K   (3) 222 K   (4) -222 K

6   The graph at the right represents the relationship between the temperature of a gas and the
average kinetic energy (KE) of the molecules
of the gas. The temperature represented at
point X is approximately
(1) 273°C
(2) 0°C
(3) -273°C
(4) -373°C

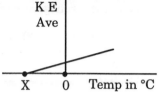

7   Which line on the graph at the right represents the relationship between the average
kinetic energy of the molecules of an ideal gas
and absolute temperature?
(1) 1
(2) 2
(3) 3
(4) 4

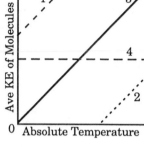

8   Block A, at 100°C, and block B, at 50°C, are brought together in a well-
insulated container. The internal energy of block A will
(1) decrease and the internal energy of block B will decrease
(2) decrease and the internal energy of block B will increase
(3) increase and the internal energy of block B will decrease
(4) increase and the internal energy of block B will increase

9   Equal masses of copper, iron, lead, and silver are heated from 20°C to
100°C. Which substance absorbs the *least* amount of heat?
(1) lead   (2) iron   (3) copper   (4) silver

10   Which graph best represents the relationship between the heat absorbed
(Q) by a solid and its temperature (T)? (Assume the specific heat of the
solid to be a constant.)

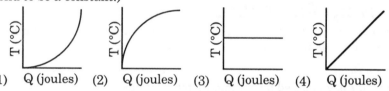

(1)   Q (joules)   (2)   Q (joules)   (3)   Q (joules)   (4)   Q (joules)

11   How much heat is required to raise the temperature of 1.00 kilogram of
liquid alcohol from its melting point to 0°C?
(1) 2.43 kJ   (2) 196 kJ   (3) 284 kJ   (4) 476 kJ

12   If 73 kilojoules of heat energy is added to 1.50 kilograms of ethyl alcohol
initially at 20. °C, what will be the final temperature of the liquid?
(1) 20. °C   (2) 22°C   (3) 40. °C   (4) 60. °C

13   Compared to the freezing point of pure water, the freezing point of a salt-
water solution is
(1) lower   (2) higher   (3) the same

14   After a hot object is placed in an insulated container with a cold object,
the hot object changes temperature and the cold object changes phase.
The total amount of internal energy in the system will
(1) decrease   (2) increase   (3) remain the same

Base your answers to questions 15 through 17 on the graph at the right which represents the variation in temperature as 1.0kilogram of a gas, originally at 200°C, loses heat at a constant rate of 2.0kilojoules per minute and eventually becomes a solid at room temperature.

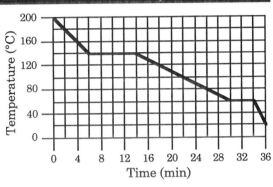

15 What is the heat of fusion of this substance?
    (1) 18 kJ/kg   (3) 8 kJ/kg
    (2) 9 kJ/kg   (4) 4 kJ/kg
16 What is the boiling temperature of this substance?
    (1) 200°C   (2) 140°C   (3) 90°C   (4) 60°C
17 Which phase has the highest specific heat?
    (1) solid   (2) liquid   (3) gas

Base your answers to questions 18 through 21 on the information below.

A 0.20-kilogram sample of ethyl alcohol is at a temperature of 28°C.

18 If the ethyl alcohol at 28°C loses 15 kilojoules of heat, it will be
    (1) liquid, only   (3) solid, only
    (2) gas, only   (4) liquid and solid
19 How much heat is needed to raise the temperature of the ethyl alcohol from 28°C to its boiling point?
    (1) 16 kJ   (2) 25 kJ   (3) 38 kJ   (4) 51 kJ
20 At its boiling point, how much heat is needed to change the ethyl alcohol from a liquid to a vapor?
    (1) 1.8×10^1 kJ   (3) 8.6×10^2 kJ
    (2) 1.7×10^2 kJ   (4) 4.4×10^3 kJ
21 The phase of ethyl alcohol at a temperature of 300 K and 1 atmosphere is
    (1) solid, only   (3) gas, only
    (2) liquid, only   (4) liquid and gas
22 For object A to have a higher absolute temperature than object B, object$A$ must have a
    (1) higher average internal potential energy
    (2) higher average internal kinetic energy
    (3) greater mass
    (4) greater specific heat
23 The sum of the kinetic and potential energies of an object's molecules is called the object's
    (1) temperature   (3) internal energy
    (2) heat of fusion   (4) specific heat
24 The random motion of small particles through a liquid is called
    (1) evaporation   (3) vaporization
    (2) condensation   (4) diffusion

25 If the pressure of a fixed mass of an ideal gas is doubled at a constant
    temperature, the volume of this gas will be
    (1) the same   (2) doubled   (3) halved   (4) quartered
26 Which graph best represents the relationship between volume and
    absolute temperature for an ideal gas at constant pressure?

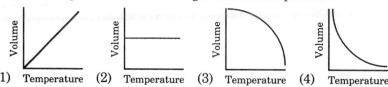

    (1)   Temperature   (2)   Temperature   (3)   Temperature   (4)   Temperature

27 The absolute temperature of a fixed mass of ideal gas is tripled while its
    volume remains constant. The ratio of the final pressure of the gas to its
    initial pressure is
    (1) 1 to 1   (2) 1.5 to 1   (3) 3 to 1   (4) 9 to 1
28 In an ideal gas, entropy is measure of the
    (1) volume of the molecules
    (2) mass of the molecules
    (3) forces of attraction between the molecules
    (4) disorder of the molecules
29 When the volume of a gas is reduced and no heat enters or leaves the
    system, the air temperature will
    (1) increase   (2) decrease   (3) remain the same
30 Two identical pieces of metal, at 15.°C and 25.°C respectively, are placed
    in contact with each other. The final block temperatures could not be
    5.0°C and 35.°C because it would violate which Law of Thermodynamics?
    (1) the First Law   (3) the Third Law
    (2) the Second Law   (4) none

3

Unit

Electricity And Magnetism

Important Terms To Be Understood

positive charge	electric potential	ampere
negative charge	electric field line	volt
elementary charge	equipotential lines	ion
conservation of charge	potential difference	plasma
charging by contact	electron volt	resistance
charging by induction	Millikan oil drop	ohm
grounding	series circuit	resistivity
coulomb	parallel circuit	power
Coulomb's Law	ammeter	ohmic conductor
electroscope	voltmeter	watt
electric field	galvanometer	Ohm's Law
intensity	flux density	induced emf
permeability	Left Hand Rules	alternating current
domains	electromagnetic	generator
flux	solenoid	weber
electromagnetic waves	tesla	Lenz's Law

I. Static Electricity

Static electricity deals with electrical charges at rest. The term "at rest" indicates that there is no net transfer of charge in any given direction. The term "at rest" does not mean that the charges themselves are not in motion. Static charges are most evident in dry environments such as the clothes dryer or indoors during winter months.

1.1 Microstructure Of Matter

The atom consists of three main particles. The proton has a positive charge and is found in the nucleus. The **neutron** has no charge and is also located in the nucleus. The **electron** has a negative charge and orbits the nucleus. The magnitude of the positive and negative charges is the same. The unit of charge in the *SI system* is the **coulomb**. One coulomb (C) represents the charge of 6.25×10^{18} electrons. Each electron has a charge of 1.6×10^{-19} coulomb. The electron and the proton are among a class of particles called **charged carriers**.

The electron has the smallest negative charge, and the proton has the smallest positive charge. These charges are equal in magnitude and opposite in sign. One **negative elementary charge** is defined as the charge on an electron. One **positive elementary charge** is the charge on one proton. The symbol for an elementary charge is **e**.

Atom

Neutral atoms have an equal number of positive and negative charges. They can be used to detect charged particles. The protons are difficult to remove because they are held together by very strong nuclear forces. Electrons are much easier to remove, because they are more loosely bound by the electrical force. Particles are said to be charged when they have an **excess** or a **deficiency** of electrons.

When electrons are lost or gained by a neutral atom, the resulting particle is electrically charged and is called an **ion**. The charge on an ion depends on the excess elementary charges. For example, an atom with 10 protons and 10 electrons is neutral. If two electrons (two elementary units of negative charge) are removed from this atom, the resulting ion is said to have a charge or +2 elementary units.

1.2 Charge Detection

A positively charged strip sets up an electric field around it. This field attracts a neutral pithball because the electrons migrate to one side of the pithball, causing one side of the pithball to be more negative than the other side. The positively charged strip attracts the negative side of the pithball. The pithball moves towards the charged strip.

A negatively charged strip will also attract a pithball. The electrons are repelled by the electric field of the charged strip leaving one side of the pithball more positive than the other side. Positive and negative charges attract each other. The pithball moves towards the strip.

When a neutral strip is in the region of a neutral pithball, there is no attraction.

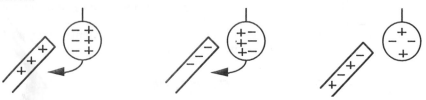

Note: The only proof that an object is charged is repulsion. If two objects are repelled, they have the same charge.

1.3 Types Of Charge

Certain neutral objects can become **charged by rubbing** them with wool or fur. During the rubbing, electrons are removed from one object and placed on the other object. When rubber is rubbed with wool, electrons are removed from the wool and left on the rubber. The rubber has an excess of electrons. This gives the rubber a negative charge. The wool has a deficiency of electrons so it has a positive charge.

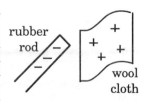

**Transference
of Electrons**

The negative and positive charge was defined by Benjamin Franklin. The charge gained by one object exactly equals the charge lost by the other object. This principle is known as the **conservation of charge**. It states that *the net charge in a closed system is constant.*

```
Practical Applications
•  photo copy machines/printers
•  lightning
•  lightning rods
•  electrostatic precipitators
•  ground wires in electrical apparatus
```

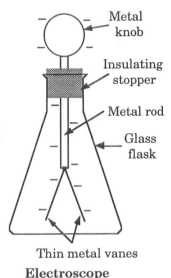

Metal knob

Insulating stopper

Metal rod

Glass flask

1.4 The Electroscope

The **electroscope** is used to detect the presence of charge. A neutral electroscope behaves the same towards positive and negative charges. A charged electroscope can be used to detect both types of charges. An electroscope may be charged in two ways:

Thin metal vanes

Electroscope

a) **Contact** is touching the electroscope with the charged object. The electroscope will have the **same** charge as the charging body, due to the transfer of electrons to or from the electroscope. When one charged object touches an identical uncharged object, the charges redistribute evenly on both objects.

b) **Induction** is a process by which a charged object causes a redistribution of the charges in another object without contact. It produces an **opposite** charge in a four step process:

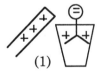 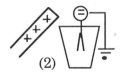

(1)   (2)   (3)   (4)

1) Bring a positively charged rod near the neutral electroscope. The leaves will diverge. Electrons have moved up to the knob leaving both leaves positively charged. The leaves repel each other.

2) Ground the electroscope with the charged object still in the region. A **ground** is an object that can give up or accept a large number of electrons without becoming charged enough to measure. (i.e. the Earth). This causes the leaves to fall. Electrons move into the electroscope from ground while the positive charge is near. The leaves now have an equal number of positive and negative charges in them and have a neutral charge.

3) Remove the ground connection. The leaves are still together.

4) Remove the rod, and the charges redistribute evenly over the electro-scope. The leaves diverge. The electroscope is now negatively charged.

Note: *The process is the same for charging with a negative object. The only difference is that in step 2 electrons leave the electroscope and return to the ground. The electroscope is now positively charged. Remember that only the electrons move.*

Lightning Produced by Opposite Charged Clouds

1.5 Coulomb's Law

Charges exert forces on each other. These forces have been shown experimentally to be directly proportional to the magnitude of the charge and inversely proportional to the distance between the charges. *Coulomb's Law is limited to point charges that are small in size.*

$$F = \frac{kq_1\, q_2}{r^2}$$

Where:

k   = constant of proportionality = 9.0×10^9 N·m²/C²
q_1, q_2 = charges measured in coulombs
r   = distance between charged particles in meters

Example

What is the force between two point charges, that each have a charge of -0.003 coulombs at a separation of 10 meters?

Given: q_1 = q_2 = -0.003 C
           r = 10.0 m

Find: F

Solution: $F = \dfrac{kq_1\, q_2}{r^2}$

$= \dfrac{(9 \times 10^9 \text{ N·m}^2/\text{C}^2)(-0.003 \text{ C})(-0.003 \text{ C})}{(10 \text{ m})^2}$

= **810 newtons**

A positive value for force indicates repulsion.

Questions

1   A neutral rubber rod is rubbed with fur and acquires a charge of -2×10^{-6} coulomb. The charge on the fur is
   (1) $+1 \times 10^{-6}$ C   (3) -1×10^{-6} C
   (2) $+2 \times 10^{-6}$ C   (4) -2×10^{-6} C

2   When an object is placed near a negatively charged electroscope, the leaves of the electroscope diverge farther. Which statement about the object is true?
   (1) It must be neutral.
   (2) It must be positively charged.
   (3) It must be negatively charged.
   (4) It may be either positively or negatively charged.

3   A body will maintain a constant negative electrostatic charge if the body
   (1) maintains the same excess of electrons
   (2) maintains the same excess of protons
   (3) continuously receives more electrons than it loses
   (4) continuously receives more protons than it loses

4   When two neutral materials are rubbed together, there is a transfer of electrical charge from one material to the other. The total electrical charge for the system
   (1) increases as electrons are transferred
   (2) increases as protons are transferred
   (3) remains constant as protons are transferred
   (4) remains constant as electrons are transferred

5   Which diagram correctly represents the charge distribution on a neutral electroscope in the presence of a negatively charged object?

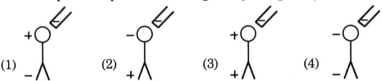

6   As a charged electroscope is grounded, the net charge on the electroscope will
   (1) decrease   (2) increase   (3) remain the same

7   An electroscope has been charged by induction with a positively charged rod. Which diagram best shows the charge distribution on the electroscope?

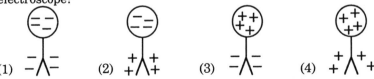

8   Which is the most massive positively charged particle?
   (1) positron   (3) neutron
   (2) electron   (4) proton

9   The ratio of the magnitude of charge on an electron to the magnitude of charge on a proton is
   (1) 1:2   (2) 1:1   (3) $1:1.6 \times 10^{18}$ (4) 1:1,840

10 An electrically neutral object can be attracted by a positively charged body because
(1) the net charge in a closed system varies
(2) the neutral body becomes charged by contact
(3) the charges on a neutral body can be redistributed
(4) like charges repel each other

11 Sphere A has a charge of +2 units and sphere B, which is identical to sphere A, has a charge of -4 units. If the two spheres are brought together and then separated, the charge on each sphere will be
(1) -1 unit   (2) -2 unit   (3) +1 unit   (4) +4 unit

12 Which magnitude of charge could NOT be found on an object?
(1) -0.8 x10^{-19} coul   (3) +1.6x10^{-19} coul
(2) -1.6x10^{-19} coul   (4) +3.2x10^{-19} coul

13 After a neutral object loses 2 electrons, it will have a net charge of
(1) -2 elementary charges
(2) +2 elementary charges
(3) -3.2 x 10^{-19} elementary charges
(4) +3.2 x 10^{-19} elementary charges

14 A small, uncharged metal sphere is placed near a larger, negatively charged sphere. Which diagram best represents the charge distribution on the smaller sphere?

(1)   (2)   (3)   (4)

15 Which is equivalent to three elementary charges?
(1) 2.4 x10^{-19} coulombs   (3) 4.8 x10^{-19} coulombs
(2) 2.0 x10^{-19} coulombs   (4) 5.4 x10^{-19} coulombs

16 How many excess electrons are contained in a charge of -8.0 x 10^{-19} coulomb?
(1) 5   (2) 2   (3) 8   (4) 4

17 A sphere has a negative charge of 6.4 x 10^{-7} coulomb. Approximately how many electrons must be removed to make the sphere neutral?
(1) 1.6 x 10^{-8} (3) 6.4 x 10^{26}
(2) 9.8 x 10^{5} (4) 4.0 x 10^{12}

18 As an electron approaches a proton, the electrostatic force acts on
(1) the electron, only   (3) both the proton and the electron
(2) the proton, only   (4) neither the proton or the electron.

19 The diagram represents two charges at a separation of d. Which would produce the greatest increase in the force between the two charges?
(1) doubling charge q_1, only
(2) doubling d, only
(3) doubling charge q_1 and d, only
(4) doubling both charges and d

20 The electrical force of attraction between two point charges is F. The charge on one of the objects is quadrupled and the charge on the other object is doubled. The new force between the objects is
(1) 6 F   (2) 2 F   (3) 8 F   (4) 4 F

II. The Electric Field

An electric field is said to exist in any region of space in which an electric force acts on a charge. The field exists around every charged object. A similarity exists between the electric field around a uniformly charged spherical body and the gravitational field around a sphere. The electric field intensity is a vector quantity. The **SI** unit for electric field strength is the **newton/ coulomb**. In comparison, the gravitational field strength can be expressed in newtons/kilogram ($g = F/m$). The magnitude of the field at any point is equal to:

$$E = F/q$$

Where:
  q   is the charge in coulombs on the test charge
  F   is the force in newtons on the test charge **q**
  E   is the electric field intensity in newtons/coulomb.

The direction of the electric field is defined as the direction of force on a positive test charge placed in the field.

An **electric field line** is the line along which a charged particle would move as a result of its interaction with the electric field. The direction of a field line is the direction of the force exerted on a positively charged particle placed at that location on the field line.

Electric field lines leave the surface of the charge at right angles and extend away from a positive charge. The lines around a negative charge point toward the charge. Charges interact through their fields. Below are several examples of interacting fields:

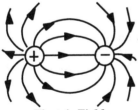

Electric Field
between opposite charges

Electric Field
between two positive charges

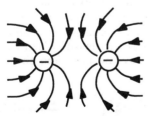

Electric Field
between two negative charges

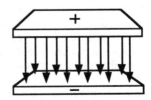

Electric Field
oppositely charged parallel plates

Notice that the field lines do not cross each other.

As charges distribute on the surface of a conductor, the field lines are normal to the surface. The field around the charged sphere acts as though all of the charge were concentrated at the center. *The field within the charged conducting sphere is zero.* The intensity of the electric field around a point charge varies *inversely with the square* of the distance from the point charge. If you double the distance from the charge, the field intensity is quartered.

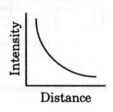

Example

Calculate the force exerted on an electron in an electric field, whose intensity is 2.0 x 10^2 N/C

Given: E = 2.0 x 10^2 N/C   q = 1.6 x 10^{-19} C

Find: F

Solution: F = Eq
               F = (2.0 x 10^2 N/C)(1.6 x 10^{-19} C)
               F = 3.2 x 10^{-17} N

The electric field around a point charge is calculated from:

$$E_{point} = \frac{kq}{r^2}$$

The electric field around a uniformly charged rod is radially directed and its intensity varies inversely with the distance from the rod. *If you double the distance from the rod the electric field with be halved.*

The electric field between parallel charged plates is essentially *uniform* if the distance between the plates is small compared to the dimensions of the plates. This uniform field produces a constant force on a given charge placed anywhere in the field.

2.1 Electric Potential

The **electric potential** at any point in an electric field is the work required to bring a unit positive charge from infinity to that point. When a charge is moved against the force of an electric field, work is done on the charge and the potential energy of the charge is increased. When the charge moves in response to the field, work is done by the field and the potential energy of the charge is decreased.

In order to move a proton from point B to point A, work must be done on the charge. The work becomes potential energy. Energy is conserved.

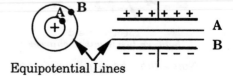

Equipotential Lines

Equipotential lines cross electric field lines at right angles. They represent positions of equal potential energy. As a charge moves on an equipotential line, there is no change in potential energy. As the charge crosses equipotential lines, the potential energy changes.

2.2 Potential Difference or Potential Drop

The potential difference between two points in an electric field is the change in energy per unit charge, as the charge is moved from one point to another. The **SI** unit of electric potential is the **volt**.

$$V = W/q$$

When: V is potential difference (measured in volts)
         W is work (measured in joules)
         q is charge (measured in coulombs)

1 volt = 1 joule/1 coul

One volt is the potential difference that exists between two points, if one joule of work is required to transfer one coulomb of charge from one point to the other against the electric force.

> **Practical Applications**
> - batteries
> - household electricity
> - power transmission lines
> - Van de Graaff generators

As an electron moves along an equipotential line, there is no change in potential energy. As the electron is moved from the negative plate to the positive plate it crosses a potential measured in volts. The electron gains kinetic energy as it crosses the gap between the plates. The energy can be calculated in the unit of electron volts. An **electron volt (eV)** is the energy required to move one elementary charge (that is, 1.6×10^{-19} coulombs) through a potential of one volt.

$$W = Vq$$

$$1.6 \times 10^{-19}\,J = (1\ Volt)(1.6 \times 10^{-19}\ coulombs)$$

Therefore, **1 eV = 1 V x 1 elementary charge = 1.6×10^{-19} J**

> **Practical Applications Of Electron Volts**
> - energy of photons
> - measuring band gaps in semiconductors
> - energy levels in atoms

At location *A*, the electron has 100 eV of electric potential energy. At location *B*, the electron has 100eV of kinetic energy.

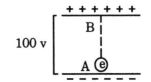

The intensity of an electric field is the rate at which electric potential changes with position. It may be expressed in terms of the electric potential and distance.

$$V = \frac{W}{q} = \frac{\text{joules}}{\text{coulomb}} = \left[\frac{\text{newton}}{\text{coulomb}}\right]\text{meter}$$

Therefore, $E = \dfrac{V}{d}$ (measured in volts/meter or newtons/coulomb)

$$\frac{\text{volt}}{\text{meter}} = \frac{\text{joules/coulomb}}{\text{meter}} = \frac{\text{joules}}{\text{coulomb·meter}} = \frac{\text{newton·meter}}{\text{coulomb·meter}} = \frac{\text{newton}}{\text{coulomb}}$$

In 1911, **Millikan** won the **Nobel Prize** for determining the fundamental unit of charge. He measured the forces on a charged oil drop in a uniform electric field. The field between two parallel plates can be adjusted so the oil drop remains suspended in the space between the plates. At that time the forces acting on the oil drop are balanced.

$F \text{ (electric)} = F \text{ (gravitational)}$

Once balanced, the amount of charge on the oil drop can be calculated. Millikan found that the charge was always an integral multiple of a small constant. He determined the elementary charge to be equal to 1.6×10^{-19} coulomb. It is the charge of an electron or proton. It is the smallest charge possible.

Example

Two large parallel metal plates are separated by a distance of 2.0×10^{-3} meters. The plates are attached to a 4.0×10^3 volt source.

a) Calculate the field intensity between the plate.
b) Determine the force on an electron in the field.
c) Calculate the work necessary to move an electron from the positive plate to the negative plate in electron volts and in joules.

Given:　　$d = 2.0 \times 10^{-3}$ m
　　　　　$V = 4.0 \times 10^3$ V
　　　　　$q = e = 1.6 \times 10^{-19}$ C

Find:　a) Field intensity
　　　　b) Force
　　　　c) Work

Solution:

(a) E = V/d

         = $(4.0 \times 10^3 \text{ V}) / (2.0 \times 10^{-3} \text{ m})$

         = 2.0×10^6 V/m

(b) F = qE

         = $(1.6 \times 10^{-19} \text{ C})(2.0 \times 10^3 \text{ N/C})$

         = 3.2×10^{-16} N

(c) W = $qV = 1$ electron $\times (4.0 \times 10^3 \text{ V})$

         = 4.0×10^3 eV

Since: 1 eV   = 1.6×10^{-19} J

              W = $(1.6 \times 10^{-19} \text{ J/eV})(4.0 \times 10^3 \text{ eV})$

                 = 6.4×10^{-16} J

Questions

1   If you charge a comb by running it through your hair, the size of the charge on the comb might most reasonably be
(1) 1 C   (2) 10^3 C   (3) 10^{-3} C   (4) 10^{-9} C

2   A metallic sphere is positively charged. The field at the center of the sphere due to this positive charge is
(1) positive
(2) negative
(3) zero
(4) dependent on the magnitude of the charge

3   Two charged spheres are shown in the diagram. Which polarities will produce the electric field.
(1) A and B both negative
(2) A and B both positive
(3) A positive and B negative
(4) A negative and B positive

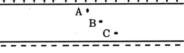

4   The diagram represents two charged parallel plates. What is the intensity of the electric field relative to locations A, B, and C?

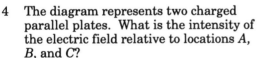

(1) greater at A than at B (3) greater at B than at C
(2) greater at C than at A (4) the same at A, B, and C

5   If 8.0 joules of work is required to transfer 4.0 coulombs of charge between two points, then the potential difference between the two points is
(1) 6.4 V   (2) 2.0 V   (3) 32 V   (4) 40. V

6   Which quantity is equivalent to 3.2×10^{-17} joule?
(1) 8.00×10^{-3} eV   (3) 3.20 eV
(2) 3.20×10^{-17} eV   (4) 200 eV

7   Gravitational force is to mass as electrical force is to
(1) weight   (2) charge   (3) gravity   (4) electricity

8   A charged particle is placed in an electric field E. If the charge on the particle is doubled, the force exerted on the particle by the field E is
(1) unchanged   (2) doubled   (3) halved   (4) quadrupled

9   The electric field between two parallel plates connected to a 45-volt battery is 500. volts per meter.   The distance between the plates is closest to
    (1) 11. m   (2) 22. m   (3) 50. m   (4) 0.090 m
10 What is the charge on an object that experiences a force of 5 newtons in and electric field of 50 newtons per coulomb?
    (1) 10 coulombs   (3) 0.2 coulomb
    (2) 2.0 coulombs   (4) 0.1 coulomb

Base your answer to questions 11 through 15 on the diagram below which represents a source connected to two large, parallel metal plates. The electric field intensity between the plates is 3.75×10^4 newtons per coulomb.

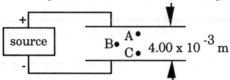

11 What is the potential difference of the source?
    (1) 9.38×10^5 volts   (3) 3.75×10^2 volts
    (2) 4.00×10^3 volts   (4) 1.50×10^2 volts
12 What would be the magnitude of the electric force on a proton at point A?
    (1) 1.60×10^{-19} newton   (3) 0 newtons
    (2) 6.00×10^{-15} newton   (4) 3.75×10^4 newtons
13 Compared to the work done in moving an electron from point A to point B to point C, the work done in moving an electron directly from point A to point C is
    (1) less   (2) greater   (3) the same
14 If the source is replaced with one having twice the potential difference and the distance between the plates is halved, the electric field intensity between plates will
    (1) decrease   (2) increase   (3) remain the same
15 As a proton moves from A to B to C, the electric force on the proton
    (1) decreases   (2) increases   (3) remains the same

Base your answers to questions 16 through 19 on the diagram below, which shows a positive charge placed at A.

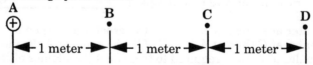

16 The electric field intensity at point B is E. At point D the field intensity will be equal to
    (1) $\frac{1}{9}E$ (2) $\frac{1}{3}E$ (3) $3E$ (4) $9E$
17 If a positive charge is placed at point B, the force exerted on this charge by charge A, will be directed toward
    (1) the top of the page   (3) A
    (2) the bottom of the page   (4) C
18 If the charge is moved from point B to point C, the force between the two charges will
    (1) decrease   (2) increase   (3) remain the same

19 The electric field surrounding charge *A* is best represented by which diagram?

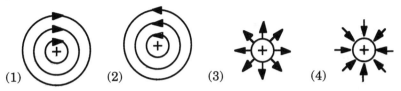

(1) (2) (3) (4)

III. Electrons In Motion

Electrons move through conductors from negatively charged objects to positively charged objects. The amount of current is the number of charges per unit time. Electron current flows from negative to positive. Conventional current flows from positive to negative. Note: *Current references in this review book refer to electron current.*

The *SI* unit for current is the **ampere** (or **amp**), which is the flow of one coulomb of charge per second or 6.25×10^{18} electrons per second.

$$I = \Delta q / \Delta t$$

Where:
I = current measured in amperes
Δq = charge measured in coulombs
Δt = time measured in seconds

An **ammeter** is a device used to measure current. The symbol for an ammeter is shown at the right.

Conductivity In Solids

Solids vary in their ability to conduct current. The conductivity of solids depends on the number of free charges per unit volume and the mobility of the charges. In general, *since **metals** have a large number of free electrons they are **good conductors**. **Nonmetals** are poor conductors of electricity.* Substances with few free electrons are called **insulators**. No solid is a perfect insulator, but in some solids, such as glass and fused quartz, the conductivity is so low that they are good insulators. Some materials whose resistivities lie between metals and insulators are called **semiconductors**.

Conductivity In Liquids

Liquids also vary in their ability to conduct an electric current. Pure water is not a good conductor. Many chemical compounds, called **electrolytes**, dissociate in aqueous solution into positively and negatively charged particles, called **ions**. In such solution both positive and negative ions are free to move. The solution can conduct an electric current. Both positive and negative charges move in solution.

The motion of positive charges in one direction is equivalent to motion of negative charges in the other direction.

Conductivity In Gases

Ionized gases conduct electric current. Gases which are normally composed almost entirely of neutral molecules may become ionized by high energy radiation, electric fields, and collisions with particles. **Ionized gases**, also known as **plasma**, are the fourth state of matter.

Plasma is the most common phase of matter in the universe. Plasma is found in space, outside of our protective atmosphere. The stars, the streams of ions (that radiate from the stars), and the Van Allen belts around our planet, are examples of this fourth phase of matter. Plasma, or an ionized gas, **may consist of positive ions, negative ions**, and **electrons which are free to move.**

Conditions Necessary For Current Flow

A **potential difference** is required to create a flow of charge between two points in a conductor. The conductor must form a complete circuit to maintain a flow of charge. Work must be done on electrons to move them from a positive object to a negative object. The work per unit charge is called the volt. One volt equals one joule per coulomb.

$$V = \frac{W}{q}$$

Where: V = electric potential measured in volts
               W = work measured in joules
               q = charge measured in coulombs

The **voltmeter** is represented in the circuit by the symbol shown at the right.

3.1 Ohm's Law

At constant temperature, the current in a metallic conductor is directly proportional to the potential difference between its ends. **Ohm's Law** is specific for certain materials and not a general law of electricity.

$$R = \frac{V}{I}$$

Where: V = potential measured in volts
            I = current measured in amperes
           R = resistance measured in ohms

The resistance is the constant of proportionality in Ohm's Law.

3.2 Resistance

Resistance is the ratio of the potential difference across a conductor to the current in it. The **SI** unit of resistance is the ohm. The symbol " Ω " is used to represent the ohm.

$$\frac{V \text{ (volts)}}{I \text{ (amps)}} = R \text{ } (\Omega)$$

The slope of a potential difference - current graph is resistance for a metallic conductor at constant temperature.

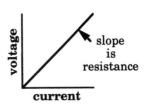

The resistance in a conducting wire can be expressed by the following formula:

$$R = \frac{\rho l}{A}$$

Where: R =   resistance in ohms
ρ =   resistivity in ohms · m
l =   length in meters
A =   cross sectional area in square meters

In the metric system, the resistivity of a substance is defined as the resistance of a cube, with edges 1 meter long, at a given temperature (usually 0 degrees or 20 degrees Celsius).

The resistance is represented in a circuit diagram by the symbol shown at the right.

3.3 Laws Of Resistance

A. **Length**: As the length of the wire increases, the resistance also increases.

B. **Cross-sectional area**: As the wire diameter increases, the resistance decreases. Since $A = \pi r^2$, the area increases as the radius squared (r^2).

C. **Temperature**: As the metal wire temperature increases, the resistance increases. The resistance of nonmetals and solutions usually decreases with increasing temperature. At extremely low temperatures, some materials have no measurable resistance. This phenomenon is known as **superconductivity**.

Questions

1   If the length of a copper wire is reduced by half, then the resistance of the wire will be
(1) halved   (2) doubled   (3) quartered   (4) quadrupled
2   As the temperature of the metal filament of an electric light bulb increases, the resistance of the filament
(1) decreases   (2) increases   (3) remains the same

3   If the cross-sectional area of a fixed length of wire were decreased, the
    resistance of the wire would
    (1) decrease   (2) increase   (3) remain the same
4   The ratio of the potential difference across a conductor to the current in
    the conductor is called
    (1) current   (3) resistance
    (2) conductance   (4) electric potential
5   What is the current in a conductor if 6.25×10^{18} electrons pass a given
    point each second?
    (1) 1 ampere   (3) 2.6 amperes
    (2) 1.6×10^{-19} amp   (4) 6.25×10^{18} amp
6   Most metals are good conductors because they have
    (1) molecules that are close together
    (2) high melting points
    (3) many intermolecular spaces through which the current can flow
    (4) a large number of free electrons
7   When 20. coulombs of charge pass a given point in a conductor in
    4.0seconds, the current in the conductor is
    (1) 80. amperes   (3) 16 amperes
    (2) 0.20 ampere   (4) 5.0 amperes
8   If the voltage across a 12 -ohm resistor is 4.0 volts, the current through
    the resistor is
    (1) 0.33 A   (2) 48 A   (3) 3.0 A   (4) 4.0 A
9   Which condition must exist between two points in a conductor in order to
    maintain a flow of charge?
    (1) a potential difference   (3) a low resistance
    (2) a magnetic field   (4) a high resistance
10 If 60. electrons pass a given point in a conductor in one second, the
    current in this conductor is
    (1) 9.6×10^{-18} ampere   (3) 1.6×10^{-20} ampere
    (2) 1.6×10^{-19} ampere   (4) 2.7×10^{-21} ampere
11 A uniform copper wire has a resistance of 100 ohms. If the wire is cut
    into 10 equal lengths, the resistance of each piece will be
    (1) 1 ohm   (2) 10 ohms   (3) 100 ohms   (4) 1,000 ohms
12 Which graph best represents the relationship between the resistance (R)
    of a solid metallic conductor of constant cross section and its length (L)?

    (1)   L   (2)   L   (3)   L   (4)   L

13 Which graph best represents the relationship between the current in a
    metallic conductor and the applied potential difference?

    (1) potential   (2) potential   (3) potential   (4) potential
         difference   difference   difference   difference

14 At room temperature, which segment of copper wire has the highest resistance?
 (1) 1.0 m length, 1.0 x 10^{-6} m^2 cross-sectional area
 (2) 2.0 m length, 1.0 x 10^{-6} m^2 cross-sectional area
 (3) 1.0 m length, 3.0 x 10^{-6} m^2 cross-sectional area
 (4) 2.0 m length, 3.0 x 10^{-6} m^2 cross-sectional area

Base your answers to questions 15 through 19 on the graph at the right which represents data obtained by applying potential differences to a metallic conductor at a constant temperature.

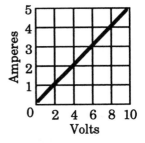

15 The resistance of the conductor is approximately
 (1) 1.0 ohm
 (2) 2.0 ohms
 (3) 0.5 ohm
 (4) 4.0 ohms
16 If the temperature of the conductor is increased, the amount of the current at 10 volts would be
 (1) less   (2) greater   (3) the same
17 If the length of the conductor were increased, the amount of current at 10volts would be
 (1) less   (2) greater   (3) the same
18 Compared with a conductor of the same material with a larger cross-sectional area, the resistance of this conductor is
 (1) less   (2) greater   (3) the same
19 Which mathematical relationship exists between the current and the voltage?
 (1) direct linear   (3) inverse
 (2) direct square   (4) inverse square

3.4 Circuits

A **circuit** is a closed loop consisting of an *energy source, connecting wire and a resistor*. The voltage supplied can be by many different sources. Chemical, mechanical, heat, and light energy can be changed into electrical energy. The connecting wire is assumed to have zero resistance.

Voltmeter is a high resistance meter placed in parallel in a circuit to measure the voltage drops across a device.

Ammeter is a low resistance meter placed in series to record the current flowing in a circuit.

The resistance can be determined by the ammeter - voltmeter method using Ohm's Law.

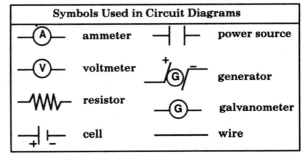

Symbols Used in Circuit Diagrams		
—(A)— ammeter	—┤├—	power source
—(V)— voltmeter	+/(G)/—	generator
—WW— resistor	—(G)—	galvanometer
┤├ cell	——	wire

Example

Using the circuit at the right, calculate the value of the unknown resistance R.

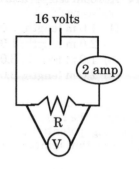

Given: V = 16 volts I = 2.0 amps

Find: **R**

Solution: Using Ohm's Law $V = IR$ and solving for **R**

$$R = V/I$$
$$R = 16 \text{ volts}/2.0 \text{ amps}$$
$$= 8.0 \ \Omega$$

3.5 Series Circuits

A **series circuit** is one in which there is only one current path. The current is the same in all the components of a series circuit. The sum of the potential drops in a series circuit is equal to the total applied potential difference. The total resistance of the series circuit is equal to the sum of the resistance of its components and can be derived as follows.

$$V_t = V_1 + V_2 + V_3 + \dots$$

Where, $V = IR$

Thus, $I_T R_T = I_1 R_1 + I_2 R_2 + I_3 R_3 + \dots$

But, $I_T = I_1 = I_2 = I_3 = \dots$

So, $R_T = R_1 + R_2 + R_3 = \dots$

$$I_{total} = I_1 = I_2 = I_3 = \dots$$
$$V_{total} = V_1 + V_2 + V_3 = \dots$$
$$R_{total} = R_1 + R_2 + R_3 = \dots$$

Example

Three resistors of 20, 30, and 40 ohms are connected in series to an applied potential of 120 volts.

Calculate: a) total resistance
               b) current through each resistor
               c) potential drop across each resistor

Given: R_1 = 20 Ω R_3 = 40 Ω
           R_2 = 30 Ω V_T = 120 V

Solution:

a) $R_T = R_1 + R_2 + R_3 = 20\,\Omega + 30\,\Omega + 40\,\Omega = 90\,\Omega$

b) $I_T = \dfrac{V_T}{R_T} = \dfrac{120\text{ V}}{90\,\Omega} = 1.3\text{ A}$

In the series circuit, current is the same in all parts of the circuit.

$$I_T = I_1 = I_2 = I_3$$

c) $V_1 = I_1 R_1 = (1.3\text{ A})(20\,\Omega) = 26\text{ V}$

$\quad V_2 = I_2 R_2 = (1.3\text{ A})(30\,\Omega) = 39\text{ V}$

$\quad V_3 = I_3 R_3 = (1.3\text{ A})(40\,\Omega) = 52\text{ V}$

In a series circuit, voltage total is equal to the sum of the voltage drops across each resistor.

Check: $\quad V_T = V_1 + V_2 + V_3$
$$= 26\text{ V} + 39\text{ V} + 52\text{ V} = 117\text{ V}$$
$$V_T = 120\text{ V (two significant digits)}$$

Practical Applications
- holiday tree lights
- fuses
- circuit breakers
- thermal switches in a hair dryer

3.6 Parallel Circuits

A **parallel circuit** is one in which there is more than one current path. The potential drop is the same across each branch of a parallel circuit. The total current in a parallel circuit is equal to the sum of the branch currents. The reciprocal of the equivalent resistance of a parallel circuit is equal to the sum of the reciprocals of the branch resistances and can be derived as follows:

$V_{total} = V_1 = V_2 = V_3 = \ldots$

$I_{total} = I_1 + I_2 + I_3 = \ldots$

$\dfrac{1}{R_{total}} = \dfrac{1}{R_1} + \dfrac{1}{R_2} + \dfrac{1}{R_3} = \ldots$

$I_T = I_1 + I_2 + I_3 + \ldots$

Where, $\quad I = V/R$

Thus, $\quad \dfrac{V_T}{R_T} = \dfrac{V_1}{R_1} + \dfrac{V_2}{R_2} + \dfrac{V_3}{R_3} + \ldots$

But, $\quad V_T = V_1 = V_2 = V_3 = \ldots$

So, $\quad \dfrac{1}{R_T} = \dfrac{1}{R_1} + \dfrac{1}{R_2} + \dfrac{1}{R_3} = \ldots$

Example

Three resistors of 20, 30, and 40 ohms are connected in parallel to an applied potential of 120 volts.

Calculate: a) the total resistance
b) the potential difference across each resistor
c) the current through each resistor

Given: $R_1 = 20\,\Omega$ $R_3 = 40\,\Omega$

$R_2 = 30\,\Omega$ $V_T = 120\,V$

Find: a) R_{total}
b) V_1, V_2, V_3
c) I_1, I_2, I_3

Solution:
a)

$$\frac{1}{R_T} = \frac{1}{R_1} + \frac{1}{R_2} + \frac{1}{R_3}$$

$$\frac{1}{R_T} = \frac{1}{20\,\Omega} + \frac{1}{30\,\Omega} + \frac{1}{40\,\Omega}$$

$$\frac{1}{R_T} = \frac{6}{120\,\Omega} + \frac{4}{120\,\Omega} + \frac{3}{120\,\Omega}$$

$$\frac{1}{R_T} = \frac{13}{120\,\Omega} \quad , \quad R_T = \frac{120\,\Omega}{13} = 9.2\,\Omega$$

b) In the parallel circuit, voltage is the same across each resistor in the circuit.

$$V_T = V_1 = V_2 = V_3 = 120\,V$$

c)

$$I_1 = \frac{V_1}{R_1} = \frac{120\ volts}{20\,\Omega} = 6.0\ amp$$

$$I_2 = \frac{V_2}{R_2} = \frac{120\ volts}{30\,\Omega} = 4.0\ amp$$

$$I_3 = \frac{V_3}{R_3} = \frac{120\ V}{40\,\Omega} = 3\ amp$$

Note: In parallel circuits, the sum of the currents in the resistors is equal to the total current from the source

$$I_T = I_1 + I_2 + I_3$$
$$= 6.0\,A + 4.0\,A + 3.0\,A$$
$$I_T = 13\,A$$

Check:

$$I_T = \frac{V_T}{R_T} = \frac{120\,V}{9.2\,\Omega} = 13\,A$$

3.7 Kirchkoff's Laws

First law - Conservation of Charge. The algebraic sum of the current entering any current junction is equal to zero. **A** represents a junction in an electric circuit. Nine amps are entering **A**; therefore, according to Kirchkoff's First Law, nine amps must come out of junction **A**.

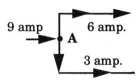

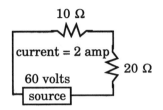

Second law - Conservation of Energy. The algebraic sum of all the potential drops, and the applied voltage around a complete circuit is equal to zero. According to Kirchkoff's Second Law, the applied voltage (+60 volts) minus the potential drops around the complete circuit, must equal zero.

60 volts - (20 Ω x 2 amp) - (10 Ω x 2 amp) = 0
60 volts - 40 volts - 20 volts = 0

Questions

1 The diagram at the right represents a segment of a circuit. The current in wire *X* may be
   (1) 1 ampere
   (2) 2 amperes
   (3) 3 amperes
   (4) 4 amperes

2 In the diagram shown, how many amperes is the reading on ammeter *A*?
   (1) 5 amperes
   (2) 2 amperes
   (3) 3 amperes
   (4) 7 amperes

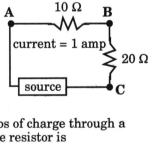

3 If the potential difference between points *A* and *B* in the electric circuit shown is 10 volts, what is the voltage between points *A* and *C*?
   (1) 5 volts
   (2) 10 volts
   (3) 20 volts
   (4) 30 volts

4 If 4 joules of work are required to move 2 coulombs of charge through a 6-ohm resistor, the potential difference across the resistor is
   (1) 1 volt   (3) 6 volts
   (2) 2 volts   (4) 8 volts

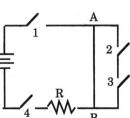

5 In the circuit represented, which switch or switches must be closed to produce a current in conductor *AB*?
   (1) 1 and 4, only   (3) 1, 2, and 3
   (2) 2 and 3, only   (4) 4, only

6   Which quantity must be the same for each component in a series circuit?
   (1) voltage   (2) power   (3) resistance   (4) current

7   What is the current in the circuit represented in the diagram?
   (1) 1 amp
   (2) 2 amp
   (3) 3 amp
   (4) 4 amp

8   The diagram represents a circuit with two resistors in series. If the total resistance of R_1 and R_2 is 24 ohms, the resistance of R_2 is
   (1) 1.0 ohm
   (2) 0.50 ohm
   (3) 100 ohm
   (4) 4.0 ohm

9   A 10-ohm and a 20-ohm resistor are connected in parallel to a constant voltage source. If the current through the 10-ohm resistor is 4.0 amperes, then the current through the 20-ohm resistor is
   (1) 1 ampere   (2) 2 amperes   (3) 8 amperes   (4) 4 amperes

10   As the number of resistors connected in parallel to a constant voltage source is increased, the potential difference across each resistor
   (1) decreases   (2) increases   (3) remains the same

11   Compared to the current in the 20. ohm resistance in the circuit diagram shown, the current in the 5.0 ohm resistor is
   (1) one-half as great
   (2) one-fourth as great
   (3) the same
   (4) four times as great

Base your answers to questions 12 through 15 on the diagram in which the source voltage is 26 volts.

12   What is the reading of voltmeter V_2?
   (1) 52 volts   (2) 26 volts   (3) 13 volts   (4) 8 volts

13   What is the total resistance of the circuit?
   (1) ¾ ohm   (2) 4/3 ohm   (3) 10 ohm   (4) 13 ohm

14   The reading of ammeter A_1 is
   (1) 6 amperes   (3) 3 amperes
   (2) 2 amperes   (4) 52 amperes

15   If additional resistances are added in series and the applied voltage is kept constant, the reading of voltmeter V_3 will
   (1) decrease   (2) increase   (3) remain the same

Base your answers to questions 16 through 20
on the diagram below which represents an
electrical circuit.

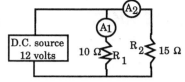

16 The equivalent resistance of the circuit is
   (1) 25 Ω   (2) 6.0 Ω   (3) 5.0 Ω   (4) 0.17Ω
17 The potential difference across R_2 is
   (1) 1.0 V   (2) 2.0 V   (3) 10 V   (4) 12 V
18 The magnitude of the current in ammeter A_1 is
   (1) 120 A   (2) 2.0 A   (3) 1.2 A   (4) 0.83 A
19 Compared to the current in A_1, the current in A_2 is
   (1) less   (2) greater   (3) the same
20 If another resistor were added to the circuit in parallel, the equivalent re-
    sistance of the circuit would
   (1) decrease   (2) increase   (3) remain the same

IV. Energy And Power In Electrical Problems

Electric power is the time rate at which electrical energy is expended.
The **watt** is the **SI** Unit of power. It is a derived unit. Power is equal to the
product of current and potential difference for any general electrical device.
For **ohmic conductors**, power can be obtained from the equation:

$$P = VI = I^2R = V^2/R$$

By substituting units into the formula: $\mathbf{P} = \mathbf{VI}$

$$P = \frac{joules}{coulomb} \cdot \frac{coulomb}{second}$$

$$P = \frac{joules}{second}$$

$$P = watts$$

The **energy** used in an electric circuit is the product of the power developed and the time during which the charges flow. The work done (or the energy expended) can be calculated with the following formula:

$$W = Pt = VIt = I^2Rt$$

Where:
   $\mathbf{W}$ = work in joules $\mathbf{I}$ = current in amperes
   $\mathbf{V}$ = potential difference in volts $\mathbf{t}$ = time in seconds

Since the joule is a small unit of energy, commercial electricity is usually
measured in **kW - hr**. One kW - hr = 3.6×10^6 J. Both the watt - sec and the
kW - hr are units of energy. Energy is a scalar quantity.

Questions

1   Which combination of current and electromotive force (emf) would use energy at the greatest rate?
    (1) 10 amperes at 110 volts   (3) 3 amperes at 220 volts
    (2) 8 amperes at 110 volts   (4) 5 amperes at 110 volts

2   Two resistors are connected in parallel to a 12-volt battery as shown in the diagram. If the current in resistance R is 3.0 amperes, the rate at which R consumes electrical energy is
    (1) 1.1×10^2 watts
    (2) 36 watts
    (3) 24 watts
    (4) 4.0 watts

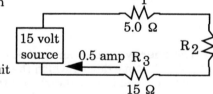

3   The potential difference across a 100.-ohm resistor is 4.0 volts. What is the power dissipated in the resistor?
    (1) 0.16 watt   (3) 4.0×10^2 watts
    (2) 25 watts   (4) 4.0 watts

4   An electrical heater raises the temperature of a measured quantity of water. Six thousand joules of energy is absorbed by the water from the heater in 30.0 seconds. What is the minimum power rating of the heater?
    (1) 5.00×10^2 W   (3) 2.00×10^3 W
    (2) 2.00×10^2 W   (4) 1.80×10^5 W

5   What is the current in a 1,200-watt heater operating on 120 volts?
    (1) 0.10 ampere   (3) 10. amperes
    (2) 5.0 amperes   (4) 20. amperes

6   A 10 -volt potential difference maintains a 2 -ampere current in a resistor. The total energy expended by this resistor in 5 seconds is
    (1) 10 J   (2) 20 J   (3) 50 J   (4) 100 J

7   How long must a 100-watt light bulb be used in order to dissipate 1,000 joules of electrical energy?
    (1) 10 sec   (2) 100 sec   (3) 1,000 sec   (4) 1000,000 sec

8   As the resistance of a lamp operated at a constant voltage increases, the power used by the lamp
    (1) decreases   (2) increases   (3) remains the same

Base your answers to questions 9 through 13 on the diagram below which shows 3 resistors connected to a 15-volt source.

9   The equivalent resistance of the circuit is
    (1) 10 ohms   (3) 30 ohms
    (2) 20 ohms   (4) 40 ohms

10 The potential difference across R_2 is
    (1) 2.5 volts   (2) 5.0 volts   (3) 7.5 volts   (4) 10 volts

11 The total power developed in the circuit is
    (1) 2.5 watts   (2) 5.0 watts   (3) 7.5 watts   (4) 10 watts

12 If resistor R_3 is removed and replaced by a resistor of lower value, the resistance of the circuit will
    (1) decrease   (2) increase   (3) remain the same

13 Compared to the heat developed in resistor R_1, the heat developed in resistor R_3 is
   (1) one-third as great
   (2) two times as great
   (3) three times as great
   (4) one-fourth as great

Base your answers to questions 14 through 18 on the diagram below which represents three resistors connected in parallel across a 24-volt source. The ammeter reads 3 amperes.

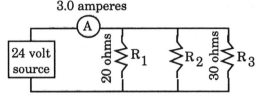

14 The equivalent resistance in the circuit is
   (1) 0.13 ohms   (2) 8.0 ohms   (3) 58 ohms   (4) 72 ohms
15 The current in R_1 is
   (1) 0.83 amperes
   (2) 1.5 amperes
   (3) 3.0 amperes
   (4) 1.2 amperes
16 The potential difference across R_3 is
   (1) 8.0 volts   (2) 24 volts   (3) 48 volts   (4) 72 volts
17 If the ratio of the current in R_3 to the current in R_2 is 4:5, the resistance of R_2 is
   (1) 5.0 ohms   (2) 8.0 ohms   (3) 24 ohms   (4) 60. ohms
18 The power supplied to the circuit is
   (1) 220 watts   (2) 190 watts   (3) 72 watts   (4) 24 watts

V. Magnetism
5.1 Magnetic Force

Magnetic force is a force that exists between charges in motion. All magnetic properties of any kind of material are due to the electrons orbiting the nucleus of the atoms. The electrons generate magnetic fields by spinning on their axis.

All substances exhibit magnetic properties. Diamagnetic substances reduce the flux density; paramagnetic and ferromagnetic increase the flux density. **Permeability** is the property of a material which changes the flux density in a magnetic field from its value in a vacuum. The permeability of air is nearly the same as that of a vaeuum which is one.

Modern theory of magnetism indicates that ferromagnetic materials such as iron, nickel, and cobalt tend to align the atoms in clusters, called domains. Within a domain the **magnetic fields** produce a relatively strong field but normally, the axes of the domains are randomly arranged so the fields cancel each other out. *These fields are due to atomic currents caused by spinning electrons.* When some domains are enlarged in respect to others a net field is produced and the object is said to be magnetized. In a magnetic field, some atoms line up with the field causing some domains to grow larger at the expense of others getting smaller. This results in boundary shifts between these domains. If the boundaries persist after removal from the field, the substance is a permanent magnet. Certain natural substances are magnets.

5.2 Magnetic Fields

A **magnetic field** is a region where magnetic force may be detected. It can be near a magnet, or a current carrying wire. The direction of a magnetic field is, according to convention, the direction in which the N - pole of a compass would point in the field. The direction of the magnetic field at any point is tangent to the field line at that point.

The magnetic field around a bar magnet is mapped by drawing magnetic flux lines (lines of force). The flux lines are imaginary and always form closed paths.

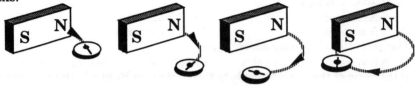

The **SI** unit for flux is the **weber**. The *magnetic lines never cross each other*. The field is *strongest near the poles when the lines of force are closest together*. The compass will point from the N (North) pole and towards the S (South) pole.

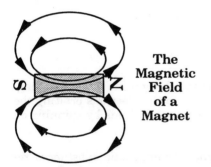

The
Magnetic
Field
of a
Magnet

Inside the magnet the lines of force go from south to north. If a magnet were broken into pieces, each piece would have a north and south pole.

A magnetic field also exists around a current carrying wire. Compasses surrounding the wire would point in a circle. The magnetic field direction can be determined by the first left-hand rule for electron current.

The **first left-hand rule** states that *when you grab a current carrying wire with your left hand, such that the thumb points in the direction of the electron flow, the curved fingers will point to the direction of the magnetic field.*

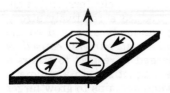

Using the first left-hand rule, one can determine the direction of the field around a loop of current carrying wire. The field is such that the faces of the loop show polarity. When the number of loops is increased, a **solenoid** is

formed. The lines of magnetic flux around a solenoid emerge from the north pole of the solenoid
and enter the south pole. Inside the lines of force
are nearly parallel to its axis and perpendicular to
its faces.

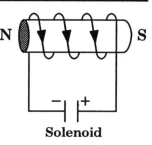

A magnetic field exists around a current carrying coil. The direction of the magnetic field
around a coil can be determined by the second
left-hand rule.

Solenoid

The **second left-hand rule** states *if you grab the
coil with your left hand so that your fingers point in the
direction of the electron flow, your thumb will point to
the N pole.* A solenoid consists of many coils wrapped
around a core, and is known as an **electromagnet**.

The magnetic field near a current carrying coil can
be made stronger by:

1) inserting an iron core (increasing permeability)
2) increasing the current in the coil
3) increasing the number of turns

The field strength of a solenoid is also affected by its shape.

Practical Applications
- electromagnets
- electric motor
- circuit breaker
- galvanometer

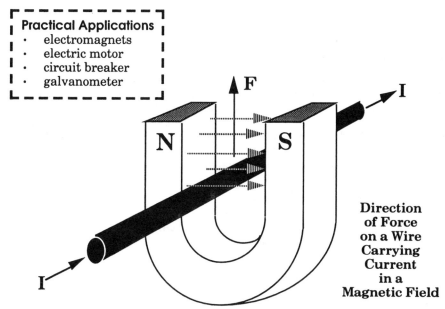

**Direction
of Force
on a Wire
Carrying
Current
in a
Magnetic Field**

When a current carrying wire is placed in a magnetic field, a force is exerted on the conductor, (if the conducting wire is not parallel to the magnetic
flux). A magnetic field exerts a force on any moving charge. The standard
unit test object (current element) for a magnetic field consists of a wire of
length *l* carrying a current of one ampere.

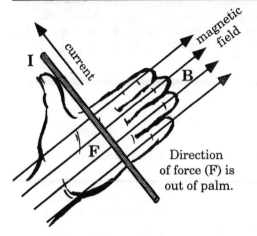

Direction
of force (F) is
out of palm.

The force is perpendicular to both the field and the current and can be determined by the **third left-hand rule.** *Placing the left hand so the thumb points to the direction (I) of the electron flow, the fingers point in the direction (B) of the magnetic flux (N to S), and the palm of the hand (F) indicates the direction of the force.*

The **first left-hand rule** can be used to determine the direction of the force in all of the previous examples.

A · (dot) indicates that current or magnetic field is traveling out of the page. An X indicates into the page. The magnetic field generated by the current in the wire is circular. Below the wire the magnetic flux lines add to each other. Above the wire the fields subtract, causing a weaker field above the wire than below the wire. Therefore the wire would move up.

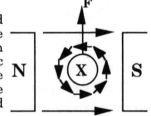

5.3 Flux Density

It is convenient to describe the interaction between charges by means of electrical fields, and the interaction between currents flowing in two parallel wires by defining a magnetic field. We can think of one of the currents as producing a magnetic field, and the field as exerting a force on the other current. The magnetic field is measured by the force it exerts on a current. The force depends on the direction of the current and varies to a maximum value. The strength of the magnetic field is defined as:

$$B = \frac{F_{max}}{I\,L}$$

Where:

F	=	maximum force measured in newtons
I	=	current in amperes
L	=	length of wire in magnetic field
B	=	the strength of the magnetic field, known as magnetic flux density.

The **flux density** (B) is the number of flux lines per unit area and is proportional to the intensity of the field. Flux density is a vector quantity. The field strength of magnets is commonly measured in **webers/meter^2, newtons/amp·m,** or **tesla.** Flux density between the poles of magnets is often given in Gauss, as

$$1.0\ \text{tesla} = 1.0 \times 10^4\ \text{Gauss}$$

Flux density units are:

$$B = \frac{\text{webers}}{\text{meter}^2} = \frac{\text{newtons}}{\text{amp} \cdot \text{m}} = \text{tesla}$$

Flux can be calculated from flux density.

Flux = BA = Webers

Where: **B** is the magnetic density (or magnetic field intensity)
A is the area

The magnetic field strength **B** is analogous to the electric field strength **E** and the gravitational field strength **g**. All of these fields are vector quantities, which means that they have magnitude and direction.

Gravitational field strength (g) = force per unit mass
Electric field strength (E) = force per unit charge
Magnetic field strength (B) = force per unit current element

When a current carrying wire is bent into a loop and placed in a magnetic field, a **torque** is experienced. This torque applied to a coil provides the basis of operation of the **galvanometer** and electric motors.

The force is a maximum when the angle between the field and the conductor is 90°. There is no force when the angle between the field and the conductor is 0°.

Two current carrying wires exert a force on each other through their magnetic fields. If the current is flowing in the same direction, the wires attract each other. Current flowing in the opposite direction of two parallel wires causes them to repel each other.

Wires Attract

Wires Repel

Legend
Current into Page = $\otimes$
Current Out of Page = $\odot$

At point **P** in both examples above, the fields are subtractive. Therefore, the wires move together or attract each other. At point **M**, the fields add together and the wires move apart.

The **ampere** is defined in terms of this relationship. An ampere is that amount of unvarying current, which, if present in each of two parallel conductors of infinite lengths and one meter apart in free space, will produce a force of exactly 2×10^{-7} newtons per meter of length.

Questions

1　The diagram represents a current-carrying loop of wire. The direction of the magnetic field at point *P* is

    (1) toward the page   (3) into the page
    (2) to the right   (4) out of the page

2　In the diagram, what is the direction of the magnetic field at point *A*?

    (1) to the left
    (2) to the right
    (3) toward the top of the page
    (4) toward the bottom of the page

3　The arrows in the diagram indicate the direction of the electron flow. The south pole of the electromagnet is located closest to point

    (1) *A*   (3) *C*
    (2) *B*   (4) *D*

4　When the iron core is removed from the center of a direct current carrying coil, the magnetic field strength of the coil

    (1) decreases   (2) increases   (3) remains the same

5　Which diagram best represents the magnetic field around a material of high permeability placed between unlike magnetic poles?

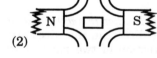

6　An electromagnet is shown in the diagram. Its north pole will be nearest which point?

    (1) 1
    (2) 2
    (3) 3
    (4) 4

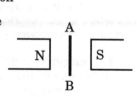

7　The diagram shows a current carrying wire in a magnetic field. If the current flows out of the page, the magnetic force on the wire will be in which direction?

    (1) 1
    (2) 2
    (3) 3
    (4) 4

8　When electrons flow from point *A* to point *B* in the wire shown in the diagram, there will be a force produced on the wire

    (1) toward N
    (2) into the page
    (3) toward S
    (4) out of the page

9 The existence of a magnetic field around a current - carrying conductor
can be demonstrated by placing the conductor near
(1) a pith ball   (3) a battery
(2) an electroscope   (4) a compass needle

10 Which arrow in the diagram represents the direction of
the flux inside the bar magnet? 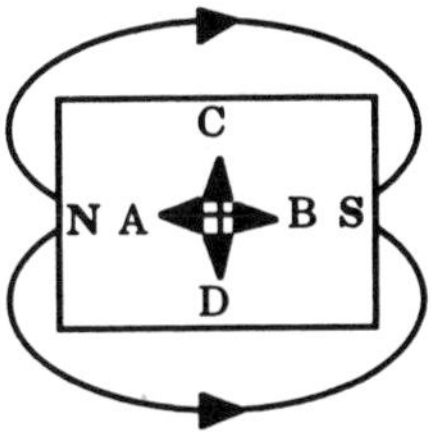
(1) *A*
(2) *B*
(3) *C*
(4) *D*

11 The field around a permanent magnet is caused by the
motions of
(1) nucleons   (2) protons   (3) neutrons   (4) electrons

12 Which diagram best represents the direction of the magnetic field around
a wire conductor in which the electrons are moving as indicated? (The x's
indicate that the field is directed into paper and the dots indicate that
the field is directed out of the page.)

(1)    (2) 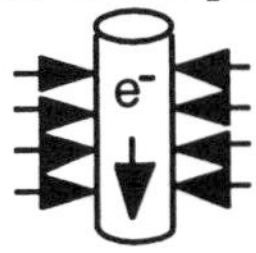   (3) 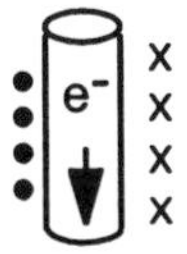   (4)

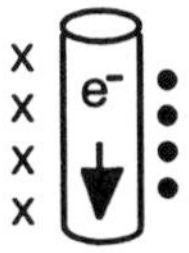

13 Which diagram below best represents a magnetic field?

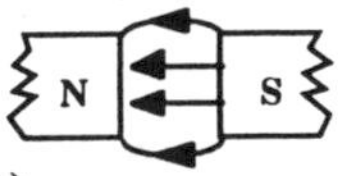

(1) 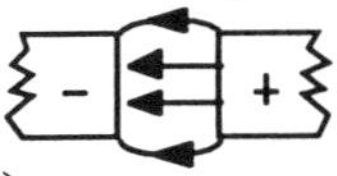   (2) 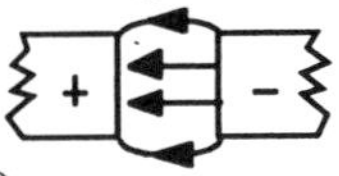   (3) 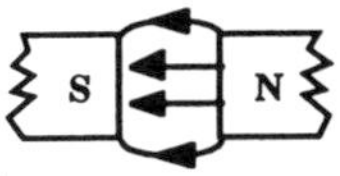   (4)

14 The accompanying diagram represents a wire carrying electrons
into the page. The direction of the magnetic field above the wire is 
(1) toward the left   (3) up from the page
(2) toward the right   (4) into the page

15 Magnetic fields are produced by
(1) motion of electric charges   (3) photon motion
(2) static electric charges   (4) gamma radiation

16 In the diagram, what is the direction of the magnetic
field at point *P*? 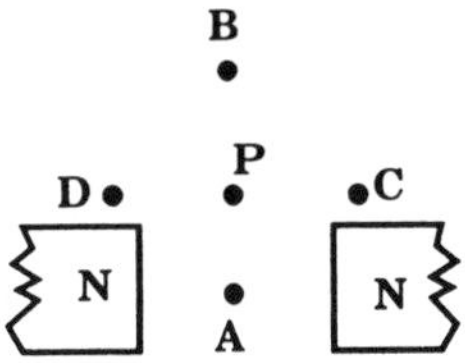
(1) toward *A*
(2) toward *B*
(3) toward *C*
(4) toward *D*

17 Which diagram best illustrates the direction of the magnetic field
between the unlike poles of two bar magnets?

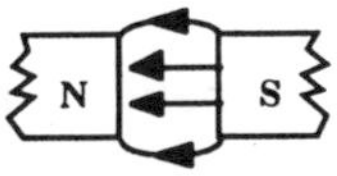

(1)   (3)

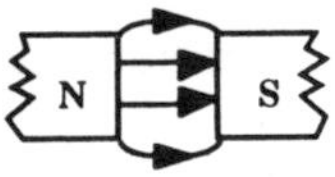

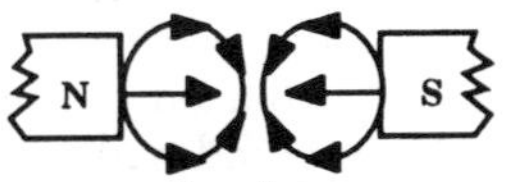

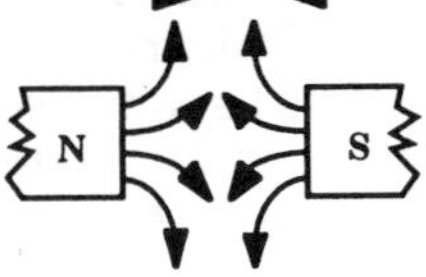

(2)   (4)

18 The magnetic lines of force near a long straight current-carrying wire are
  (1) straight lines parallel to the wire
  (2) straight lines perpendicular to the wire
  (3) circles in a plane perpendicular to the wire
  (4) circles in a plane parallel to the wire

19 In the diagram, A, B, C, and D are points in the magnetic field near a current-carrying loop. At which points is the direction of the magnetic field into the page?
  (1) A and B
  (2) B and C
  (3) C and D
  (4) A and D

20 As two parallel conductors with currents in the same direction are moved apart, their force of
  (1) attraction increases
  (2) attraction decreases
  (3) repulsion increases
  (4) repulsion decreases

21 Wires x and y experience a force when electrons pass through them as shown in the diagram. The force on wire y will be toward
  (1) A
  (2) B
  (3) C
  (4) D

Base your answers to question 22 through 26 on the diagram, which represents a circuit containing a solenoid on a cardboard tube, a variable resistor R, and a source of potential difference.

22 The north pole of the solenoid is nearest to point
  (1) A   (3) C
  (2) B   (4) D

23 Due to the current in the FE section of the circuit, the direction of the magnetic field at point X is
  (1) into the page   (3) to the left
  (2) out of the page   (4) to the right

24 If the resistance of the variable resistor R is increased, the magnetic field strength of the solenoid will
  (1) decrease   (2) increase   (3) remain the same

25 If a soft iron rod is placed in the cardboard tube, the magnetic field strength of the solenoid will
  (1) decrease   (2) increase   (3) remain the same

26 If the number of turns in the solenoid is increased and the current is kept constant, the magnetic field strength of the solenoid will
  (1) decrease   (2) increase   (3) remain the same

5.4 Electromagnetic Induction

If a straight conductor is moved across a magnetic field, the electrons in the conductor will be acted upon by a magnetic force. This force will tend to move the electrons along the conductor from one end to another. As the motion of the electrons cause an excess of negative charge at one end and a deficiency of charge at the other end, a potential difference is established between the ends of the conductor. This effect can also be observed if the conductor is stationary and the field is moving.

Therefore, any time a conductor **cuts lines of flux**, a potential is induced in the conductor. If the conductor is connected to a circuit, a current will flow around the circuit. The direction of the induced current is in such a direction that its magnetic field opposes that of the field that induced it. This relationship is known as **Lenz's Law** and is *an example of conservation of energy.*

The magnitude of an induced electromotive force is directly proportional to the flux density, the length of the conductor, and the speed of the conductor relative to the flux. This can be written as:

$$\textbf{EMF} = \textbf{B l v}$$

Where: **EMF**   =   induced electromotive force in volts
             B   =   magnetic field strength in N/A · m
             l   =   length of conductor in meters
             v   =   velocity in m/s

5.5 The Generator Principle

A conducting loop rotating in a uniform magnetic field experiences a continual change in the total number of flux lines crossing the loop. This change induces a potential across the ends of the loop, which alternates in direction and varies in magnitude, between zero and a maximum. When the plane of the loop is perpendicular to the field, the induced potential is zero. When the plane of the loop is parallel to the field, the induced potential is a maximum.

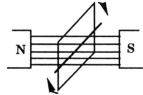

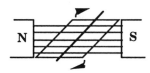

Loop is perpendicular.
Velocity is parallel. } Emf = 0

Loop is parallel.
Velocity is perpendicular. } Emf = max

The magnitude of the induced potential is proportional to the component of the velocity perpendicular to the field and the intensity of the magnetic field. When the loop is part of a complete circuit, the induced potential causes a current in the loop. Since the induced potential is alternating, the current is an alternating current. An **alternating current** is a current that reverses its direction with regular frequency.

Practical Applications
- generators
- transformers

5.6 Electromagnetic Radiation

As alternating voltages and current are produced in rotating loops by electromagnetic induction, electrons are moved back and forth in the conductor. This motion means that the electrons are accelerating alternately in opposite directions. Oscillating charges produce electric and magnetic fields that radiate outward in the form of waves. This combined electric and magnetic wave is called an **electromagnetic wave** and is propagated by interchanging electric and magnetic fields. The waves make up the electromagnetic spectrum. (See Unit 4, Section 5.4 on the Electromagnetic Spectrum.)

The illustration (right) shows interchanging electric and magnetic fields at right angles to each other.

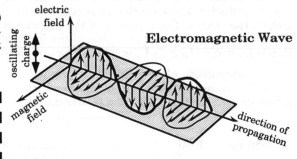

Electromagnetic Wave

```
r --------------------- 1
| Practical Applications |
|   •  electromagnetic   |
|      spectrum          |
|   •  antenna           |
L --------------------- J
```

Questions

1 A conducting loop is rotated one full turn (360°) in a uniform magnetic field. Which graph best represents the induced potential difference across the ends of the loop as a function of the angular rotation?

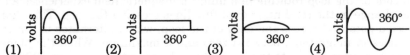

(1)   (2)   (3)   (4)

Base your answers to questions 2 through 6 on the diagram which shows a cross section of a wire *A* moving down through a uniform magnetic field (*B*) towards *y*.

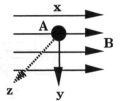

2 What is the direction of the magnetic force on the electrons in the wire?
   (1) toward *x*   (3) into the page
   (2) toward *y*   (4) out of the page

3 What is the direction of the magnetic force on the wire due to the induced current in the wire?
   (1) toward *x*   (2) toward *y*   (3) into page   (4) out of page

4 If the cross section of wire *A* moved at an angle with the field towards *z*, what would be the direction of the magnetic force on the electrons in the wire?
   (1) towards *x*   (2) towards *y*   (3) into page   (4) out of page

5 The maximum potential difference will be induced across the wire when the angle between the direction of motion of the wire and the direction of the magnetic field is
   (1) 0°   (2) 45°   (3) 90°   (4) 180°

6   If the velocity of the wire is increased, the induced potential difference across the wire will
(1) decrease   (2) increase   (3) remains the same

7   A charged particle is moving in a magnetic field, with its velocity parallel to the direction of the field. If the magnetic flux density is doubled, the force on the moving charge
(1) is halved   (3) remains the same
(2) is doubled   (4) is quadrupled

8   As the speed of a conducting loop rotating in a magnetic field decreases, the magnitude of the induced current in the loop
(1) decreases   (2) increases   (3) remains the same

9   Electromagnetic radiation can be generated by
(1) a neutron moving with constant velocity
(2) an electron moving with constant velocity
(3) an accelerating neutron
(4) an accelerating electron

10   The diagram shows a copper wire located between the poles of a magnet. Maximum electric potential will be induced in the wire when it is moved at constant speed toward which point?
(1) *A*   (3) *C*
(2) *B*   (4) *D*

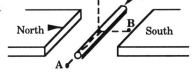

11   The magnitude of the electric potential induced across the ends of a conductor moving in a magnetic field may be increased by
(1) increasing the diameter of the conductor
(2) increasing the speed of the conductor
(3) decreasing the resistance of the conductor
(4) decreasing the length of the conductor

Base your answers to questions 12 through 15 on the diagram which shows a conductor that is moving toward point *A*, through a uniform magnetic field with a speed of 2 meters per second. The direction of the magnetic field is downward and the conductor is moving at right angles to the direction of the magnetic field.

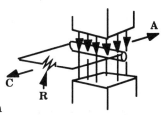

12   If the speed of the conductor is increased, the induced electromotive force
(1) decreases   (2) increases   (3) remains the same

13   If the resistor *R* is decreased, the force needed to move the conductor at the same speed
(1) decreases   (2) increases   (3) remains the same

14   If the motion of the conductor is changed to 2 meters per second upward, the magnitude of the induced electromotive force
(1) decreases   (2) increases   (3) remains the same

15   If the motion of the conductor is changed to 2 meters per second towards point *C*, the induced electromotive force will
(1) remain the same and its direction will be reversed
(2) decrease and its direction will remain the same
(3) increase and its direction will be reversed
(4) increase and its direction will remain the same

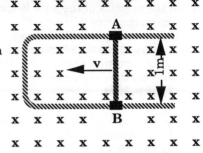

Base your answers to questions 16 through 19 on the diagram below which represents a U-shaped wire conductor positioned perpendicular to a uniform magnetic field which acts into the page. *AB* represents a second wire which is free to slide along the U-shaped wire.

16 If wire *AB* is moved to the left at a constant speed, the direction of the induced electron motion in the wire will be
(1) toward *A, only*   (3) first toward *A* and then toward *B*
(2) toward *B, only*   (4) first toward *B* and then toward *A*

17 Wire *AB* is moved at constant speed to the left. The current induced in the conducting loop will produce a force on wire *AB* which acts
(1) to the right   (2) to the left   (3) into page   (4) out of page

18 The resistance of wire *AB* is increased, and the wire is moved to the left at a constant speed. Compared to the induced potential difference before the resistance was increased, the new potential difference will be
(1) less   (2) greater   (3) the same

19 If wire *AB* is accelerating to the left, the potential difference induced across *AB*
(1) decreases   (2) increases   (3) remains the same

20 The diagram shows the cross section of a wire which is perpendicular to the page and a uniform magnetic field directed to the right. Toward which point should the wire be moved to induce maximum electric potential? (Assume the same speed would be used in each direction.)
(1) 1   (3) 3
(2) 2   (4) 4

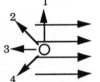

Thinking Physics

Explain the following:

1 Wintergreen Life Savers glow in the dark when you bite into them.

2 As you walk over a rug, do you become positively or negatively charged?

3 Lightning rods attract lightning and are considered a protection device.

4 Can you set up a circuit in which a 30 watt light bulb would be more powerful than a 60 watt light bulb?

5 Iron flag poles become magnetic after remaining in the same position for many years.

6 A car made of iron and steel passes over a wide closed loop embedded in the road surface. How can this trigger an automatic traffic light?

Free Response Questions

1   A proton is placed between two parallel charged plates in a vacuum. The electric field is adjusted such that the particle will remain suspended.

   a) Draw a labeled diagram indicating the forces acting on the proton.

   b) The proton is replaced by an electron. Describe how the forces on the electron will be different from those shown on the proton.

2   A charged particle gains 1.0×10^{-3} joule of energy by moving through a potential difference of 500. volts. The same particle experiences an attractive electrostatic force of 5.4 newtons when it is 1. meter from a second charged particle.

   Describe a step-by-step procedure, using complete sentences, for determining the charge of the second particle. Do not solve any equations. You may refer to equations in your step-by-step procedure. If you do, state how these equations are to be used.

3   A conductor is connected to a source of variable potential difference. Ammeter readings are taken as the potential difference is increased. These readings are listed in the table below.

Voltage (volts)	0	20	40	60	80	100
Current (amps)	0	5	10	15	20	25

   a) Sketch the graph with properly labeled axes.

   b) What is the resistance of the conductor (in ohms) when the potential difference is 40. volts?

   c) What power (in watts) will be used when the potential difference applied to the conductor is 60. volts?

   d) Does the conductor obey Ohm's Law, assuming constant temperature?

   e) The conductor is replaced with two identical resistors in parallel with each other. What must be the resistance of each one (in ohms) for the graph to remain unchanged?

4　Base your answers to parts *a* through
　d on the diagram which represents a
　circuit containing a 120-volt power
　supply with switches S_1 and S_2 and
　two 60.-ohm resistors.

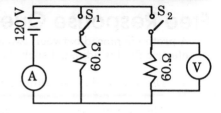

a)　If switch S_1 is kept open and
　　switch S_2 is closed, what is the
　　circuit resistance?

b)　If switch S_2 is kept open and switch S_1 is closed, how much current
　　will flow through the circuit? {Show all calculations, including
　　equations and substitutions with units.]

c)　When both switches are closed, what is the current in the ammeter?

d)　When both switches are closed, what is the reading of the voltmeter?

5　Suppose you are provided with the following apparatus:

　　one 100 ohm resistor　　　　　an ammeter
　　one 50 ohm resistor　　　　　　a voltmeter
　　a power source of 100 volts　　wires of negligible resistance.

　You are asked to design a circuit that will heat 100 g of water at 25° C
　the fastest.

a)　Diagram the circuit.
b)　Determine the total current in the circuit.
c)　Place the ammeter in the circuit to read the total current.
d)　Place the voltmeter in the circuit to read the smallest V.
e)　What is the power of the circuit?

Self–Help Questions

1　When two neutral materials are rubbed together, there is a transfer of
　electrical charge from one material to another. The total electrical charge
　for the system
　　(1)　increases as electrons are transferred
　　(2)　increases as protons are transferred
　　(3)　remains constant as protons are transferred
　　(4)　remains constant as electrons are transferred

2　As shown in the diagram, a charged rod is held
　near, but does not touch a neutral electroscope.
　The charge on the knob becomes
　　(1)　positive and the leaves become positive
　　(2)　positive and the leaves become negative
　　(3)　negative and the leaves become positive
　　(4)　negative and the leaves become negative

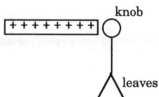

3 When hair is combed with a hard rubber comb, the hair becomes
   positively charged because the comb
   (1) transfers protons to the hair
   (2) transfers electrons to the hair
   (3) removes protons from the hair
   (4) removes electrons from the hair

4 One of two identical metal spheres has a charge of +q, and the other
   sphere has a charge of -q. The spheres are brought together and then
   separated. Compared to the total charge on the two spheres before
   contact, the total charge on the two spheres after contact is
   (1) less   (2) greater   (3) the same

5 The electrostatic force of attraction between two small spheres that are
   1.0 meter apart is F. If the distance between the spheres is decreased to
   0.5 meters, the electrostatic force will then be
   (1) F/2   (2) 2F   (3) F/4   (4) 4F

6 If the charge on one of two small charged spheres is doubled while the
   distance between them remains the same, the electrostatic force between
   the point sources will be
   (1) halved   (2) doubled   (3) quadrupled (4) unchanged

+12 coulombs

Questions 7 through 10 refer to the diagram.
Sphere *B* is neutral initially.

$\left(\begin{array}{c}A\end{array}\right)$ $\left(\begin{array}{c}B\end{array}\right)$

7 When spheres *A* and *B* come into contact, sphere *B* will
   (1) gain 6 coulombs of protons   (3) gain 6 coulombs of electrons
   (2) lose 6 coulombs of protons   (4) lose 6 coulombs of electrons

8 When spheres *A* and *B* are in total contact, the total charge of the system
   is
   (1) neutral   (3) +12 coulombs
   (2) +6 coulombs   (4) +24 coulombs

9 When spheres *A* and *B* are separated, the charge on *A* will be
   (1) +12 coulombs   (3) ½ the original amount
   (2) ¼ the original amount   (4) 4 times the original amount

10 After spheres *A* and *B* are separated, which graph best represents the
   relationship of the force between the spheres and their separation?

(1)   d   (2)   d   (3)   d   (4)   d

11 What is the charge to mass ratio, e/m, on an electron?
   (1) 9×10^{-9} C/kg   (3) 1.10×10^{30} C/kg
   (2) 1.76×10^{11} C/kg   (4) 6.86×10^{48} C/kg

12 Which diagram best illustrates the electric field around two unlike
   charges?

(1)   (2)   (3)   (4)

13 The diagram at the right represents a positive test   •A   •B
    charge located near a positively charged sphere.
    The greatest increase in the electric potential ener- ⊕ •C
    gy of the test charge relative to the sphere would •D ⊕ sphere
    be caused by moving the charge to point
    (1) *A*   (3) *C*
    (2) *B*   (4) *D*

14 What is the maximum amount of kinetic energy that may be gained by a
    proton accelerated through a potential difference of 50 volts?
    (1) 1 eV   (2) 10 eV   (3) 50 eV   (4) 100 eV

15 If 10 joules of work must be done to move 2.0 coulombs of charge from
    point *A* to point *B* in an electric field, the potential difference between
    points *A* and *B* is
    (1) 5.0 V   (2) 10 V   (3) 12 V   (4) 20 V

Base your answers to questions 16 through 19 on the + ————————
accompanying diagram which represents two large   A•   •B
parallel plates which are oppositely charged. *A, B, C*   •C
are reference points.   − ————————

16 If an electron moves from point *A* to point *B*, the electron's electric
    potential energy will
    (1) decrease   (2) increase   (3) remain the same

17 If an electron is moved by the electric field from point *C* to point *A*, what
    happens to the total energy of the electron? (Assume no gravitational
    effects)
    (1) It decreases   (2) It increases   (3) It remains the same

18 Compared to the amount of electric potential energy that an electron has
    at point *C*, the amount of electric potential energy that a proton has at
    point *B* is
    (1) less   (2) greater   (3) the same

19 If an electron is moved from point *A* to point *C*, the potential energy of
    the electron will
    (1) decrease   (2) increase   (3) remain the same

Base your answers to questions 20 through 24 on
the diagram which represents an electron project- + + + + + + + +
ed into the region between two parallel charged ⊖→
plates which are 10^{-3} meter apart. The electric – – – – – – – –
field between the plates is 10^6 volts per meter.

20 In which direction will the electron be deflected?
    (1) into the page   (3) toward the bottom of the page
    (2) out of the page   (4) toward the top of the page

21 What is the potential difference across the two plates?
    (1) 10^{-3} volt   (3) 10^6 volts
    (2) 10^3 volts   (4) 10^9 volts

22 What is the magnitude of the force acting on the electron, when it is in
    the electric field?
    (1) 1.6×10^{-25} N   (3) 1.0×10^6 N
    (2) 1.6×10^{-13} N   (4) 1.6×10^{25} N

23 As an electron moves from the negatively charged plate to the positively
charged plate, the force on the electron due to the electric field
(1) decreases   (2) increases   (3) remains the same

24 The electron is replaced by a proton. Compared to the magnitude of the
force on the electron, the magnitude of the force on the proton will be
(1) less   (2) greater   (3) the same

25 The potential difference between a pair of charged parallel plates
0.50meter apart is 50 volts. What is the electric field intensity between
the plates?
(1) 1.0×10^2 N/C   (3) 5.0×10^2 N/C
(2) 2.5×10^2 N/C   (4) 8.0×10^2 N/C

26 What is the magnitude of the electric field intensity at a point in the field
where an electron experiences a 1.0-newton force?
(1) 1.0 N/C   (3) 6.3×10^{18} N/C
(2) 1.6×10^{-19} N/C   (4) 9.1×10^{-31} N/C

27 A beta particle with a charge of -1 elementary charges is accelerated by a
potential difference of 1.0×10^6 volts. The energy acquired by the particle
is
(1) 0.5×10^6 eV   (3) 1.6×10^{-19} eV
(2) 1.0×10^6 eV   (4) 3.2×10^{-13} eV

28 How much energy is required to move 3.2×10^{-19} coulomb of charge
through a potential difference of 5 volts?
(1) 5 eV   (3) 10 eV
(2) 2 eV   (4) 20 eV

29 The diagram shows the electric field in the vicinity of two charged conducting spheres, *A* and
B. What is the static electric charge on each of
the conducting spheres?
(1) *A* is negative and *B* is positive.
(2) *A* is positive and *B* is negative.
(3) Both *A* and *B* are positive.
(4) Both *A* and *B* are negative.

30 A negatively charged object is brought near the knob of a negatively
charged electroscope. The leaves of the electroscope will
(1) move closer together   (3) become positively charge
(2) move farther apart   (4) become neutral

31 Which is a vector quantity?
(1) electric charge   (3) electric power
(2) electric field intensity   (4) electrical energy

32 An oil drop has a charge of -4.8×10^{-19} coulomb. How many excess
electrons does the oil drop have?
(1) 1.6×10^{-19} (3) 3.0
(2) 2.0   (4) 6.3×10^{18}

33 At room temperature, which segment of copper wire has the *lowest*
resistance?
(1) 1.0 m length, 1.0×10^{-6} m^2 cross-sectional area
(2) 2.0 m length, 1.0×10^{-6} m^2 cross-sectional area
(3) 1.0 m length, 3.0×10^{-6} m^2 cross-sectional area
(4) 2.0 m length, 3.0×10^{-6} m^2 cross-sectional area

34 A copper wire has a resistance of 100 ohms. A second copper wire with twice the cross-sectional area and the same length would have a resistance of
    (1) 50 Ω   (2) 100 Ω   (3) 200 Ω   (4) 400 Ω

35 A charge of 5.0 coulombs moves through a circuit in 0.50 second. How much current is flowing through the circuit?
    (1) 2.5 A   (2) 5.0 A   (3) 7.0 A   (4) 10. A

36 The graph shows the relationship between current and potential difference for four resistors, A, B, C, and D. Which resistor has the *least* resistance?
    (1) A
    (2) B
    (3) C
    (4) D

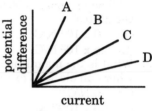

37 If 20 coulombs of charge passes a given point in a conductor every 4 seconds, the current at that point is
    (1) 2.0 A   (2) 5 A   (3) 10 A   (4) 20 A

38 The diagram represents currents flowing in branches of an electric circuit. What is the reading on ammeter A?
    (1) 13 A-
    (2) 17 A
    (3) 3 A
    (4) 33 A

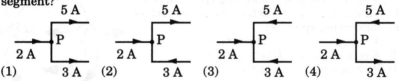

39 Which diagram below shows correct current direction in a circuit segment?

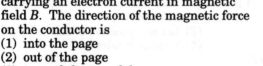

40 The diagram at the right represents a conductor carrying an electron current in magnetic field B. The direction of the magnetic force on the conductor is
    (1) into the page
    (2) out of the page
    (3) toward the top of the page
    (4) toward the bottom of the page

41 Which circuit segment has an equivalent resistance of 6 ohms?

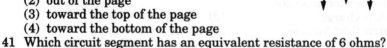

42 Three resistors of 10. ohms, 30. ohms, and 60. ohms, respectively, are connected in series with a battery. A current of 2.0 amperes will flow through this circuit when the potential difference of the battery is
(1) 20 V   (2) 55 V   (3) 110 V   (4) 200 V

43 If the voltage across a 4.0-ohm resistor is 12 volts, the current through the resistor is
(1) 0.33 A   (2) 48 A   (3) 3.0 A   (4) 4.0 A

44 In the circuit shown at the right, what is the potential difference of the source?
(1) 33 V
(2) 100 V
(3) 300 V
(4) 10,000 V

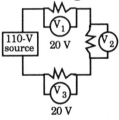

45 A toaster connected to a 120-volt outlet draws a current of 6.0 amperes. How much electrical energy does the toaster use in 5.0 seconds?
(1) 1.4×10^2 J   (2) 7.2×10^2 J   (3) 3.6×10^3 J   (4) 2.2×10^4 J

46 Which circuit below would have the *lowest* volt-meter reading?

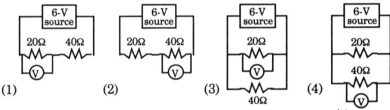

47 In the circuit diagram at the right, which is the correct reading for meter V_2?
(1) 20 V
(2) 70 V
(3) 90 V
(4) 10 V

48 If the power developed in an electric circuit is doubled, the energy used in one second is
(1) halved   (2) doubled   (3) quartered   (4) quadrupled

49 Which diagram best represents the lines of magnetic flux between the ends of two bar magnets?

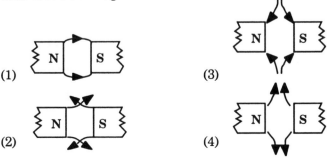

50 In which diagram below is the magnetic flux density at point P greatest?

(1)   (2)   (3)   (4)

51 Which diagram best represents the magnetic field near the poles of a
    horseshoe magnet?

(1)   (2)   (3)   (4)

52 Electrons are moving to the right in the conductor represented in the
    diagram. What is the direction of the magnetic field above the wire at
    point P?
    (1) into the page
    (2) out of the page
    (3) toward the top of the page
    (4) toward the bottom of the page

53 The wire loop shown at the right has a clockwise electron current
    What is the direction of the magnetic field at point P?
    (1) to the right   (3) into the page
    (2) to the left   (4) out of the page

54 A bar magnet is dropped through a wire loop as
    shown in the diagram at the right. As the south pole
    approaches the loop, the electron flow induced in the
    loop
    (1) is clockwise
    (2) is counterclockwise
    (3) attracts the south pole
    (4) speeds up the magnet

55 The diagram shows an end view of a current
    carrying wire between the poles of a magnet. The wire
    is perpendicular to the magnetic field. If the direction
    of the electron flow is out of the page, which arrow
    correctly shows the direction of the magnetic force F
    acting on the wire?

    (1)   (2)   (3)   (4)

56 In the diagram at the right, A, B, C, and D are
    points near a current carrying solenoid. Which
    point is closest to the north pole of the solenoid?
    (1) A
    (2) B
    (3) C
    (4) D

3 Electromagnetic Applications

Optional Unit

Important Terms To Be Understood

galvanometer	Lenz's law	step up
ammeter	tesla	step down
voltmeter	cathode ray tube	eddy currents
electric motor	mass spectrograph	laser
split ring commutator	Millikan oil drop	induction coil
back EMF	particle accelerator	coherent light
	transformer	

I. Torque on a Current Carrying Loop

A current carrying loop of wire experiences a torque in a magnetic field. A **torque** is a force that tends to cause a body to rotate. Torque is usually produced by a force acting at a distance from the axis of rotation. The direction of the torque tends to turn the loop so that the field due to the current is parallel to the existing field. The torque is proportional to the current in the loop.

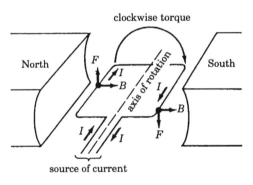

The above figure shows the torque on a current-carrying loop when the loop is in a magnetic field. I represents electron flow.

1.1 Meters

1. Galvanometer. A galvanometer is a device used to measure weak electric current. A galvanometer has a coil placed so that its field is perpendicular to a uniform field produced by a permanent horseshoe magnet. This coil is free to rotate against a spring. The degree of deflection of the coil is directly proportional to the current in the coil. The figure at the left illustrates the construction of a galvanometer.

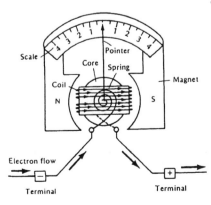

Galvanometer

2. Ammeter. An ammeter is a modified galvanometer used to measure larger currents. In an ammeter, a shunt is placed in parallel with the coil allowing most of the current to bypass the coil.

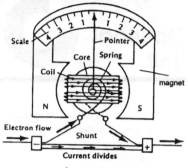

Ammeter

The resistance of the shunt is very small compared to the resistance of the coil. Decreasing the shunt resistance increases the maximum readings on the meter.

The low resistance of the shunt makes the internal resistance of the ammeter very small so that the potential drop across the meter has a negligible effect on the circuit being measured. The ammeter is placed in series in the circuit being measured.

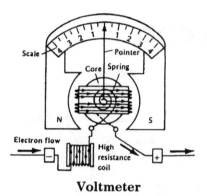

Voltmeter

3. Voltmeter. A voltmeter is a modified galvanometer used to measure potential differences. In a voltmeter a high resistance wire is connected in series with the coil. Increasing the series resistance increases the maximum readings on the meter.

The high resistance of the series resistor makes the internal resistance of the voltmeter very high so that it draws very little current and has a negligible effect on the circuit being measured.

A voltmeter is placed in parallel across the element whose voltage is being measured.

1.2 Motors

An **electric motor** is a device which converts electrical energy into rotational mechanical energy. Its operation is based on the principle that a current carrying coil experiences a torque in a magnetic field.

1. Iron Core. In a practical motor, the coil is wound on an **iron core** causing the torque to become very large. Since iron is highly permeable to magnetic flux, it concentrates and strengthens the magnetic field produced in the coil.

2. Split-ring Commutator and Brushes. A split-ring commutator is used to reverse the current in the armature coil after each half rotation, so that the torque is always in the same direction. **Brushes,** usually made of graphite or copper, ride on the split-ring commutator and connect the armature to the external circuit.

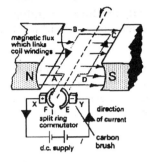

(a) *DC Electric Motor*

3. How a DC Motor Works. In **(a)** the coil begins in a horizontal position. The force on **AB** and **CD** causes the coil to rotate counterclockwise. As the coil passes through the vertical position **(b)**, no current is flowing, but the coil's momentum keeps it rotating. Just past the vertical position **(c)**, the force on **AB** and **CD** continues to rotate the coil counterclockwise. In position **(d)**, the coil is horizontal again and experiences a maximum amount of force and torque.

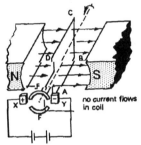

(b) *DC Electric Motor*

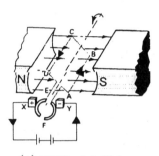

(c) *DC Electric Motor*

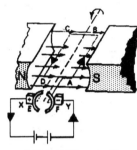

(d) *DC Electric Motor*

How A DC Electric Motor Works

Position of coil	Split - ring and Carbon brush in contact	Direction of current	Force on AB	Force on CD	Movement of Coil
horizontal	F with X E with Y	DCBA	downwards	upwards	starts to rotate counterclockwise
vertical	no contact between split - rings and brushes	no current flowing	zero	zero	coil's momentum keeps it rotating
just past the vertical	F with Y E with X	ABCD	upwards	downwards	continues to rotate counterclockwise
horizontal	F with Y E with X	ABCD	upwards	downwards	counterclockwise
vertical	no contact between split - rings and brushes	no current flowing	zero	zero	coil's momentum keeps it rotating
just past the vertical	F with X E with Y	DCBA	downwards	upwards	counterclockwise

and so on ...

4. Back EMF. An operating electric motor will produce an induced EMF in the armature coil which will oppose the applied potential difference and reduce the current in the armature.

The back EMF arises because the armature coil is rotating in an external magnetic field and is therefore acting as a generator, whose induced voltage must oppose the applied voltage that provides the driving current. This is an example of **Lenz's Law** and can be explained by conservation of energy.

Questions

1 As the armature of a motor turns, an EMF is induced which is opposite to the applied voltage. The existence of this back EMF can best be accounted for by
(1) the conservation of momentum
(2) the conservation of energy
(3) Coulomb's laws
(4) Newton's laws of motion

2 In the diagram of a DC motor, the arrows represent the direction of electron current. In what direction will end A of the armature spin?
(1) into the page
(2) out of the page
(3) clockwise
(4) counterclockwise

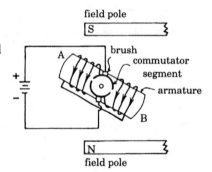

3 An ammeter is made by placing the current carrying wire loop of a galvanometer in
(1) series with a high resistance
(2) series with a low resistance
(3) parallel with a high resistance
(4) parallel with a low resistance

4 When a galvanometer is used, deflection of the needle occurs as a result of
(1) electrostatic force   (3) gravitational force
(2) magnetic force   (4) photoelectric effect

5 To convert a galvanometer into a voltmeter, it is necessary to connect a
(1) resistor in series with the coil
(2) resistor in parallel with the coil
(3) shunt wire in series with the coil
(4) shunt wire in parallel with the coil

6 The purpose of the shunt in an ammeter is to provide
(1) electrostatic deflection of the coil
(2) magnetic deflection of the coil
(3) resistance to current flow
(4) a path for some current to bypass the coil

7   The current in the armature of an electric motor switches direction with each half rotation. Which motor part produces this phenomenon?
(1) magnet   (3) armature
(2) split-ring commutator   (4) stator

8   If the amount of current flowing in a current carrying loop is doubled, the strength of the torque (twisting force) is
(1) halved   (2) doubled   (3) the same   (4) quadrupled

9   As a twisting torque causes the current carrying loop in an electric motor to begin rotating, the current in that loop
(1) decreases   (2) increases   (3) remains the same

10 If the shunt resistance of an ammeter is decreased, the maximum possible reading on the meter
(1) decreases   (2) increases   (3) remains the same

11 If the internal resistance of a voltmeter is increased, the maximum possible reading on the meter
(1) decreases   (2) increases   (3) remains the same

12 When a permeable substance is inserted into a current carrying coil of wire, the strength of the magnetic field
(1) decreases   (2) increases   (3) remains the same

II. Electron Beams

When metallic objects are heated to high temperatures, electrons are emitted from the filament. This phenomena is called **thermionic emission.** A space charge will be developed around an incandescent object which will impede the continued emission of electrons. The rate of the electron emission can be controlled by the temperature of the filament.

Thermionic emission is used to produce electron beams and currents between the electron emitter or the **cathode** and the positive plate, called the **anode.** Electron beams can be deflected by electric and magnetic fields. In an electric field the beam is deflected by a force which is parallel to the field and directed toward the positive plate.

In a magnetic field, the beam is deflected by a force, which is perpendicular to both the beam and the field. Its direction can be determined by the appropriate left hand rule. Electron beams will pass straight through crossed electric and magnetic fields when the electric force exactly balances the magnetic force. At that time:

$$F(\text{electric}) = F(\text{magnetic})$$
$$Eq = Bqv$$
$$E = Bv$$
$$v = E/B$$

In the diagram at the right of crossed electric and magnetic fields, the magnetic field is directed out of the page.

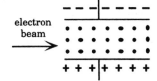

2.1 Magnetic Effect On A Moving Charge

If a charged particle moves at right angles to a magnetic field, the magnitude of the magnetic force on the particle is directly proportional to the charge on the particle, the velocity of the particle, and the strength of the magnetic field.

$$F = qvB$$

Where: F is the force in newtons,
        q is the charge in coulombs,
        v is the velocity in meters per second,
        B is the magnetic field in N/amp-m (or) Weber/m^2 or tesla

A positive charged particle can be treated as a negative charged particle moving in the opposite direction. For example, protons moving to the right, act the same as electrons moving to the left with respect to the forces involved.

The direction of the force can be determined by the **Third Left Hand Rule**. If the thumb is pointed in the direction of the electron flow, the outstretched fingers pointed to the direction of the magnetic field from North to South, the palm will indicate the direction of the force.

If the charge moves parallel to the field and cuts no lines of magnetic flux, then no force is experienced. The maximum force occurs when motion is perpendicular to the field.

Example

Calculate the force exerted on an electron as it moves through a magnetic field as indicated below.

Given: B = 100.0 N/amp-m
           v = 3.0×10^2 m/s.
           q = 1.6×10^{-19} coul

Find: Force

Solution: F = qvB
            F = $(1.6 \times 10^{-19}$ coul$)(3.0 \times 10^2$ m/s$)(100$ N/amp-m$)$
            F = 4.8×10^{-15} N

2.2 Charge To Mass Ratio Of An Electron

The charge to mass ratio of an electron was discovered in 1897, by an English scientist, **J. J. Thomson**. He was one of the first users of combined electric and magnetic fields on electron beams. When the fields were adjusted so the net force on the electrons was zero, the theory showed that the e/m ratio was readily measured using the values of **B** and **E**.

The numerical value of this ratio was later used by **R. A. Millikan** in the "oil drop" experiment. The value of the charge to mass ratio for an electron is 1.76×10^{11} coul/kg.

III. Charged Particle Beams
3.1 Cathode Ray Tube

A **cathode ray tube** is an evacuated glass tube containing a source of electrons at one end, a fluorescent screen inside the surface at the other end and two pairs of deflecting plates in between.

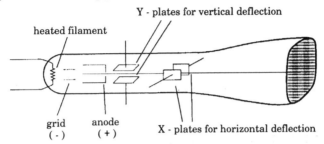

The uniform electric field between the oppositely charged plates controls the direction of the electron beam. The electron beam produces a fluorescent spot of light when it strikes the screen. The brightness of this spot is directly related to the intensity of the electron beam striking the screen. Magnetic fields can also be used for deflection.

3.2 Mass Spectrograph

A **mass spectrograph** is a device used to determine the masses of individual atoms. An element in the gaseous phase is bombarded with electrons causing one or more electrons to be removed from the atoms of the element. The resulting positive ions are subjected to magnetic and electric fields.

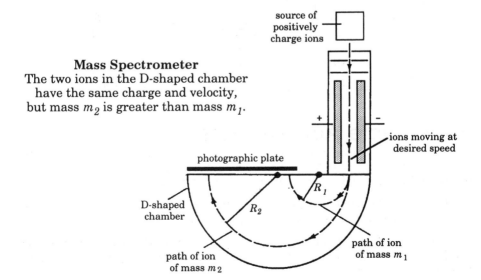

Mass Spectrometer
The two ions in the D-shaped chamber have the same charge and velocity, but mass m_2 is greater than mass m_1.

The charge to mass ratio of ions may be determined by measuring the radius of the circular path that ions travel. If the charge on the ion is known, then its mass can be calculated. This is a common method of separating the isotopes of an element and determining their masses.

To determine: $\dfrac{q}{m}$

From the electric field: $W = Vq = \frac{1}{2}mv^2$

From the magnetic field: $F = qvB = \dfrac{mv^2}{r}$ $v = \dfrac{rqB}{m}$

Substituting for v: $Vq = \frac{1}{2}m \left(\dfrac{rqB}{m} \right)^2$ and $\dfrac{q}{m} = \dfrac{V}{B^2 r^2}$

3.3 Mass Of The Electron

The Millikan oil drop experiment determined that an electric charge is always an integral multiple of an indivisible unit of charge called the elementary charge. Since the charge to mass ratio of the electron was known, the mass of the electron could be calculated. Thomson determined e/m for electrons in 1897. Millikan performed his series of measurements between 1909 and 1913.

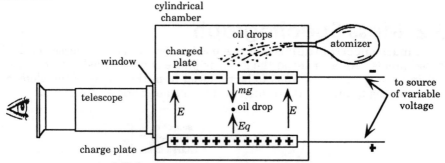

Millikan's Oil-drop Experiment
The oil drop, charged by friction, remains suspended between the charged plates when the upward electric force equals the downward force of the weight. The oil drops can have an excess or deficiency of electrons.

3.4 Particle Accelerators

A particle accelerator is a device used to accelerate *charged* particles to speeds approaching the speed of light by subjecting the particles to a large potential drop or a series of repeated smaller potential drops.

The charged particles bombarded nuclei to provide information about the structure of the nucleus and subatomic particles.

Examples of particle accelerators include the **Van de Graaff generator** which uses a single large potential drop, the **cyclotron**, **synchrotron** and **linear accelerator** which require repeated smaller potential drops.

Questions

Base your answers to questions 1 through 5 on the diagram which represents an electron beam in a vacuum. The beam is emitted by cathode C, accelerated by anode A, and passes through electric and magnetic fields.

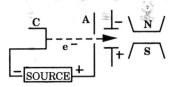

1   If an electron in the beam is accelerated to a kinetic energy of 4.8×10^{-16} joule, the potential difference between the cathode and the anode is
(1) 7.7×10^3 V   (2) 4.8×10^{-3} V   (3) 3.0×10^3 V   (4) 3.0×10^{-3} V

2   In which direction will the electron beam be deflected by the electric field?
(1) into the page   (3) towards the top of the page
(2) out of the page   (4) towards the bottom of the page

3   In which direction will the force of the magnetic field act on the electron beam?
(1) into the page   (3) towards the top of the page
(2) out of the page   (4) towards the bottom of the page

4   If an electron in the beam moves at 2.0×10^8 meters per second between the magnetic poles where the flux density is 0.20 tesla, the force on the electron is
(1) 6.4×10^{-12} N   (2) 6.4×10^{-10} N   (3) 4.0×10^7 N   (4) 4.0×10^9 N

5   If the plates producing the electric field are 0.020 meter apart and have a potential difference across them of 1,000 volts, what is the electric field intensity between them?
(1) 1,000 volts/meter   (3) 5,000 volts/meter
(2) 2,000 volts/meter   (4) 50,000 volts/meter

Base your answers to questions 6 through 10 on the diagram which represents a helium ion with a charge of +2 elementary charges, moving toward point A with a constant speed v of 2.0 meters per second perpendicular to a uniform magnetic field between the poles of a magnet. The strength of the magnetic field is 0.10 tesla.

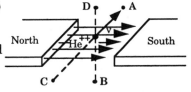

6   The direction of the magnetic force on the helium ion is toward point
(1) A (2) B (3) C (4) D

7   The magnitude of the magnetic force exerted on the helium ion is
(1) 3.2×10^{-20} N   (2) 0.20 N   (3) 0.10 N   (4) 6.4×10^{-20} N

8   If the strength of the magnetic field and the speed of the helium ion are both doubled, the force on the helium ion will be
(1) halved   (2) doubled   (3) the same   (4) quadrupled

9   If the polarity of the magnet is reversed, the magnitude of the magnetic force on the helium ion will
(1) decrease   (2) increase   (3) remain the same

10   The helium ion is replaced by an electron moving at the same speed. Compared to the magnitude of the force on the helium ion, the magnitude of the force on the electron is
(1) less   (2) greater   (3) the same

11 An electron traveling at a speed v in the plane of this paper enters a uniform magnetic field. Which diagram best represents the condition under which the electron will experience the greatest magnetic force as it enters the magnetic field?

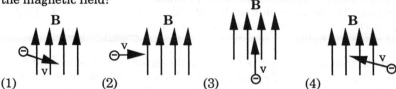

(1)   (2)   (3)   (4)

Base your answers to questions 12 through 16 on the diagram which represents an electron moving at 2.0×10^6 meters per second into a magnetic field which is directed into the paper. The magnetic field has a strength of 2.0 newtons per ampere meter.

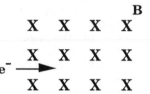

12 Which vector best indicates the direction of the force on the electron?

(1)   (2)   (3)   (4)

13 What is the magnitude of the force on the electron?
    (1) 6.4×10^{-13} N   (3) 6.4×10^6 N
    (2) 4.0×10^6 N   (4) 8.0×10^6 N

14 If the strength of the magnetic field were increased, the force on the electron would
    (1) decrease   (2) increase   (3) remain the same

15 If the velocity of the electron were increased, the force on the electron would
    (1) decrease   (2) increase   (3) remain the same

16 The electron is replaced with a proton moving with the same velocity. Compared to the magnitude of the force on the electron, the force on the proton would be
    (1) less   (2) greater   (3) the same

17 The charge to mass ratio of an electron is approximately 1.76×10^{11} coulombs per kilogram. This value indicates that the
    (1) electron's charge is about equal to its mass
    (2) charge on an electron is extremely small compared to its mass
    (3) mass of an electron is the same as that of an atom
    (4) mass of an electron is extremely small compared to its charge

18 An electron and a proton are projected into a magnetic field at right angles to the field. Both particles have the same velocity. Which statement best describes the resultant motion.
    (1) Both particles will travel in identical circles.
    (2) Only one of the particles will travel in a circle.
    (3) Both travel in circles in the same direction but with a different radii.
    (4) Both travel in circles, but both the direction and the radii will be different.

IV. Electromagnetic Induction

If a straight conductor is moved across a magnetic field, the electrons in the conductor will be acted upon by a magnetic force. This force will tend to move the electrons along the conductor from one end to another. As the motion of the electrons cause an excess of negative charge at one end and a deficiency of charge at the other end, a potential difference is established between the ends of the conductor. This effect can also be observed if the conductor is stationary and the field is moving.

Therefore, any time a conductor **cuts lines of flux**, a potential is induced in the conductor. If the conductor is connected to a circuit, a current will flow around the circuit. The direction of the induced current is in such a direction that its magnetic field opposes that of the field that induced it. This relationship is known as **Lenz's Law** and is *an example of conservation of energy.*

```
┌ ─ ─ ─ ─ ─ ─ ─ ─ ─ ─ ─ ┐
│ Practical Applications │
│  ·  generators        │
│  ·  transformers       │
└ ─ ─ ─ ─ ─ ─ ─ ─ ─ ─ ─ ┘
```

As a conductor is moved through a magnetic field, the direction of the induced current can be determined by using the left hand rule in reverse. Place the palm of the left hand to oppose the motion of the conductor. Line the fingers up with the magnetic field, pointing from N to S, and the thumb will point to the direction of the induced current.

The magnitude of an induced electromotive force is directly proportional to the flux density, the length of the conductor, and the speed of the conductor relative to the flux. This can be written as:

$$\text{EMF} = B \, l \, v$$

Where: **EMF**   =   induced electromotive force in volts

              B   =   magnetic field strength in $N/(A{\cdot}m)$

              l   =   length of conductor in meters

              v   =   velocity in m/s

4.1 The Generator Principle

A conducting loop rotating in a uniform magnetic field experiences a continual change in the total number of flux lines crossing the loop. This change induces a potential across the ends of the loop, which alternates in direction and varies in magnitude, between zero and a maximum. When the plane of the loop is perpendicular to the field, the induced potential is zero. When the plane of the loop is parallel to the field, the induced potential is a maximum.

The magnitude of the induced potential is proportional to the component of the velocity perpendicular to the field and the intensity of the magnetic field. When the loop is part of a complete circuit, the induced potential causes

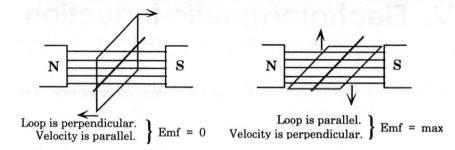

$$\left.\begin{array}{l}\text{Loop is perpendicular.}\\ \text{Velocity is parallel.}\end{array}\right\}\text{Emf} = 0$$ $$\left.\begin{array}{l}\text{Loop is parallel.}\\ \text{Velocity is perpendicular.}\end{array}\right\}\text{Emf} = \text{max}$$

a current in the loop. Since the induced potential is alternating, the current is an alternating current. An **alternating current** is a current that reverses its direction with regular frequency. Below is a graph of the induced potential difference in a generator coil rotating at constant speed.

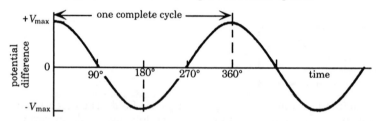

V. Changing Voltage

The basic principle for the operation of transformers and induction coils is electromagnetic induction. A changing magnetic field can induce a potential difference in a conductor.

5.1 Transformers

A **transformer** is a device used to change the voltage of an alternating current into a larger or smaller voltage of alternating current.

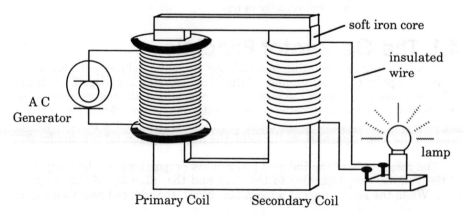

A transformer consists of a primary and secondary coil wound on an iron core. A continually changing current in the primary coil produces a continually changing magnetic field that induces an alternating voltage in the secondary coil. The induced **EMF** depends on the ratio of turns on the two coils.

The illustration at the bottom of the opposite page is a **step-down** transformer. The primary has more turns than the secondary. Step-down transformers are used in electric toy racing cars and electric toy trains.

Step-up transformers have a larger number of turns on the secondary than on the primary coil. They are used in fluorescent lights.

The ratio of the number of turns on the primary coil to the number of turns on the secondary coil of a transformer is equal to the ratio of the voltage across the primary coil to the voltage induced in the secondary coil.

$$\frac{N_p}{N_s} = \frac{V_p}{V_s}$$

Where: N_p =   number of turns on primary
N_s =   number of turns on secondary
V_p =   voltage of primary
V_s =   voltage of secondary

A transformer will transfer energy from the primary coil to the secondary coil if it is connected in a complete circuit. Power output from the secondary can never exceed the power input in the primary coil.

In a 100% efficient transformer, the power input to the primary coil is equal to the power output from the secondary coil.

$$Power_{input} = Power_{output}$$

$$V_p I_p = V_s I_s$$

Where: V is voltage,   I is current

The **efficiency** of the transformer is equal to the ratio of the power output from the secondary coil to the power input of the primary coil times 100.

$$\% \text{ efficiency} = \frac{V_s I_s}{V_p I_p} \times 100$$

Transformers get hot when used. This heating effect created within the metal core of the transformer is caused by **eddy currents**. This is the name given to the induced currents set up within a mass of conducting material (solid, liquid, or gas). The currents are visualized as being like swirling eddy currents seen in water.

If the transformer core is laminated, this reduces the eddy currents and hence the heating effect within the core. This is also true for motor and generator armatures.

Example

An alternating emf of 110 volts is applied to a step-up transformer having 300 turns on its primary and 6000 turns on its secondary. The secondary current is 0.100 A.

a) What is the secondary EMF?
b) What is the primary current?
c) What is the power input?
d) Assuming 90% efficiency, what is the power output?

Given: V_p = 110. V   **efficiency** = 90%
              N_p = 300. I_s = 0.100 A
              N_s = 6000.

Find: a) V_s = ?   c) Power input
              b) I_p = ?   d) Power output

Solution a): $\dfrac{V_s}{V_p} = \dfrac{N_s}{N_p}$

$V_s = \dfrac{6000}{300} \times 110 \text{ V}$

$V_s = 20 \times 110 \text{ V} = 2200 \text{ V}$

Solution b): $I_s = 0.100 \text{ A}$

$\dfrac{I_p}{I_s} = \dfrac{V_s}{V_p} = \dfrac{N_s}{N_p}$ or $\dfrac{I_p}{I_s} = \dfrac{N_s}{N_p}$, $I_p = \dfrac{N_s}{N_p} I_s$

$I_p = \dfrac{6000}{300} \times 0.100 \text{ A}$

$I_p = 20 \times 0.100 \text{ A} = 2.00 \text{ A}$

Solution c):   The power input is calculated by the product of the voltage and the current.

$V_p I_p = 110 \text{ V} \times 2.00 \text{ A}$

$V_p I_p = 220 \text{ W}$

Solution d):   Since the transformer is 90% efficient

$\%\text{ efficiency} = \dfrac{\text{power output}}{\text{power input}} \times 100$

power output = power input (90%)
             = (220 W) (.90)
$V_s I_s$ = 198 W

5.2 Induction Coil

A transformer will not work when a DC supply is passed through it, but an induction coil, a device which is similar in principle to a step-up transformer, can be used as a high voltage DC supply.

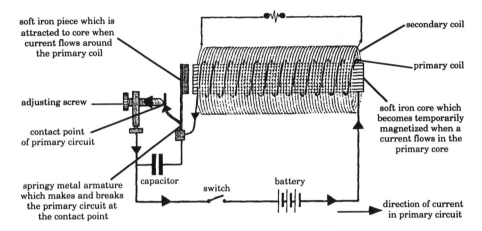

soft iron piece which is attracted to core when current flows around the primary coil

secondary coil

primary coil

adjusting screw

contact point of primary circuit

soft iron core which becomes temporarily magnetized when a current flows in the primary core

springy metal armature which makes and breaks the primary circuit at the contact point

capacitor   switch   battery

direction of current in primary circuit

A continuously changing current is produced by a mechanism in the primary coil which completes and breaks the circuit. The making and breaking occurs in the region of 50 times per second and on average a 12 volt input will produce an output in the region of 100,000 V. Such an induction coil is used within the ignition system of automobiles.

5.3 Laser

A **laser** is an acronym for light amplification by stimulated emission of radiation. Laser light is nearly **coherent**. That means that the light is in a narrow band of frequencies and also "in step" or "in phase." Lasers produce an intense, narrow, beam of parallel light. Many types of lasers exist. However, they all operate in the following basic way.

1) An outside agent (an electrical discharge, for example) excites atoms within the laser causing electrons to go to higher energy states.

2) With proper selection of the laser material, certain atoms in it are excited and remain in the excited state for an appreciable time.

3) One excited electron falls to a lower state and emits a photon. This photon's wave travels past the other excited atoms and triggers more electrons to fall to a lower state. In doing so, they emit photons with waves in step with the original photon wave.

4) The laser material is in the form of a tube or rod. The ends of the material are closed by accurately parallel mirrors. The beam travels back and forth between two mirrors. A very intense, coherent beam develops as more and more excited electrons emit photons in phase.

5) Only parallel rays will remain in the tube after many reflections.

6) One of the end mirrors of the laser leaks light slightly allowing a small fraction of the beam to pass through. This is what is referred to as a laser beam.

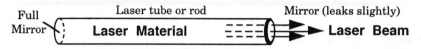

The laser light is very intense because all of the waves are in phase. Waves which are identical and in step are said to be coherent. The many reflections cause the beam to be very parallel. The beam retains its pencil-like quality over many miles.

Because of these unique properties, lasers have become widely used. Their high intensity makes them useful for drilling through the hardest metals or welding a loosened retina in place in an eye. Their high coherence and precision wavelength has led to the use of laser light in communications. Their pencil-like quality is useful in surveying and equipment alignment. The present and probable future of the laser is endless. One future use of the laser is to ignite the nuclear fusion reaction which could influence the course of civilization for years into the future.

Questions

1   A wire is moved through a magnetic field. The magnitude of the induced potential current cannot be increased by
    (1) moving the wire faster   (3) using a stronger field
    (2) using a better conductor   (4) using a longer wire

2   Current is induced in the secondary of an induction coil when the primary current is
    (1) direct and steady   (3) large
    (2) changing   (4) small

3   In a step-down transformer with a turn ratio of 5 to 1, a primary current of 10amperes will produce a current of
    (1) 1 ampere   (3) 5 amperes
    (2) 2 amperes   (4) 50 amperes

4   In a 100 % efficient transformer, both the primary and the secondary circuits would always have the same
    (1) voltage   (3) current
    (2) wattage   (4) resistance

5   An essential part of an induction coil is (are) the
    (1) slip rings   (3) brushes
    (2) interrupter   (4) magnets

6   A step-up transformer with a turn ratio of 5 to 1 will change a primary voltage of 100 volts into a secondary voltage of
    (1) 500 volts   (2) 20 volts   (3) 10 volts   (4) 5 volts

7   If the output of a 90 % efficient transformer is 180 watts, the input must be
    (1) 200 watts   (2) 90 watts   (3) 180 watts   (4) 233.3 watts

Base your answers to questions 8 and 9 on the diagram and the information below. The diagram represents a 100% efficient transformer connected to an AC source. This transformer has two turns in the secondary coil for each turn in the primary coil.

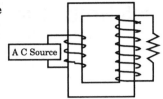

8   The voltage across the secondary coil is
   (1) the same as that in the primary
   (2) twice that in the primary
   (3) half that in the primary
   (4) zero

9   Compared to the power in the primary coil, the power in the secondary coil is
   (1) less   (2) greater   (3) the same

10 If a closed loop is pulled out of a magnetic field, it should be expected that
   (1) there will be no net force on the loop
   (2) the magnetic field will exert a force tending to pull it back
   (3) the loop will tend to rotate
   (4) the magnetic field will push it out

11 Compared to the voltage in the coil of a transformer with more turns of wire, the voltage in the coil with fewer turns is
   (1) smaller   (2) greater   (3) the same

12 A wire 1 meter long moves at 5 meters per second perpendicularly to a uniform magnetic field of intensity of 0.02 tesla. What is the potential difference induced across the ends of the wire?
   (1) 1.0 volt   (2) 0.01 volt   (3) 0.1 volt   (4) 10 volts

13 A wire 0.10 meter long is pushed through a magnetic field of strength 4.0 tesla in a direction perpendicular to the field. If the speed of the wire is 2.0 meters per second, what is the magnitude of the induced voltage across the ends of the wire?
   (1) 0.20 volt   (2) 2.0 volts   (3) 0.80 volt   (4) 0.5 volt

Thinking Physics

1   Why do American homes run on AC current rather than DC current?

2   Electrical forces between charges are huge compared to gravitational forces. Why don't we sense electrical forces within our environment while we do sense gravitational forces with the Earth?

3   In supermarkets, lasers "read" bar code information. What components make up a scanning system? Where is the laser located?

4   The picture tube of your television is a cathode–ray tube. The electron beam is controlled by magnetic fields. What effect would you expect from a magnet held in front of the tube?

5   Electric generators are used on some bicycles. As the wheel turns, the generator provides the energy for the bike's lamp. Does it make any difference if the light is off or on when the bike is coasting?

Self–Help Questions

1   In the diagram at the right, a solenoid that is
    free to rotate around an axis at its center, *C*, is
    placed between the poles of a permanent
    magnet. As an electron current starts through
    the solenoid in the direction shown, the
    solenoid will
    (1) remain motionless
    (2) vibrate back and forth
    (3) start turning clockwise
    (4) start turning counterclockwise

2   A galvanometer can be modified to function as
    (1) a voltmeter, only
    (2) an ammeter, only
    (3) either a voltmeter or an ammeter

3   As the shunt resistance of an ammeter is decreased, the maximum
    current that the ammeter can measure
    (1) decreases   (2) increases   (3) remains the same

4   A galvanometer with a low-resistance shunt in parallel with its moving
    coil is
    (1) a motor   (3) a voltmeter
    (2) a generator   (4) an ammeter

5   Which statement about ammeters and voltmeters is correct?
    (1) The internal resistance of both meters should be low.
    (2) Both meters should have a negligible effect on the circuit being
        measured.
    (3) The potential drop across both meters should be made as large as
        possible.
    (4) The scale range on both meters must be the same.

6   In a practical motor, the coil is wound around a soft iron core. The
    purpose of the soft core is to
    (1) strengthen and concentrate the magnetic field through the coil
    (2) cause the torque on the coil to remain in the same direction
    (3) convert alternating current to direct current
    (4) oppose the applied potential difference and reduce the current in the
        coil

7   What is the purpose of the split-ring commutator in a direct current
    motor?
    (1) to eliminate the external magnetic field
    (2) to increase the current in the armature
    (3) to maintain the direction of rotation of the armature
    (4) to decrease the induced back EMF

8   The physical structure of an electric motor most closely resembles that of
    (1) a mass spectrograph   (3) a transformer
    (2) a cathode ray tube   (4) an electric generator

9   An electron is moving at a velocity of 4.0×10^6 meters per second
    perpendicular to a magnetic field with a flux density of 6.0 teslas. The
    magnitude of the magnetic force acting on the electron is
    (1) 1.6×10^{-13} N   (3) 3.8×10^{-12} N
    (2) 6.4×10^{-13} N   (4) 2.4×10^7 N

10 Which could *not* be accelerated using an electric field?
    (1) electron   (2) positron   (3) photon   (4) alpha particle

Base your answers to question 11 through 14 on the diagram at the right, which represents an electron beam entering the space between two parallel, oppositely charged plates. A uniform magnetic field, directed out of the page, exists between the plates.

11 If the magnitude of the electric force on each electron and the magnetic force on each electron are the same, which diagram best represents the direction of the *vector sum* of the forces acting on one of the electrons?

    (1)   (2)   (3)   (4)

12 In which direction would the magnetic field have to point in order for the magnetic force on the electrons to be opposite in direction from the electric force on the electrons?
    (1) toward the bottom of the page
    (2) toward the top of the page
    (3) out of the page
    (4) into the page

13 If the electric force were equal and opposite to the magnetic force on the electrons, which diagram would best represent the path of the electrons as they travel in the space between the plates?

    (1)   (2)   (3)   (4)

14 If only the potential difference between the plates is increased, the force on the electron will
    (1) decrease   (2) increase   (3) remain the same

Base your answers to questions 15 through 18 on the diagram at the right which shows a cathode ray tube. The electrons in the tube are emitted from a heated cathode and travel in a beam to the face of the tube, causing a bright spot to appear where they hit.

15 The process by which the electrons are emitted from the cathode as it becomes hot is called
    (1) thermionic emission   (3) photoelectric emission
    (2) alpha particle emission   (4) photon emission

16 If an upward *magnetic* field is applied to the neck of the tube, the bright spot will move toward point
    (1) *A*   (2) *B*   (3) *C*   (4) *D*

17 If an upward *electric* field is applied to the neck of the tube, the bright spot will move toward point
    (1) *A*   (2) *B*   (3) *C*   (4) *D*

18 A 5.0×10^{-4} tesla magnetic field is placed so its direction is perpendicular to the path of the electrons. The electrons move toward the tube face at a speed of 4.0×10^6 meters per second. What is the magnitude of the force exerted on each electron by this magnetic field?
(1) 1.6×10^{-9} N   (3) 3.2×10^{-16} N
(2) 2.0×10^3 N   (4) 8.0×10^{-23} N

Base your answers to questions 19 through 21 on the diagram at the right which shows an apparatus for demonstrating the effect of a uniform magnetic field on a beam of electrons moving in the direction shown.

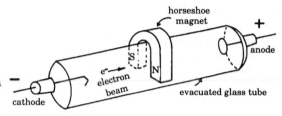

19 As the electron beam enters the magnetic field of the horseshoe magnet, the beam will be deflected
(1) toward the south pole of the magnet
(2) toward the north pole of the magnet
(3) downward, toward the bottom of the tube
(4) upward, toward the top of the tube

20 The velocity of the electron beam is 3.0×10^6 meters per second, perpendicular to the 5.0×10^{-3} tesla magnetic field. What is the magnitude of the force acting on each electron in the beam?
(1) 8.0×10^{-22}N   (2) 2.4×10^{-15}N   (3) 1.7×10^{-9}N   (4) 1.5×10^4N

21 If the speed of the electrons traveling through the magnetic field increases, the magnetic force on the electrons will
(1) decrease   (2) increase   (3) remain the same

Base your answers to questions 22 through 26 on the diagram which represents a beam of electrons entering a magnetic field between the poles of a magnet. The magnetic flux density between the poles of the magnet is 5.0×10^{-5} tesla.

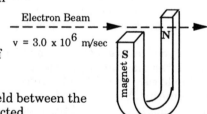

22 As the electrons enter the magnetic field between the poles of the magnet, they will be deflected
(1) toward the south pole   (3) downward
(2) toward the north pole   (4) upward

23 The magnitude of the force acting on each electron as it travels through the magnetic field is
(1) 1.5×10^{-13} N   (3) 2.4×10^{-17} N
(2) 8.0×10^{-17} N   (4) 4.8×10^{-24} N

24 The electrons are replaced by helium ions with a charge of +2. The ions enter the magnetic field between the poles of the magnet with the same velocity as the electrons. Compared to the force on each electron, the force on each helium ion will be
(1) the same   (3) sixteen times as great
(2) twice as great   (4) four times as great

25 If the velocity of the helium ions entering the magnetic field between the poles is increased, the force on each helium ion will
(1) decrease　　(2) increase　　(3) remain the same

26 Both helium ions and helium atoms enter the magnetic field with the same velocity. Compared to the force exerted by the magnetic field on the helium ions, the force exerted on the helium atoms will be
(1) less　　　(2) greater　　　(3) the same

27 What is the potential difference induced in a wire 0.10 meter long as it moves with a speed of 50. meters per second perpendicular to a magnetic field that has a magnetic flux density of 0.050 tesla?
(1) 0.25 V　　(2) 25 V　　(3) 250 V　　(4) 2500 V

28 A wire loop is rotating between the poles of a magnet as represented in the diagram at the right. As the loop rotates 90 degrees from the position shown, the magnitude of the induced current in resistor R
(1) decreases
(2) increases
(3) remains the same

29 The diagram at the right shows a wire loop rotating between magnetic poles. During 360° of rotation from the position shown, the induced potential difference changes in
(1) direction, only
(2) magnitude, only
(3) both magnitude and direction
(4) neither magnitude nor direction

30 The diagram at the right shows conductor C between two opposite magnetic poles. Which procedure will produce the greatest induced potential difference in the conductor?
(1) holding the conductor stationary between the poles
(2) moving the conductor out of the page
(3) moving the conductor toward the right side of the page
(4) moving the conductor toward the N-pole

31 The loop shown in the diagram at the right rotates about an axis which is perpendicular to a constant uniform magnetic field. If only the direction of the field is reversed, the magnitude of the maximum induced potential difference will
(1) decrease
(2) increase
(3) remain the same

32 The input to the primary of a transformer is 2.0 amperes at 400 volts. If the output at the secondary is 1.0 ampere at 600 volts, the efficiency of the transformer is
(1) 25%　　(2) 50%　　(3) 75%　　(4) 133%

33 A transformer is designed to step 220 volts up to 2,200 volts. There are
    200 turns on the primary. How many turns are there on the secondary?
    (1) 20   (2) 200   (3) 1,000   (4) 2,000
34 As the resistance of a constant-voltage circuit is increased, the power
    developed in the circuit
    (1) decreases   (2) increases   (3) remains the same

Base your answers to questions 35 and 36 on the diagram and the
information below.

The diagram represents a 100% efficient transformer connected to an AC
source. This transformer has two turns
in the secondary coil for each turn in
the primary coil.

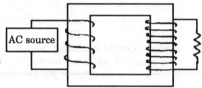

35 The voltage across the secondary coil is
    (1) the same as that in the primary
    (2) twice that in the primary
    (3) half that in the primary
    (4) zero
36 Compared to the power in the primary coil, the power in the secondary
    coil is
    (1) less   (2) greater   (3) the same
37 An ideal transformer has a current of 2.0 amperes and a potential differ-
    ence of 120 volts across its primary coil. If the current in the secondary
    coil is 0.50 ampere, the potential difference across the secondary coil is
    (1) 480 V   (2) 120 V   (3) 60. V   (4) 30. V
38 A transformer has 150 turns of wire in the primary coil and 1,200 turns
    of wire in the secondary coil. The potential difference across the primary
    is 100 volts. What is the potential difference induced across the secondary
    coil?
    (1) 14 V   (2) 110 V   (3). 150 V   (4) 800 V
39 A potential difference of 50. volts
    is required to operate an electri-
    cal device. The potential differ-
    ence of the source is 120 volts.
    The table shows the primary and
    secondary windings for four avail-
    able transformers. Which trans-
    former is suitable for this applica-
    tion?
    (1) A   (2) B   (3) C   (4) D

Transformer	Primary	Secondary
A	250	600
B	600	250
C	240	150
D	150	240

Unit 4
Wave Phenomena

Important Terms To Be Understood

pulse	angle of incidence	angle of reflection
wave	standing wave	virtual image
longitudinal wave	dispersion	index of refraction
transverse wave	reflection	polarization
wavelength	refraction	coherent
frequency	interference	diffraction grating
velocity	Doppler Effect	resolution
amplitude	diffraction	electromagnetic spectrum
period	resonance	continuous spectra
hertz	open air tube	line spectra
node	closed air tube	band spectra
antinode	phase difference	red shift and blue shift

I. Waves

A **pulse** is a single vibratory disturbance which moves from point to point through a medium. A **wave** is several pulses generated at regular time intervals. *Waves are important in physics because waves can transfer energy without transferring mass.*

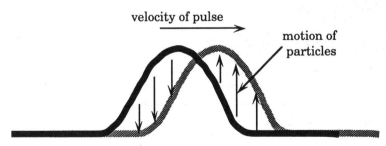

velocity of pulse

motion of particles

wave pulse — passing to the right

Some waves need material mediums in which to travel. Examples of waves that travel in mediums include water waves, sound waves or waves in a rope. The particles of the medium vibrate around a rest position. Light, radio and other electromagnetic waves are periodic disturbances in an electromagnetic field. They need no material medium and can travel through a vacuum.

Waves are classified by the way they vibrate. The most common types are **longitudinal** and **transverse** waves.

Particles in longitudinal waves vibrate *parallel to the direction of the wave* motion. Each particle moves back and forth parallel to the wave direction. Sound is an example of a longitudinal wave. Seismic waves (P - waves) are longitudinal waves.

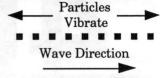

Longitudinal Wave

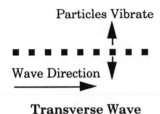

Transverse Wave

Particles in transverse waves vibrate *perpendicular to the direction of the waves*. Each particle moves perpendicular to the wave direction. Electromagnetic waves, such as light and radio are examples of transverse waves. Seismic waves (S–waves) are transverse waves. Transverse waves can travel in the same direction but in different planes.

The electromagnetic wave illustration at the right has a magnetic field that vibrates in the X direction and has an electric field that vibrates in the Y direction. The velocity of an electromagnetic wave is the speed of light in the Z direction.

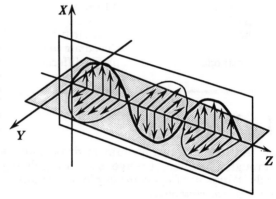

II. Wave Characteristics

A. The **frequency** is the number of waves passing a point per unit time, determined by the vibrating source. The unit of frequency is **hertz** (or s^{-1}).

Practical Applications
- radio frequencies   • sonar
- tuning forks   • dog whistles

B. The **period** of the wave (**T**) is the time for one complete cycle to pass a point. Period is the reciprocal of frequency and is calculated from the formula, **T = 1/f**, where **T** is the period in seconds and **f** is the frequency in hertz.

C. The **amplitude** of a wave is related to energy of a wave. In a transverse wave, it is defined as the maximum distance above or below the wave axis (equilibrium position).

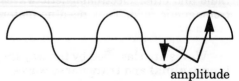

amplitude

In a longitudinal wave, the amplitude is determined by the separation of particles in the compression and rarefactions.

As the amplitude of a light wave increases, the brightness of the light increases. As the amplitude of a sound wave increases, the loudness of the sound increases.

> **Practical Applications**
> - "boom box" stereo
> - threshold of pain
> - dimmer switch for a lamp
> - rock concerts
> - "walkman" radios

D. **Phase.** Points on a periodic wave having the same displacement (amplitude) from their equilibrium position and moving in the same direction, are said to be *inphase* (for example, *A* and *B* or *C* and *D*).

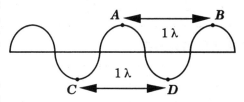

E. The **wavelength** (λ) is the distance between corresponding points in phase on successive waves. Wavelength is measured in meters.

F. The **velocity** is the speed of the wave. It is determined by the type of wave medium. It is measured in meters per second. These characteristics are related to each other by the wave equation:

$$\text{Velocity} = \text{frequency x wavelength}$$
$$v = f\lambda$$

G. **Wave Propagation — Wave Fronts.** A wave front is the locus of adjacent points of a wave that are in phase.

H. **Doppler Effect** is observed when the source or the observer is moving. There is an increase in the observed frequency, when the vibrating source approaches the observer (for example, the distance between the source and the receiver is decreasing). There is a decrease in observed frequency as the source moves away from the observer (for example, the distance between the source and the receiver is increasing).

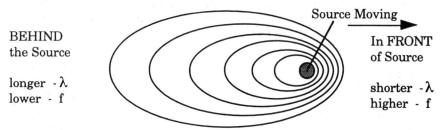

Doppler Effect of a Source Moving to the Right

If a source is moving at constant velocity, the frequency heard as it comes towards you will be a constant higher frequency than produced by the source or a constant lower frequency, going away.

If the source is accelerating towards you, the frequency will be higher and will continue to get higher. Accelerating away from you, it will be lower and continue to get lower.

Practical Applications
- train whistle
- passing car's horn
- radar
- red or blue shift of starlight
- measuring the speed of pitched baseballs

Example

1) Calculate the velocity of a wave that has a frequency of 6.0 hertz and a wavelength of 3.0 meters.

Given: frequency   = 6.0 hertz
           wavelength = 3.0 meters

Find: velocity

Solution: v = $f(\lambda)$
                   = (6.0 hertz)(3.0 m)
                   = 18 m/s

2) Calculate the period of the wave in example (1)

Given: frequency = 6.0 hertz

Find: Period

Solution: T = $1/f$
                   = ⅙ hertz
                   = 0.17 seconds

Questions

1   The diagram illustrates the wave pattern formed when a stone is dropped into still water. The locus of points on the outermost circle represents
    (1) a wave front
    (2) a wavelength
    (3) the frequency
    (4) the period

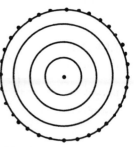

2   If the frequency of a wave is increased, its wavelength will
    (1) decrease   (2) increase   (3) remain the same

3   Which distance represents the wavelength of the wave shown?
(1) *A*
(2) *B*
(3) *C*
(4) *D*

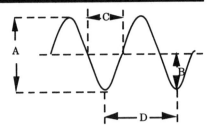

4   If the period of a wave is doubled, its wavelength will be
(1) halved   (2) doubled   (3) unchanged   (4) quartered

5   Which characteristic varies directly with the energy that produces it?
(1) amplitude   (2) speed   (3) period   (4) frequency

6   Which diagram best represents a periodic wave traveling through a
uniform medium?

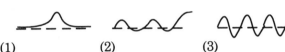

(1)   (2)   (3)   (4)

7   In the accompanying diagram, a point is
shown on a transverse wave. As the wave
pulse moves to the right, in which direction
will the point move?
(1) 1
(2) 2
(3) 3
(4) 4

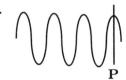

8   The diagram represents a wave traveling in
a uniform medium. Which characteristic of
the wave is constant?
(1) amplitude
(2) frequency
(3) period
(4) wavelength

9   Waves are traveling with a speed of three meters per
second toward *P* as shown in the diagram. If four
crests pass *P* in one second, the wavelength is
(1) 1 m   (3) 3 m
(2) 6 m   (4) 9 m

10 To the nearest order of magnitude, how many times greater than the
speed of sound is the speed of light?
(1) 10^4 (3) 10^{10}
(2) 10^6 (4) 10^{12}

11 What is the frequency of a wave if its period is 0.25 second?
(1) 1.0 Hz   (3) 12 Hz
(2) 0.25 Hz   (4) 4.0 Hz

12 Two points on a periodic wave in a medium are said to be in phase if they
  (1) have the same amplitude, only
  (2) are moving in the same direction, only
  (3) have the same period
  (4) have the same amplitude and are moving in the same direction

13 What is the amplitude of the wave represented in the diagram?
  (1) 1 m
  (2) 2 m
  (3) 3 m
  (4) 6 m

14 Which point on the wave shown in the diagram is 180° out of phase with point P?
  (1) 1
  (2) 2
  (3) 3
  (4) 4

15 The diagram represents a pulse traveling from left to right in a stretched heavy rope. The heavy rope is attached to light rope which is attached to a wall. When the pulse reaches the light rope, its speed will
  (1) decrease   (2) increase   (3) remain the same

16 The diagram represents a series of wave fronts produced by a wave generator that operated for 2 seconds. What was the frequency of the wave produced?
  (1) 6 hertz   (2) 2 hertz   (3) 3 hertz   (4) 12 hertz

17 The graph represents the displacement of a point in a medium as a function of time when a wave passes through the medium. What is the frequency of the wave?
  (1) 0.50 Hz
  (2) 2.0 Hz
  (3) 0.25 Hz
  (4) 4.0 Hz

18 What is the period of a wave whose frequency is 256 hertz?
  (1) 1 s   (2) $\frac{1}{256}$ s   (3) 256 s   (4) 512 s

19 A light wave has a frequency of 5.4×10^{14} hertz and a wavelength of 5.5×10^{-7} meter. What is the approximate speed of the wave?
  (1) 1.0×10^8 m/s   (3) 3.0×10^8 m/s
  (2) 2.0×10^8 m/s   (4) 4.0×10^8 m/s

20 A wave has a frequency of 2.0 hertz and a velocity of 3.0 meters per second. The distance covered by the wave in 5.0 seconds is
  (1) 30. meters   (2) 15 meters   (3) 7.5 meters   (4) 6.0 meters

III. Periodic Wave Phenomena

Periodic waves respond to different conditions in predictable ways.

3.1 Interference

Interference is the effect produced by two or more waves which are passing simultaneously through a region.

Superposition. The resultant disturbance at any point is the algebraic sum of displacements due to individual waves.

Constructive interference occurs at points where path distances to the two sources differ by an even number of half wavelengths. A phase difference of 0° or 360° is a wavelength. **Destructive interference** occurs when a crest meets a trough or compressions meet rarefactions. Destructive interference occurs at points where path differences to the two sources differ by an odd number of half wavelengths. A phase difference of 180° is a half wavelength. The placement of sound system speakers should take into account interference phenomena. Beats are usually heard when two sound frequencies are within 10 hertz of each other. Band members listen for beats when tuning their instruments.

Standing waves are produced when two waves of the same frequency and amplitude travel in opposite directions in the same medium. Some points, called **nodes**, appear to be standing still. Other positions called **antinodes**, vibrate with the maximum amplitude above and below the axis. *When a wave strikes a boundary, some of the wave's energy will be **reflected**, some will be **transmitted** and some will be **absorbed**.* Standing waves are most often produced by reflection of a wave train at a fixed boundary of a medium.

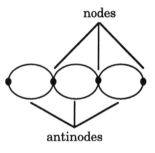

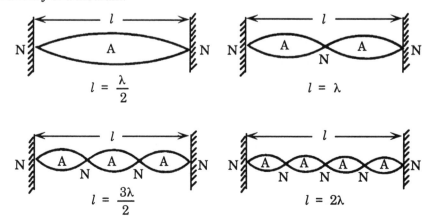

Standing Waves — *Different Wavelengths Along a Rope Bridge*
(N = node   A = antinode)

Resonance occurs when one vibrating body sympathetically causes the other to vibrate. This occurs when both objects have the same natural vibration frequency. Most vibrating systems will vibrate at a particular frequency if disturbed. Air columns, both open and closed, provide a good example of resonance (see the College Board Section).

Practical Applications
- Tacoma Narrows Bridge
- behavior of atoms
- music and acoustics
- opera singer shattering a crystal goblet
- marchers breaking step when crossing a bridge

Questions

1   The musical instrument shown in the diagram has four strings of identical metal under identical tension. Which string will produce the highest pitch when plucked?

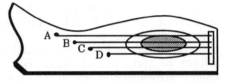

   (1) *A*   (3) *C*
   (2) *B*   (4) *D*

2   Sound waves may form standing waves as shown in this apparatus. Which points represent a complete wavelength?

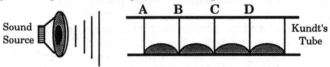

   (1) *AB*   (2) *AC*   (3) *BC*   (4) *CD*

3   It takes 1 second for a sound wave to travel from a source to observer *A*. How long does it take the same sound wave to travel in the same medium to observer *B*, who is located twice as far from the source as observer *A*?
   (1) ¼ s   (2) 2 s   (3) ½ s   (4) 4 s

4   The diagram shows an observer *A* moving through a series of waves traveling the direction shown. If the speed of observer *A* increases, the frequency of the waves that he observes

   (1) decreases
   (2) increases
   (3) remains the same

Wave Direction

5   Resonance occurs when one object causes a second one to vibrate. The second object must have the same natural
   (1) frequency   (3) loudness
   (2) speed   (4) amplitude

6   If two identical sound waves arriving at the same point are in phase, the resulting wave will have
   (1) an increase in speed   (3) a larger amplitude
   (2) an increase in frequency   (4) a longer period

7   As the phase difference between two superimposed waves changes from 180° to 90°, the amount of destructive interference
(1) decreases   (2) increases   (3) remains the same

8   The diagram represents two waves traveling simultaneously in the same medium. At which of the given points will maximum constructive interference occur?

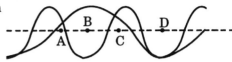

(1) *A*   (3) *C*
(2) *B*   (4) *D*

9   The diagram represents pulses traveling toward point *B* in a spring, *AB*. Which diagram best represents the motion of point *C* as the pulses pass along the spring?

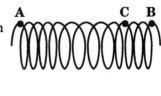

     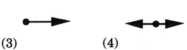

(1)   (2)   (3)   (4)

10 Which pair of moving pulses in a rope may produce destructive interference?

(1)    (3)

(2)    (4)

IV. Light

Visible light is an electromagnetic disturbance that can produce the sensation of sight. It is produced as excited electrons return to a lower energy state. Light is a small part of the electromagnetic spectrum.

The speed of light was determined to be 3.0 x 10^8 m/s over 100 years ago. Because of its great speed, light was difficult to measure. The speed of light in a vacuum or in air are approximately the same, but in any other substance, the speed of light is less than 3.0 x 10^8 m/s. In 1905, **Albert Einstein** pointed out the theoretical importance of the velocity of light. All electromagnetic radiations travel at the speed of light (c) through a vacuum, and only electromagnetic radiation can travel at the speed of light.

In a vacuum or air, all light frequencies travel at the same speed. In other materials such as glass, different frequencies travel at different speeds. These mediums are said to be **dispersive**. The speed of light in a medium is dependent on the frequency of the light and the optical density of the medium.

4.1 Reflection

The law of reflection states that the angle of incidence equals the angle of reflection, and that the incident ray, the reflected ray, and the normal all lie in the same plane. The **angle of incidence** is defined as the angle between the normal and the incident ray. The **angle of reflection** is the angle between the normal and the reflected ray.

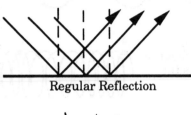

Regular Reflection

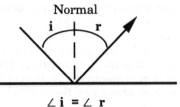

Normal

$\angle i = \angle r$

$\angle i$ — **angle of incidence**

$\angle r$ — **angle of reflection**

Diffuse Reflection

Every ray of light hitting a surface behaves according to the law of reflection. When the surface is very smooth, all the rays are reflected back parallel. This is known as **specular** or **regular reflection** and usually produces an image of the source. When the surface is irregular, (although each ray behaves according to the laws of reflection), since the normals are non-parallel, the reflected rays are non-parallel. This type of reflection is known as **diffuse reflection**. Diffuse reflection is responsible for the blue sky and the red sky at night.

Reflection in a plane mirror produces virtual, upright images in the same plane as the object. The distance the object is in front of the mirror is the same as the image distance behind the mirror.

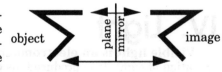

object   plane mirror   image

Practical Applications (regular reflection)
- mirrors
- pool of water
- looking out a window at night

Questions

1   A ray is reflected from a surface as shown in the diagram. Which letter represents the angle of incidence?
(1) *A*   (3) *C*
(2) *B*   (4) *D*

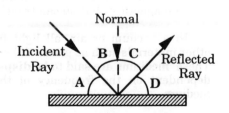

Normal

Incident Ray   B   C   Reflected Ray

A   D

2   The behavior of the light incident upon this page best illustrates the
    phenomenon of
    (1) diffraction   (3) regular reflection
    (2) refraction   (4) diffuse reflection
3   Which phenomenon of light is illustrated by the dia-
    gram?
    (1) regular reflection   (3) diffraction
    (2) diffuse reflection   (4) refraction
4   A light ray is incident upon a plane mirror. If the angle of incidence is
    increased, the angle of reflection will
    (1) decrease   (2) increase   (3) remain the same
5   An object is placed in front of a plane mirror as shown in
    the diagram. The image produced by the plane mirror is
    best represented by

    (1)   (2)   (3)   (4)
6   What is the angle of reflection represented in the
    diagram?   130°
    (1) 25°   (3) 65°
    (2) 80°   (4) 130°

7   The images formed by reflections in a plane mirror are
    (1) virtual and upright   (3) virtual and inverted
    (2) real and upright   (4) real and inverted

Base your answers to questions 8 and 9 on the diagram.   object

8   The length of the image in the mirror is
    (1) 0.5 m   (3) 8 m
    (2) 2 m   (4) 4 m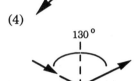
9   The distance of the image from the mirror is
    (1) 0.5 m   (2) 2 m   (3) 8 m   (4) 4 m

4.2 Refraction

Refraction is the bending of light as the
ray enters a new medium obliquely. It is due to
the change in velocity. A light traveling to a
more optically dense medium bends towards
the normal. Light going into a less optically
dense medium bends away from the normal.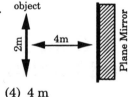

Any ray of light striking a surface along the   Ø₁ = **angle of incidence**
normal will pass straight through the new
material at a lower speed. It will not be bent.   Ø₂ = **angle of refraction**

Practical Applications
*   twinkling stars   •   straw appears bent in water
*   "water" on road on a hot day   •   one - way mirror

4.3 Absolute Index Of Refraction

The **absolute index of refraction** is the ratio between the speed of light in a vacuum or air and the speed of light in the material.

$$n = \frac{c}{v} = \frac{\text{Speed of Light in a Vacuum}}{\text{Speed of Light in the Material}}$$

The absolute index of refraction is always a number greater than one (1). **Snell's Law** relates the angle of incidence and the angle of refraction:

$$\frac{\sin \emptyset_1}{\sin \emptyset_2} = \frac{n_2}{n_1}$$

or $n_1 \sin \emptyset_1 = n_2 \sin \emptyset_2$ or $n_1 v_1 = c$ (speed of light)

also $n_1 v_1 = n_2 v_2$

Where:

n = the absolute index of refraction
n_1 = index of refraction of the first medium
$\emptyset_1$ = angle measured with the normal in the first medium
n_2 = index of refraction of the second medium
$\emptyset_2$ = angle measured with the normal in the second medium
v_1 = velocity of light in the first medium
v_2 = velocity of light in the second medium

4.4 Critical Angle
And Total Internal Reflection

As a ray of light travels from a substance with a high index of refraction (low velocity) and passes into a substance of lower index of refraction (higher velocity), the ray is refracted away from the normal. Increasing the angle of incidence will increase the angle of refraction according to Snell's Law.

The maximum angle of refraction is 90°. The angle of incidence for which the angle of refraction is 90°, is called the **critical angle**. The sine of the critical angle can be calculated from $\sin \emptyset_c = 1/n$ (when the ray is entering air). The **n** is the index of refraction of the material.

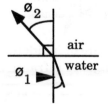 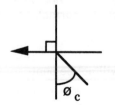 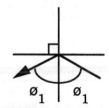

If $\emptyset_1$ is less than the critical angle, the ray will be refracted at $\emptyset_2$.

At critical angle, the angle of refraction is 90°.

Beyond critical angle, total internal reflection.

If the angle of incidence continues to increase in this situation, **total internal reflection** occurs. There is no refraction (transmission) and no absorption at the boundary.

```
┌ ─ ─ ─ ─ ─ ─ ─ ─ ─ ┐
│ Practical Applications │
│  ·   fiber - optics      │
│  ·   diamonds            │
│  ·   rainbow             │
└ ─ ─ ─ ─ ─ ─ ─ ─ ─ ┘
```

4.4 Dispersion

A **prism** separates white or polychromatic light into its component wavelengths, because each frequency has a different index of refraction.

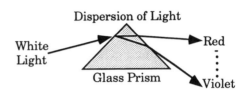

Dispersion of Light

The result is, that since violet light travels slower in the glass prism, it has a higher index of refraction and is bent more than red light, as illustrated above. The light must enter the prism obliquely to separate component wavelengths.

Under everyday classroom conditions, air is a non-dispersal medium for light. The dispersion effect of air on light can be observed in the Michelson Interferometer and, rarely, as a green flash as the sun sets.

Questions

1   As a wave is refracted, which characteristic of the wave will remain unchanged?
    (1) velocity   (2) wavelength (3) frequency   (4) direction

2   The diagram represents a light ray traveling from crown glass into air. The position of the light source is changed to vary the angle (Ø). As Ø approaches the critical angle, the angle of refraction approaches
    (1) 0°
    (2) 41°
    (3) 90°
    (4) 180°

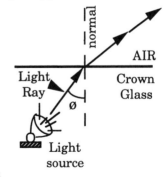

3   The accompanying diagram represents a light ray passing from one medium into another. The light ray must be traveling
    (1) from medium x into medium y
    (2) from medium y into medium x
    (3) faster in medium x than in medium y
    (4) faster in medium y than in medium x

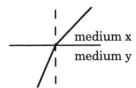

4   In the diagram, light ray *AO* is incident on a Lucite - air surface at point *O*. Through which point will the reflected ray pass?
    (1) 1
    (2) 2
    (3) 3
    (4) 4

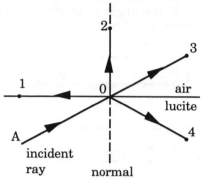

5   For a given angle of incidence, the greatest change in the direction of a light ray will be produced when the light ray passes obliquely from air into
    (1) Lucite   (3) corn oil
    (2) glycerol   (4) flint glass

6   When a wave goes from a medium of higher refractive index to one of lower refractive index, its speed
    (1) decreases   (2) increases   (3) remains the same

7   A ray of light traveling through water strikes a water-air surface with an angle of incidence equal to the critical angle. What will be the angle of refraction?
    (1) 180°   (2) 90°   (3) 45°   (4) 30°

8   The index of refraction of a transparent material is 2.0. Compared to the speed of light in air, the speed of light in this material is
    (1) less   (2) greater   (3) the same

9   As a wave travels through different dispersive media, there will be a change in the wave's
    (1) frequency   (2) velocity   (3) period   (4) phase

Base your answers to questions 10 through 14 on the diagram. The diagram shows two light rays originating from source *S* in medium *y*. The dashed line represents a normal to each surface.

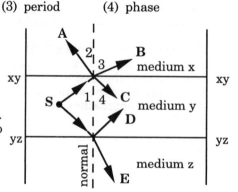

10 Which light ray would not be produced in this situation?
    (1) *A*   (2) *B*   (3) *C*   (4) *D*

11 A reflected light ray is ray
    (1) *A*   (2) *B*   (3) *C*   (4) *E*

12 Which two angles must be equal?
    (1) 1 and 2   (3) 3 and 4
    (2) 2 and 3   (4) 1 and 4

13 Light originating from source S could produce total internal reflection at
   (1) surface yz, *only*
   (2) surface xy, *only*
   (3) *neither* surface xy nor yz
14 Compared to the speed of light in medium x, the speed of light in medium z
   (1) is less   (2) is greater   (3) is the same

Base your answers to questions 15 through 20 on the diagram which represents a ray of monochromatic green light incident upon the surface of a glass prism.

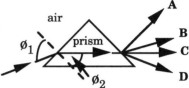

15 The index of refraction of the glass for green light equals
   (1) $\emptyset_1/\emptyset_2$ (3) $\sin \emptyset_2$
   (2) $\emptyset_2/\emptyset_1$ (4) $\sin \emptyset_1/\sin \emptyset_2$
16 After the ray leaves the prism, it will most likely pass through point
   (1) A (2) B (3) C (4) D
17 If the monochromatic green ray is replaced by a monochromatic red ray, $\emptyset_2$ will
   (1) decrease   (2) increase   (3) remain the same
18 Compared to the speed of the monochromatic green light in the prism, the speed of the monochromatic red light is
   (1) less   (2) greater   (3) the same
19 Compared to the frequency of the green light in the prism, the frequency of the red light in the prism is
   (1) less   (2) greater   (3) the same
20 Periodic waves with a wavelength of 0.05 meter move with a speed of 0.30 meter per second. When the waves enter a dispersive medium, they travel at 0.15 meter per second. What is the wavelength of the waves in the dispersive medium?
   (1) 20. m   (2) 1.8 m   (3) 0.05 m   (4) 0.025 m

V. Wave Nature Of Light

Newton's corpuscular theory of light suggested that light consisted of a stream of particles. These particles traveled in a straight line. In 1678, **Christian Huygens** proposed a model of light as a wave. He thought that each point on a circular wave front consisted of "wavelets" which were in themselves a source of circular waves. Huygens believed that the direction of the propagation of the wave could be represented by a light ray.

Reflection, refraction, and **ray optics** could easily be explained by these light rays. These two views of light persisted together until the early nineteenth century when **Thomas Young** demonstrated interference of light and explained it in terms of wave interference and diffraction. This crucial experiment led to the general acceptance of the *wave theory of light.*

Polarization of light is evidence that light is a transverse wave rather than a longitudinal one, since only transverse waves can be polarized (restricted to vibrating in one plane).

```
Practical Applications
  •  polarized sunglasses
  •  3-D films
  •  stress analysis
```

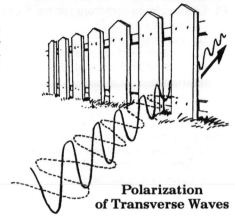

**Polarization
of Transverse Waves**

5.1 Diffraction

Diffraction is the spreading of waves in all directions behind an obstacle. The following figure illustrates the diffraction of a wave by a narrow slit.

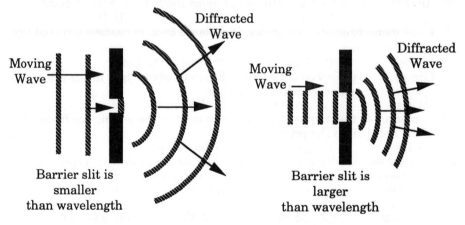

Moving Wave — Diffracted Wave — Barrier slit is smaller than wavelength

Moving Wave — Diffracted Wave — Barrier slit is larger than wavelength

The amount of diffraction depends on the ratio between the wavelength and the slit size. If the slit opening is the same size as the wavelengths or smaller, the slit acts like a point source. If the slit opening is large compared to the wavelength, the wave will pass through with only the ends rounded. This phenomena is important in the selection of hi-fi speaker systems.

If the wave hits a barrier instead of a slit, it bends around the edges, producing a shadow which is usually too small to notice.

5.2 Double Slit Apparatus

In the **Young's double slit apparatus**, the light passing through each slit was **coherent**. Sources that produce waves with a constant phase relation are said to be coherent. **Lasers** produce a beam of light with a very narrow band of frequencies in which all the waves remain in phase for a longer time than ordinary sources.

If the length of the paths from the slit to the screen differ by a number of whole wavelengths, then the crests of the waves coincide and a bright region is produced. If they differ by $\frac{1}{2}(\lambda)$, crests coincide with troughs and a dark region is produced.

The equation describing the pattern for first order is: $\lambda = \dfrac{dx}{L}$

Where:
- λ is wavelength of incident light
- d is spacing between the slits
- x is the distance between the central maximum (0 order) and the first order maximum
- L is the distance between the slits and the screen

Notice how the pattern varies with wavelength, distance from screen, and spacing between slits.

The Young's double slit pattern illustrates interference and diffraction of waves as they pass through the slits.

Constructive interference appears when at an antinode a crest meets a crest. The path differences to the two sources differ by an even number of half wavelengths. A phase difference of 0° or 360° is a wavelength. A bright line is formed at those points. **Destructive interference** occurs when the crest of one wave meets the trough of another wave. The path differences to the two sources differ by an odd number of half wavelengths. A phase difference of 180° is a half wavelength.

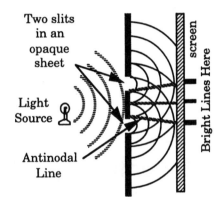

The pattern width is directly proportional to the wavelength and the distance from the screen to the slits. The pattern width **x** is inversely proportional to the separation between the slits. A central bright region is always produced because the path length from each slit is equal. As more slits are added with the same **(d)**, more light produces a brighter pattern. This device is called a **diffraction grating**.

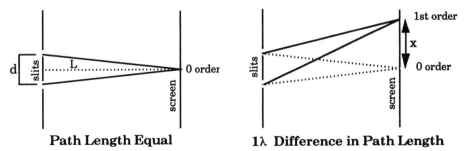

Path Length Equal   **1λ Difference in Path Length**

Illustrations of Diffraction Grating

5.3 Resolution

When light passes through an opening of limited size, it is *diffracted*. If two sources are close, their diffraction patterns may overlap. The resolution of an optical instrument is a measure of its ability to separate images of objects, that are close together. The resolution varies directly as the diameter of the opening. The larger the lens, the greater the resolution.

Example

Calculate the distance between two slits if the wavelength of the incident light is 5.0×10^{-7} m and the distance between the central maximum and the first bright line is 2.0×10^{-2} m. The distance from the slits to the screen is 4.0m.

Given: $\lambda = 5.0 \times 10^{-7}$ m, $x = 2.0 \times 10^{-2}$ m, $L = 4.0$ m

Find: d

Solution: $\lambda = \dfrac{dx}{L}$

$$d = \dfrac{\lambda L}{x}$$

$$= \dfrac{(5 \times 10^{-7} \text{ m}) (4.0 \text{ m})}{2.0 \times 10^{-2} \text{ m}}$$

$$= 1 \times 10^{-4} \text{ m}$$

5.4 Electromagnetic Radiation

Electromagnetic radiations are transverse wave disturbances generated by the acceleration of charged particles that are propagated with the speed of light. The connection between electricity and light was predicted by **James C. Maxwell** in 1870.

The **electromagnetic spectra** includes *radio waves, infrared, visible light, ultraviolet, x-rays, and gamma rays.* The divisions are not well defined but rather represent ranges of frequency which overlap. The names were derived from the sources as well as from the frequency ranges.

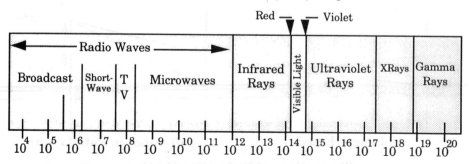

Frequency Ranges in Hertz for Electromagnetic Spectrum

Light is the best known of the electromagnetic waves. The waves originate in the motion of electrons within atoms. Light is but a small portion of the electromagnetic spectrum.

All of the waves in the electromagnetic spectrum travel in a vacuum at the same speed, $c = (3.0 \times 10^8$ m/s). The frequency and wavelength are related by the formula:

$$c = f \lambda$$

Where: λ is the wavelength
           f is the frequency

As the frequency increases, the wave nature becomes less apparent.

Types of Spectra
1. **Continuous spectra** are produced by incandescent solids, liquids and incandescent gases under extremely high pressure.

2. **Line spectra** are produced by luminous gases and vapors at low pressures. Line spectra originate in the atoms of the chemical elements.

3. **Band spectra** consisting of closely spaced spectral lines are sometimes produced. They have their origin in molecules or molecular ions, and are also produced by gases under moderate pressure.

4. **Absorption spectra** is a continuous spectrum interrupted by dark bands or lines characteristic of the cool, low temperature medium through which the radiation passes. The colors removed are the same as those which the medium would emit as a line spectrum.

5.5 Doppler Effect

Electromagnetic radiations *exhibit the Doppler Effect.* Since the speed is constant in space, the changes in the observed frequency and wavelength are such that $f\lambda = c$.

If the distance between the source and the receiver is decreasing, there is a decrease in wavelength and an increase in observed frequency. This is known as a **blue shift**.

If the distance between the source and the receiver is increasing, there is a decrease in the observed frequency and an increase in the wavelength. This is known as a **red shift**.

The radial velocity of stars may be found by their spectral shift. The speed of an earth satellite may be determined from the Doppler shift in the frequency of the radio waves it transmits. Some types of radar depend on the Doppler effect.

Questions

1   Maximum destructive interference between two waves of the same frequency occurs when their phase difference is
(1) $\lambda/4$ (2) $\lambda/2$ (3) $3\lambda/4$ (4) λ

2   Which diagram best represents the phenomenon of diffraction?

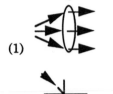

(1)

(3)
white red yellow violet

(2)

(4)

3   The difference in path length for the light from each of two slits to the first maximum is
(1) λ (2) 2λ (3) $\lambda/2$ (4) 0λ

4   The diffraction pattern produced by a double slit will show greatest separation of maxima when the color of the light source is
(1) red   (2) orange   (3) blue   (4) green

5   The double slit experiment involves
(1) refraction   (3) diffraction and polarization
(2) diffraction, only   (4) diffraction and interference

6   Which of the electromagnetic radiations has the shortest wavelength?
(1) ultraviolet   (2) visible   (3) infrared   (4) radio

7   Which phenomenon is the best evidence for the wave nature of light?
(1) reflection   (3) diffusion
(2) photoelectric emission   (4) interference

8   The diagram represents a group of light waves emitted simultaneously from a single source. The light waves would be classified as
(1) coherent, but not monochromatic
(2) monochromatic, but not coherent
(3) both monochromatic and coherent
(4) neither monochromatic nor coherent

9   Whether or not a wave is longitudinal or transverse may be determined by its ability to be
(1) diffracted   (2) reflected   (3) polarized   (4) refracted

10 Which diagram best illustrates diffraction of light incident on a barrier?

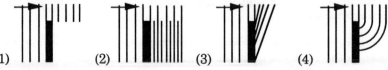

(1)   (2)   (3)   (4)

11 Light is shining through a pair of slits 3.0×10^{-3} meter apart and onto a screen one meter away. What color of light produces a first order bright band that is 2.0×10^{-4} meter from the center of the interference pattern?
(1) green   (3) red
(2) orange   (4) violet

12 Interference and diffraction can be explained by
    (1) the wave theory, only
    (2) the particle theory, only
    (3) neither the wave nor the particle theory
    (4) both the wave and the particle theory

Thinking Physics

1   How can the sound from a rifle trigger an avalanche?
2   What items are added to rooms and auditoriums to minimize noise levels?
3   The frequency of a sound wave determines the pitch. Human ears are sensitive to frequencies in the range of 20 to 20,000 hertz. Is that the same range for all organisms?
4   Why are runners told to look for the flash of the starting pistol instead of listening for the sound?
5   How is it possible to sit at one position in a room and hear almost no sound, yet if you lean to the right or the left, the sound gets louder?
6   Where would you stand on a rope bridge to cause it to vibrate in the fundamental mode?
7   The flash of lightning and the sound of the thunder do not usually occur at the same time. Is it possible to hear the sound before the "clap" of thunder?
8   Years ago "blueing" was added to clothes to make them whiter. Why did it work?
9   Meteorologist say that when we observe the sun "on the horizon" during a pretty sunset, the sun has already set. Can you explain why?
10   Why can you get a sunburn on a cloudy day?
11   Why would people who go fishing want polarized eyeglasses?

Free Response Questions

1   A beam of light (λ = 5.9 x 10^{-7} m) passes through parallel sections of four different material media as shown in the diagram below. The beam's paths through media A and D are parallel. At boundary AB, the beam bends toward the normal; at boundaries BC and CD, the beam bends away from the normal.

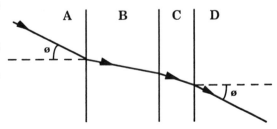

Identify a combination of four substances that could result in the path of the light ray shown in the diagram.

2   Sound generally travels faster through solids than through air. In school, the hallway is very noisy so the teacher closes the door, and less noise enters the classroom. Based on what you know about waves, describe what would cause this phenomenon to exist.

3   Base your answers to parts a and b on the diagram below of a light ray
     ($\lambda = 5.9 \times 10^{-7}$ m) in air incident on a rectangular block of Lucite.

a   Determine the angle of
     refraction, in degrees, of the
     light ray as it enters the Lucite
     from air. [Show all calculations.]

b   On the diagram (right), using a
     protractor and straight edge,
     draw the path of the light ray as
     it travels from air, through the
     Lucite, and back into the air.
     Label all angles of incidence and
     refraction with their
     appropriate numeric values.

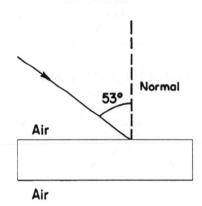

Self–Help Questions

1   A single pulse in a uniform medium transfers
     (1) standing waves   (3) mass
     (2) energy   (4) wavelength
2   The number of water waves passing a given point each second is the
     wave's
     (1) frequency   (3) wavelength
     (2) amplitude   (4) velocity
3   Which characteristic of a wave changes as the wave travels across a
     boundary between two different media?
     (1) frequency   (2) period   (3) phase   (4) speed
4   The speed of a transverse wave in a string is 12 meters per second. If the
     frequency of the source producing this wave is 3.0 hertz, what is its
     wavelength?
     (1) 0.25 m   (2) 2.0 m   (3) 36 m   (4) 4.0 m
5   What is the approximate speed of light in alcohol?
     (1) 1.4×10^8 m/s   (3) 3.0×10^8 m/s
     (2) 2.2×10^8 m/s   (4) 4.4×10^8 m/s
6   A periodic wave with a frequency of 100 hertz would have a period of
     (1) 1 s   (2) 0.01 s   (3) 0.1 s   (4) 10 s
7   What is the wavelength of the wave shown in the diagram?

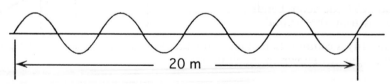

     (1) 2.5 m   (2) 5.0 m   (3) 10. m   (4) 4.0 m
8   What is the frequency of green light?
     (1) 2×10^{-15} Hz   (3) 6×10^{14} Hz
     (2) 2×10^{-14} Hz   (4) 6×10^{15} Hz

9 A ray of light strikes a mirror at an angle of incidence of 30°. What is the angle of reflection?
   (1) 0°   (2) 30°   (3) 60°   (4) 90°

10 The diagram at the right shows a light ray being reflected from a plane mirror. What is the angle of reflection?
   (1) 90°   (3) 50°
   (2) 80°   (4) 40°

mirror

11 Which diagram best illustrates wave refraction?

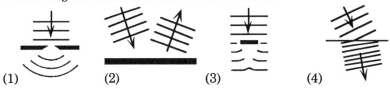

   (1)   (2)   (3)   (4)

12 Which phenomenon *must* occur when two or more waves pass simultaneously through the same region in a medium?
   (1) refraction   (2) interference (3) dispersion   (4) reflection

13 As shown in the diagram, a transverse wave is moving along a rope. In which direction will segment *X* move as the wave passes through it?

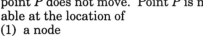

   (1) down, only
   (2) up, only
   (3) down, then up
   (4) up, then down

14 Which two wave representations in the diagram at the right have the same amplitude?

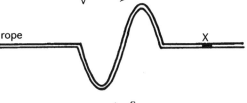

   (1) *A* and *C*   (3) *B* and *C*
   (2) *A* and *B*   (4) *B* and *D*

15 When the stretched string of the apparatus represented at the right is made to vibrate, point *P* does not move. Point *P* is most probable at the location of

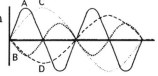

   (1) a node
   (2) an antinode
   (3) maximum amplitude
   (4) maximum pulse

16 An observer detects an apparent change in the frequency of sound waves produced by an airplane passing overhead. This phenomenon illustrates
   (1) the Doppler effect   (3) wave amplitude increase
   (2) wave intensity increase   (4) the refraction of sound waves

17 An Earth satellite in orbit emits a radio signal of constant frequency. Compared to the emitted frequency, the frequency of the signal received by a stationary observer will appear to be
   (1) higher as the satellite approaches
   (2) higher as the satellite moves away
   (3) lower as the satellite approaches
   (4) unaffected by the satellite's motion

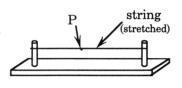
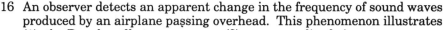

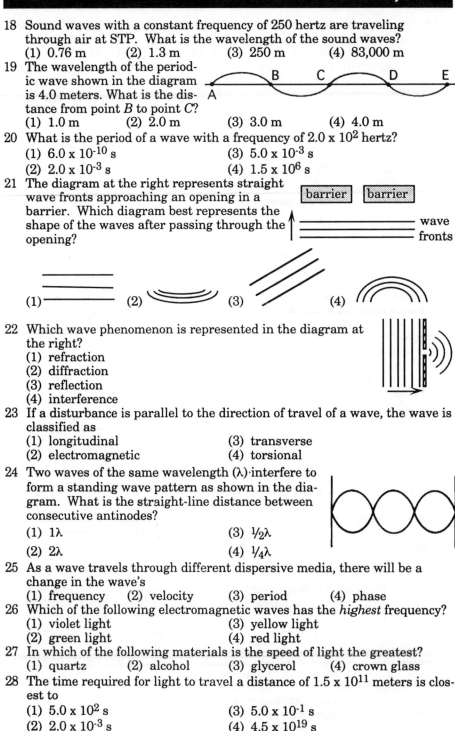

18 Sound waves with a constant frequency of 250 hertz are traveling through air at STP. What is the wavelength of the sound waves?
(1) 0.76 m   (2) 1.3 m   (3) 250 m   (4) 83,000 m

19 The wavelength of the periodic wave shown in the diagram is 4.0 meters. What is the distance from point B to point C?
(1) 1.0 m   (2) 2.0 m   (3) 3.0 m   (4) 4.0 m

20 What is the period of a wave with a frequency of 2.0×10^2 hertz?
(1) 6.0×10^{-10} s   (3) 5.0×10^{-3} s
(2) 2.0×10^{-3} s   (4) 1.5×10^6 s

21 The diagram at the right represents straight wave fronts approaching an opening in a barrier. Which diagram best represents the shape of the waves after passing through the opening?

22 Which wave phenomenon is represented in the diagram at the right?
(1) refraction
(2) diffraction
(3) reflection
(4) interference

23 If a disturbance is parallel to the direction of travel of a wave, the wave is classified as
(1) longitudinal   (3) transverse
(2) electromagnetic   (4) torsional

24 Two waves of the same wavelength (λ) interfere to form a standing wave pattern as shown in the diagram. What is the straight-line distance between consecutive antinodes?
(1) 1λ (3) $\frac{1}{2}\lambda$
(2) 2λ (4) $\frac{1}{4}\lambda$

25 As a wave travels through different dispersive media, there will be a change in the wave's
(1) frequency   (2) velocity   (3) period   (4) phase

26 Which of the following electromagnetic waves has the *highest* frequency?
(1) violet light   (3) yellow light
(2) green light   (4) red light

27 In which of the following materials is the speed of light the greatest?
(1) quartz   (2) alcohol   (3) glycerol   (4) crown glass

28 The time required for light to travel a distance of 1.5×10^{11} meters is closest to
(1) 5.0×10^2 s   (3) 5.0×10^{-1} s
(2) 2.0×10^{-3} s   (4) 4.5×10^{19} s

29 Which optical medium would have the smallest critical angle (θ_c) in the situation shown in the diagram?
(1) Lucite
(2) water
(3) crown glass
(4) diamond

30 In the diagram at the right, a ray of light enters air from a transparent medium. If angle X is 45° and angle Y is 30.°, what is the absolute index of refraction of the medium?
(1) 0.667
(2) 0.707
(3) 1.41
(4) 1.50

31 Which term describes two points on a periodic wave that are moving in the same direction and have the same displacement from their equilibrium positions?
(1) dispersed   (2) refracted   (3) polarized   (4) in phase

32 When a light ray is reflected from a surface, the ratio of the angle of incidence to the angle of reflection is
(1) less than one
(2) greater than one
(3) equal to one

33 In the diagram at the right, which wave has the *smallest* amplitude?
(1) A
(2) B
(3) C
(4) D

34 When a ray of light strikes a mirror perpendicular to its surface, the angle of reflection is
(1) 0°   (2) 45°   (3) 60°   (4) 90°

35 A prism disperses white light, forming a spectrum. The best explanation for this phenomenon is that different frequencies of visible light
(1) move at different speeds in the prism
(2) are reflected inside the prism
(3) are absorbed inside the prism
(4) undergo constructive interference inside the prism

36 When a ray of white light is refracted, the component color that has the greatest change in direction is
(1) orange   (2) red   (3) violet   (4) green

37 Refraction of a wave is caused by a change in the wave's
(1) amplitude   (3) speed
(2) frequency   (4) phase

38 Light travels from medium A where its speed is 2.5×10^8 meters per second into medium B where its speed is 2.0×10^8 meters per second. Compared to the absolute index of refraction of medium A, the absolute index of refraction of medium B is
(1) less   (2) greater   (3) the same

39 In the diagram at the right, a monochromatic
    light ray is passing from medium A into medium
    B. The angle of incidence q is varied by moving
    the light source S. When angle q becomes the
    critical angle, the angle of refraction will be
    (1) 0°
    (2) q
    (3) greater than q, but less than 90°
    (4) 90°

40 Total internal reflection can occur as light waves pass from
    (1) water to air   (3) alcohol to glycerol
    (2) Lucite to crown glass   (4) air to crown glass

41 Which phenomenon can be observed for transverse waves only?
    (1) reflection   (3) polarization
    (2) diffraction   (4) refraction

42 When the beams from two coherent light sources are projected on a
    screen, alternating bright and dark areas are produced. This
    phenomenon illustrates light
    (1) polarization   (3) dispersion
    (2) interference   (4) refraction

43 Monochromatic light is passed through two slits 2×10^{-4} meters apart.
    The pattern produced illuminates a screen 4 meters from the slits. If the
    first-order maximum is 1×10^{-2} meter from the central maximum, then
    the wavelength of the light is
    (1) 2×10^{-7} m   (2) 5×10^{-7} m   (3) 2×10^{-4} m   (4) 5×10^{-2} m

44 Coherent light of wavelength 6.4×10^{-7} meter passes through two narrow
    slits, producing an interference pattern on a screen with the first-order
    bright bands 2.0×10^{-2} meter from the central maximum. The screen is
    4.0 meters from the slits. What is the distance between the slits?
    (1) 2.6×10^{-6} m   (3) 1.3×10^{-4} m
    (2) 3.2×10^{-9} m   (4) 7.8×10^{3} m

Base your answers to questions 45
through 47 on the diagram which
represents a side view of part of
an interference pattern produced
on a screen by light passing
through a double slit. The distance (d) between the centers of
the slits is 1.0×10^{-4} meter.

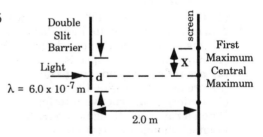

45 What is the distance x from the central maximum to the first maximum
    in the interference pattern?
    (1) 1.0×10^{-4} m   (2) 6.0×10^{-7} m   (3) 1.2×10^{-3} m   (4) 1.2×10^{-2} m

46 The difference between the distances from each of the slits to the first
    maximum is
    (1) 1 wavelength   (3) ½ wavelength
    (2) ¾ wavelength   (4) ¼ wavelength

47 If the distance between the double slits were increased, distance x would
    (1) decrease   (2) increase   (3) remain the same

Optional Unit 4

Geometric Optics

Important Terms To Be Understood

real image
light ray
virtual image
focus
law of reflection

ray diagram
diverge
principal axis
nearsightedness
converge
chromatic observation

convex
spherical aberration
concave
farsightedness
center of curvature

I. Real And Virtual Images

An **image** is formed when light rays originating from the same point intersect on a surface or appear to intersect for an observer.

Mirrors and lenses can form two types of images, **real** and **virtual**. A **real image** is formed when the rays of light from a common point pass through an optical system that causes them to converge and intersect on a surface. The light rays actually cross; the image actually exists. A real image can be projected onto a screen. Real images are formed by cameras and projectors.

Examples of lenses that produce real or virtual images.

A **virtual image** is formed when the light rays from a common point pass through an optical system that causes them to diverge and appear to originate from a point. Virtual images *cannot* be projected on a screen. This image does not exist because the rays do not actually intersect. Virtual images are formed by a plane mirror, a make-up mirror, a magnifying glass, and a concave lens. Real images are always inverted, and virtual images are always erect.

II. Reflection Images

The images formed by mirrors are caused by reflection of light. For all light rays, the angle of incidence equals the angle of reflection. From this Law of Reflection, we are able to predict the size and location of the image of an object formed by a mirror. This may be done by constructing a ray diagram to show the position of the image. The object is represented by an arrow. The image end points are located and the image is drawn.

2.1 Plane Mirror Images

The ray diagram locates the image formed by a plane mirror. **The image is virtual, erect and the same size as the object. It is exactly as far behind the mirror as the object is in front of it.** It only differs from the object in that it is laterally reversed. The image of things on the right appear to be coming from the left and is called a perverted image.

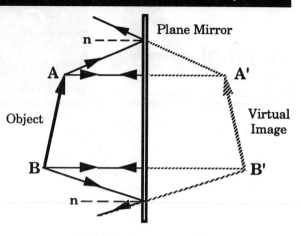

Ray diagram showing how a virtual image is formed by a plane mirror.

2.2 Spherical Mirror Images

Curved mirrors may be either concave or convex. In each case the reflecting surface is a small portion of a hollow sphere. The center of the sphere is called the **center of curvature**. A line joining the center of curvature of a curved mirror and the middle point (vertex) of the mirror is called the **principal axis**.

The **focus** (or focal point) of the mirror is where the rays of light converge (or appear to converge). The focus (**F**) is located on the principal axis midway between the mirror and the center of curvature (**C = 2F**). When the inside spherical surface is used as a reflecting surface, the mirror is said to be **concave**. When the outside spherical surface is used as the reflecting surface, the mirror is said to be **convex**.

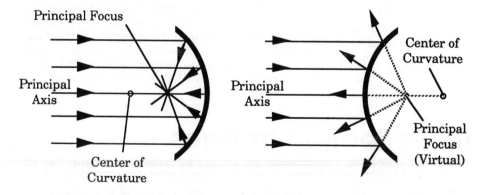

Concave mirrors cause parallel light to converge.

Convex mirrors cause parallel light to diverge.

To locate an image produced by a spherical mirror, it is possible to draw the ray diagram. First, draw a line to indicate the principal axis and place the mirror in the correct orientation to represent a concave or convex mirror. Second, indicate the center of curvature along the principal axis and insert the principal **focus** which is located halfway between the mirror and the center of curvature. Third, place the object at the correct position. Now, draw the first ray from the top of the object, parallel to the principal axis, hitting the mirror and passing through the focus (see example 1).

Example 1

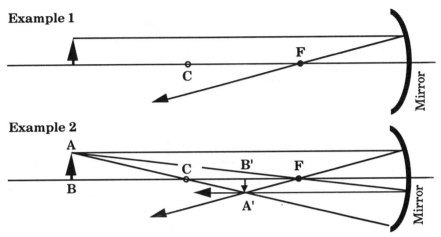

Example 2

Next, draw the second ray from the top of the object, through the center of curvature (or **2F**), so that it hits the mirror and is reflected straight back. Another ray possible goes from the top of the object, through the focus hitting the mirror and returning parallel (see example 2). There are an infinite number of rays which can be drawn from an object. Any two rays will locate the position of the image. Since the base of the object (**A,B**) rests on the principal axis, the base of the image (**A',B'**) will also rest on the same axis.

Ray Diagrams for Converging Mirrors

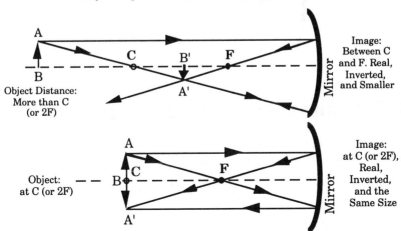

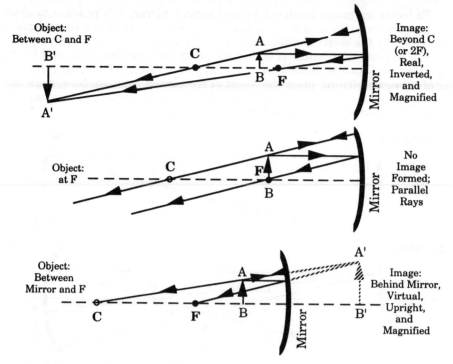

Rays of light parallel to the principal axis meet at the focus. If a light is placed at the focus, the emerging rays will be parallel to the principal focus. This is the principal of reversability.

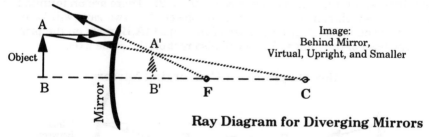

Ray Diagram for Diverging Mirrors

All ray diagrams for diverging mirrors produce the same kind of image.

2.3 Formulas For Curved Mirrors

If the distance from the object to the mirror is indicated by d_o, the distance from the image to the mirror is indicated by d_i and the focal length is indicated by f, the following relationship holds true:

$$\frac{1}{f} = \frac{1}{d_o} + \frac{1}{d_i}$$

(Sometimes referred to by students as "If I do, I die.")

Where: **f** is positive for concave (converging mirrors) and negative for convex (diverging) mirrors

d_o is always positive

d_i is positive for real images and negative for virtual images

Note: Since convex mirrors produce only virtual images, both **f** and d_i will always be negative. The size of the image and object are related as follows:

If s_o is the object size and s_i is the image size then $\dfrac{s_o}{s_i} = \dfrac{d_o}{d_i}$.

Example

A 5.0 cm object is placed 0.80 m in front of a converging mirror whose focal length is 0.20 m.

a) Describe the image.
b) Determine the position of the image.
c) Determine the size of the image.

Solution:

a) Since the object is beyond 2F and the mirror is converging, the image will be real, inverted, and smaller than the object.

b) **Given:**   f = 0.20 m d_o = 0.80 m

   Find: d_i

   Use:
$$\frac{1}{f} = \frac{1}{d_o} + \frac{1}{d_i}$$

$$\frac{1}{0.20 \text{ m}} = \frac{1}{0.80 \text{ m}} + \frac{1}{d_i}$$

$$\frac{1}{d_i} = \frac{1}{0.20 \text{ m}} - \frac{1}{0.80 \text{ m}} = 3.8 \text{ m}$$

$$d_i = 0.27 \text{ m} \qquad \text{The image is located 0.27 m}$$
from the mirror.

c) **Given:** s_o = 5.0 cm

   Find: s_i = _____

   Use:
$$\frac{s_o}{s_i} = \frac{d_o}{d_i}$$

$$\frac{5.0 \text{ cm}}{0.80 \text{ m}} = \frac{s_i}{0.27 \text{ m}}$$

$$s_i = \frac{(5.0 \text{ cm})(0.27 \text{ m})}{0.80 \text{ m}}$$

$$s_i = 1.7 \text{ cm} \qquad \text{The size of the image is 1.7 cm.}$$

Questions

1 A light ray is incident upon a plane mirror. If the angle of incidence is increased, the angle of reflection will
   (1) decrease   (2) increase   (3) remain the same

2 An object is placed in front of a plane mirror as shown in the diagram. The image produced by the plane mirror is best represented by

   (1)   (2)   (3)   (4)

3 The images formed by reflections in a plane mirror are
   (1) virtual and upright   (3) virtual and inverted
   (2) real and upright   (4) real and inverted

Base your answers to questions 4 and 5 on the diagram at the right.

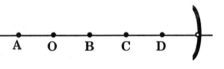

4 The length of the image in the mirror is
   (1) 0.5 m   (3) 8 m
   (2) 2 m   (4) 4 m

5 The distance of the image from the mirror is
   (1) 0.5 m   (3) 8 m
   (2) 2 m   (4) 4 m

6 Which diagram best represents the reflection of an object by a plane mirror?

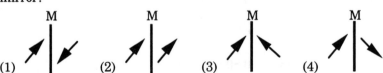

   (1)   (2)   (3)   (4)

Base your answers to questions 7 through 10 on the diagram which represents an object placed at point O, located 2.0 meters from a mirror of focal length 1.0 meters.

7 The distance between the mirror and the image is
   (1) 1.0 m   (3) 0.5 m
   (2) 2.0 m   (4) 4.0 m

8 Compared to the size of the object, the size of the image is
   (1) smaller   (2) larger   (3) same size

9 Compared to the image when the object is at point O, the image when the object is moved to point A will
   (1) be larger   (3) be changed from inverted to erect
   (2) not exist   (4) move closer to mirror

10 The original mirror is replaced with one of less curvature. If the object remains at point O, compared to the original image distance, the new image distance
   (1) is smaller   (2) is larger   (3) remains the same

11 In the diagram at the right of a convex mirror, which point represents the center of curvature?
    (1) *A*
    (2) *B*
    (3) *C*
    (4) *D*

12 The diagram represents a spherical mirror with three parallel light rays approaching. Which light ray will be reflected normal to the surface of the mirror?
    (1) *A, only*   (3) *C, only*
    (2) *B, only*   (4) all of the rays

III. Refraction Images

3.1 Lenses

Lenses form images by the refraction of light. Lenses are divided into two general types.

Converging Lens   A **Converging Lens** (convex) is a lens that is thicker at the center than at the edges. This lens converges parallel rays of light.

A **Diverging Lens** (concave) is a lens which is thinner at the center than at the edges and diverges parallel rays of light.   **Diverging Lens**

The above statements are true as long as the index of refraction of the lens is greater than that of the surrounding medium. In such cases, increasing the index of refraction of the lens reduces the focal length. Images are formed by refraction of light through the lens.

The image formed by a lens may be located by drawing a ray diagram. Two rays must be drawn from the same point on an object. The first ray is drawn parallel to the principal axis through the lens and then through the focal point.

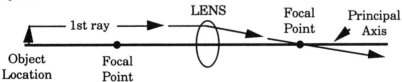

The second ray is drawn through the center of the lens. The intersection of the two rays indicates the position of the image.

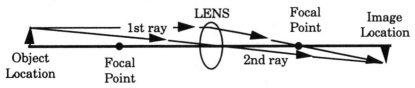

As long as the object is on the principal axis of the lens, only two rays need to be drawn. The image will also be on the principal axis. Remember that light travels in all directions. If the lens is cut so that a portion of the lens is removed, the image produced with the cut lens will be less bright than an image produced with a complete lens.

Ray Diagrams for Converging Lenses

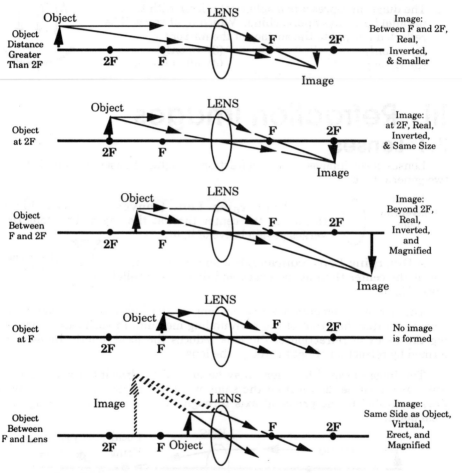

Rays of light parallel to the principal axis meet at the focus. If a light is placed at the focus, the emerging rays will be parallel to the principal axis. This is sometimes called reversability.

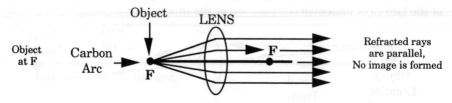

Ray Diagram for a Diverging Lens

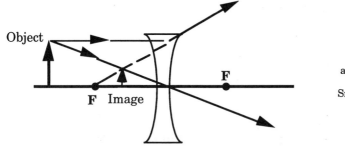

All Images
areBetween F and
Lens, Upright,
Smaller,and Virtual

Each lens has a specific focal length which is determined by the index of refraction of the lens and the radius of curvature of the two surfaces. The focal length, **f**, is related to the object distance, d_o, and image distance, d_i, by the equation:

$$\frac{1}{f} = \frac{1}{d_o} + \frac{1}{d_i}$$

Where: **f**   is positive for converging lens and negative for a diverging lens

   d_o is always positive and measured from lens

   d_i is positive for real images and negative for virtual images

The size of the image is related to the size of the object by the equation:

$$\frac{s_o}{s_i} = \frac{d_o}{d_i}$$

Where: s_o is size of object

   d_o is object distance

   s_i is size of image

   d_i is image distance

Example

A 3.0 cm tall object is placed 0.50 m in front of a diverging lens, whose focal length is 0.10 m.

(a) describe the image
(b) determine the position of the image
(c) determine the size of the image

Solution:
(a) All diverging lenses produce virtual, smaller, erect images.

(Solution (b) is continued on the next page)

(b) **Given:** f = -0.1m (negative because the lens is diverging)
                    d_o = $+0.5$m

Find: d_i

Solution: $\dfrac{1}{f} = \dfrac{1}{d_o} + \dfrac{1}{d_i}$

$$\frac{1}{d_i} = \frac{1}{f} - \frac{1}{d_o} = \frac{1}{-0.10\text{ m}} - \frac{1}{0.50\text{ m}}$$

$$\frac{1}{d_i} = -10\text{ m} - 2.0\text{ m} = -12\text{ m}$$

$$d_i = -0.083\text{ m}$$

The negative sign (-) indicates that the image is located in front of the lens and is virtual.

(c) **Given:** d_i = 0.083 m
                    d_o = 0.50 m
                    s_o = 3.0 cm

Find: s_i

Solution: $\dfrac{s_o}{d_o} = \dfrac{s_i}{d_i}$

$$s_i = \frac{(s_o)(d_i)}{d_o}$$

$$s_i = \frac{(3.0\text{ cm})(0.083\text{ m})}{0.50\text{ m}}$$

$$s_i = 0.50\text{ cm}$$

Practical Applications

Objective Location	Applications For Converging Lenses
• Very far away	*Burning glass*
• Beyond 2F	*Camera*
• At 2F	*Copiers (full size images)*
• Between F and 2F	*Slide projectors*
• At F	*Parallel light*
• Between F and lens	*Magnifying glass*

3.2 The Human Eye

In order for a person to see clearly, the image must land in focus on the light sensitive area of the eye, called the **retina**. The eye obtains focus for different object distances by changing the focal length of the lens. This is done by intrinsic eye muscles that make the lens thicker or thinner as necessary. For distant objects, the muscles stretch the lens to flatten it. For near objects, the muscles relax and the lens becomes more convex.

In a condition called **nearsightedness**, the lens is unable to become thin enough, making the focal length large enough to cause the image to land on the retina. The image is formed in front of the retina. A **diverging lens** is used to correct nearsighted vision.

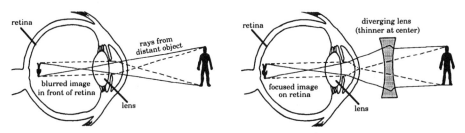

The Cause and Correction of Nearsightedness
Nearsightedness is corrected by wearing diverging lenses.

In **farsightedness**, the opposite of a nearsighted vision condition, the focal length of the eye lens cannot be made short enough to focus on nearby objects. The image is formed beyond the retina. A **converging lens** is used to correct farsighted vision.

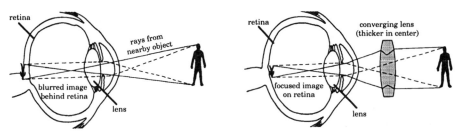

The Cause and Correction of Farsightedness
Farsightedness is corrected by wearing converging lenses.

3.3 Defects In Lenses

Chromatic aberration results when all of the colors of light do not come to a focus at the same point. Violet light focuses closer to the lens than red light. It is caused by dispersion. Chromatic aberration can be corrected by using a combination of two lenses, one converging and one diverging. Dispersion caused by one lens can be eliminated by the second lens, if the lenses are made of two different types of glass. (*Note the illustration of chromatic aberration on the top of the next page.*)

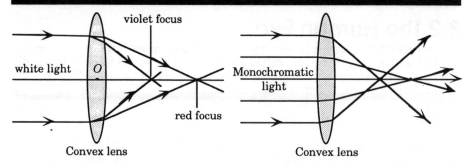

Chromatic Aberration
Violet rays are refracted through a larger
angle by a glass lens than are red rays.

Spherical Aberration
Rays passing through the outer portions of a lens
are not refracted exactly through the focus,
causing a blurring of the image.

   Spherical aberration results when all the rays do not come to focus at
the principal focus, causing a blurring of the image. It results from rays passing through the edges of the lens. They are refracted more. Spherical aberration can be corrected using a diaphragm to cover the edges of the lens. Spherical aberration can be reduced by using the smallest possible area around the
optical center of the lens.

Questions

1   Which diagram represents a type of lens that will produce only virtual
    images?

2   Rays parallel to the principal axis of a converging lens are refracted so
    that they
    (1) pass through the center of the lens
    (2) reverse their direction
    (3) pass through the focus of the lens
    (4) remain parallel after reflection
3   When a convex lens is used as a magnifying glass, the object should be
    placed
    (1) between *F* and the lens   (3) between *F* and *2F*
    (2) at *F*   (4) at *2F*
4   The image formed by a diverging lens will always be
    (1) inverted   (2) enlarged   (3) real   (4) virtual
5   As an object is moved toward the principal focus of a converging lens, the
    size of the image which is formed
    (1) decreases   (2) increases   (3) remains the same
6   An object located at twice the focal length from a converging lens will
    produce an image that is
    (1) smaller than the object   (3) the same size as the object
    (2) larger than the object   (4) no answer is correct

7   In which direction does most of the light in ray R pass?
    (1) A
    (2) B
    (3) C
    (4) D

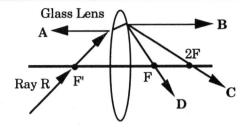

8   A spectrometer deflects the violet light to larger angles than the red. The critical element in this instrument
    (1) must be a prism
    (2) must be a diffraction grating
    (3) must be a lens
    (4) may be a prism or a diffraction grating

9   Which diagram most accurately represents the behavior of light rays passing through a glass lens in air?

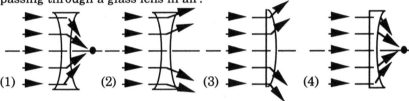

10 A magnifying glass is
    (1) a diverging lens forming a virtual image
    (2) a diverging lens forming a real image
    (3) a converging lens forming a virtual image
    (4) a converging lens forming a real image

11 The image formed by a diverging lens will always be
    (1) upright and larger   (3) inverted and larger
    (2) upright and smaller   (4) inverted and smaller

Base your answers to 12 through 17 on the diagram below which represents an object A located at O near a crown glass lens with a focal length of 1.0 meter. F is the principal focus of the lens.

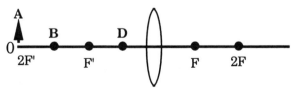

12 The distance between the image and lens is
    (1) 1.0 m   (2) 2.0 m   (3) 3.0 m   (4) 4.0 m

13 Compared to the image when the object is at point O, the image when the object is at point B will be
    (1) changed from erect to inverted
    (2) changed from inverted to erect
    (3) nearer the lens
    (4) farther from the lens

14 Moving the object from point *O* to point *D* will cause its image to
   (1) continuously decrease in size
   (2) continuously move toward the lens
   (3) change from real to virtual
   (4) change from erect to inverted

15 Compared to the original lens, another crown glass lens with less curvature would have a focal length which is
   (1) shorter   (2) longer   (3) the same

16 Compared to the focal length of the crown glass lens in the diagram, the focal length of a flint glass lens with identical curvature would be
   (1) shorter   (2) longer   (3) the same

17 Four identically shaped converging lenses are made of crown glass, flint glass, Lucite, and fused quartz. Which lens would have the shortest focal length?
   (1) flint glass   (2) crown glass   (3) Lucite   (4) fused quartz

18 If a converging lens made of Lucite is made more curved, the index of refraction of the Lucite
   (1) decreases   (2) increases   (3) remains the same

19 As an object at a very great distance from the converging lens of a camera moves up to one focal distance from the lens, the image moves
   (1) from the lens to the principal focus
   (2) from the principal focus to the lens
   (3) from the principal focus to infinity
   (4) from infinity to the principal focus

20 The cause of chromatic aberration in lenses is
   (1) the spherical shape of their surfaces
   (2) the differential refraction by the lens of different colors of light
   (3) the diffraction of light around the edges of the lens
   (4) the differential absorption of different colors of light by the lens

21 The distance of an object from a converging lens is varied until both object and image are 100 cm from the lens. The focal length of the lens is
   (1) 25 cm   (2) 50 cm   (3) 75 cm   (4) 100 cm

Thinking Physics

1 How long should a mirror be to view the entire body of a six foot person?

2 Satellite dishes are spherical "mirrors." How do they work?

3 Some automobile side-view mirrors have a warning that "objects appear to be farther away than the actually are." What kind of mirrors are these? What are the possible problems that can occur if these mirrors are not used with the warning in mind?

4 When a lens breaks in half, what happens to the image produced by the lens?

5 Slides are placed in a projector inverted. Overhead transparencies are placed on the projector right side up. What is the difference in the equipment that allows for the difference in object placement?

Self–Help Questions

1   Which diagram below best represents the image of the object formed by the plane mirror in the diagram at the right?

object

    (1)   (2)   (3)   (4)

2   A person stands in front of a vertical plane mirror 2.0 meters high as shown in the diagram at the right. A ray of light reflects off the mirror, allowing him to see his foot. Approximately how far up the mirror from the floor does this ray strike the mirror?
  (1) 1.0 m
  (2) 2.0 m
  (3) 0.25 m
  (4) 0 m

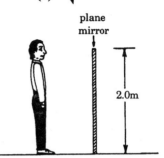

plane mirror

2.0m

3   The diagram at the right shows two rays of light striking a plane mirror. Which diagram below best represents the reflected rays?

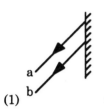   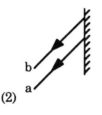   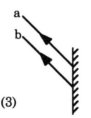   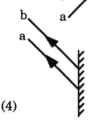

(1)   (2)   (3)   (4)

4   In the diagram at the right, a light ray leaves a light source and reflects from a plane mirror. At which point does the image of the source appear to be located?
  (1) *A*
  (2) *B*
  (3) *C*
  (4) *D*

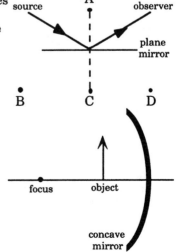

5   The diagram at the right represents an object in front of a concave mirror. The image of the object formed by the mirror is
  (1) real and larger than the object
  (2) real and smaller than the object
  (3) virtual and larger than the object
  (4) virtual and smaller than the object

focus   object

concave mirror

6   Which diagram is a correct representation of a light ray reflected from a spherical surface? (Point C is the center of curvature and point F is the focal point.)

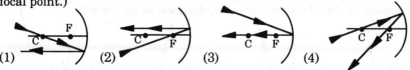

(1)   (2)   (3)   (4)

7   The diagram at the right represents a spherical mirror with its center of curvature at *C* and focal point at *F*. At which position must a point source of light be placed to produce a parallel beam of reflected light?
    (1) *A*   (3) *C*
    (2) *B*   (4) *F*

Base your answers to questions 8 through 11 on the diagram at the right which shows four rays of light from object *AB* incident upon a spherical mirror whose focal length is 0.04meter. Point *F* is the principal focus of the mirror, point *C* is the center of curvature, and point *O* is located on the principal axis.

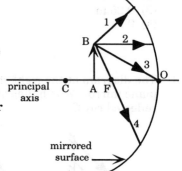

8   Which ray of light will pass through *F* after it is reflected from the mirror?
    (1) 1
    (2) 2
    (3) 3
    (4) 4

9   If object *AB* is located 0.05 meter from point *O*, its image will be located
    (1) farther from the mirror than *C* (3) between *F* and the mirror
    (2) between *C* and *F*   (4) behind the mirror

10 As object *AB* is moved from its present position toward the left, the size of the image produced
    (1) decreases   (2) increases   (3) remains the same

11 If the mirror's radius of curvature could be increased, the focal length of the mirror would
    (1) decrease   (2) increase   (3) remain the same

12 A convex (converging) lens can form images that are
    (1) real, only   (3) either real or virtual
    (2) virtual, only   (4) neither real nor virtual

13 The diagram at the right shows a thin convex (converging) lens with *F* as the principal focus. After passing through the lens, the light rays from the arrowhead of the object will
    (1) converge at *F*
    (2) converge at 2*F*
    (3) emerge as a parallel beam
    (4) diverge

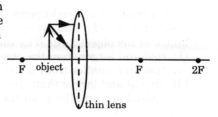

14 An object is placed 0.40 meter in front of a convex (converging) lens whose focal length is 0.30 meter. What is the image distance?

   (1) 0.17 m   (2) 0.83 m   (3) 1.2 m   (4) 5.8 m

15 In the diagram at the right, ray *XO* is incident upon the concave (diverging) lens. Along which path will the ray continue?

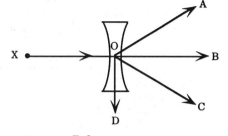

   (1) *OA*
   (2) *OB*
   (3) *OC*
   (4) *OD*

Base your answers to questions 16 through 19 on the diagram at the right which represents a converging lens with a focal length of 0.20 meter. *DE* represents a light ray parallel to the principal axis. *OD* is an object perpendicular to the principal axis.

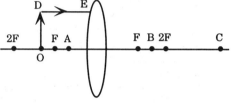

16 When ray *DE* emerges from the lens, it will most likely pass through point

   (1) *F*   (2) *B*   (3) *C*   (4) *2F*

17 This lens can *not* be used to form images which are

   (1) real and enlarged   (3) virtual and enlarged
   (2) real and reduced   (4) virtual and reduced

18 If the object is 0.25 meter from the lens, at what distance from the lens would the image be located?

   (1) 1.0 m   (2) 0.80 m   (3) 0.40 m   (4) 0.25 m

19 As the object is moved toward *F*, the size of the image will

   (1) decrease   (2) increase   (3) remain the same

Base your answers to questions 20 through 22 on the diagram at the right which shows a crown glass lens of focal length *f* in air. Monochromatic red light from an object placed on the left side of the lens passes through the lens and forms a real, inverted image on the right side of the lens. The image size is 0.04 meter, and it is smaller than the object size. The image forms at a distance of 0.1 meter from the center of the lens.

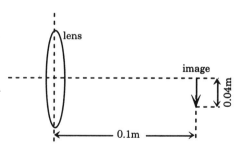

20 If the size of the object is 0.08 meter, then the distance from the object to the center of the lens is

   (1) 0.1 m   (2) 0.2 m   (3) 0.3 m   (4) 0.5 m

21 The distance from the object to the center of the lens must be

   (1) greater than $2f$ (3) between f and $2f$
   (2) equal to $2f$ (4) less than f

22 A flint glass lens of identical curvature is substituted for the crown glass lens. Compared to the focal length of the crown glass lens, the focal length of the flint glass lens is
(1) shorter   (2) longer   (3) the same

23 An object is located 0.12 meter in front of a concave (converging) mirror of 0.16-meter radius. What is the distance between the image and the mirror?
(1) 0.07 m   (2) 0.20 m   (3) 0.24 m   (4) 0.48 m

24 An object is placed in front of a convex (diverging) mirror. The image of that object will be
(1) nonexistent   (3) virtual and smaller
(2) real and smaller   (4) virtual and larger

25 An object 0.16 meter tall is placed 0.20 meter in front of a concave (diverging) lens. What is the size of the image that is formed 0.10 meter from the lens?
(1) 0.040 m   (2) 0.080 m   (3) 0.16 m   (4) 0.32 m

26 Which optical device may form an enlarged image?
(1) plane mirror   (3) converging lens
(2) glass plate   (4) diverging lens

27 A student places her eyeglasses directly on a printed page. As she raises them, the lenses cause the image of the print to remain erect while gradually decreasing in size. She should conclude from this that the lenses of the eyeglasses are
(1) polarized   (2) plane   (3) converging   (4) diverging

5 Modern Physics

Unit

Important Terms To Be Understood

quanta	scattering angle	ionization potential
Planck's Constant	Rutherford's Model	spectra
photoelectric effect	alpha particle	emission spectra
work function	scintillation	absorption spectra
threshold frequency	Bohr Model	cloud model
Compton effect	ground state	orbital
matter waves	excited state	electron cloud

I. Dual Nature Of Light

Light exhibits the characteristics of waves and particles. This duality is true for all electromagnetic radiation. It means that some phenomena are more easily explained by the use of the wave model, while other phenomena are better explained by considering light a particle. **Interference, polarization** and **diffraction** is explained only on the basis of **wave** theory. The **photoelectric effect** is explained only on the basis of the **particle** theory.

1.1 Quantum Theory

The Quantum Theory, announced by Max Planck at the close of the nineteenth century, was developed to explain certain phenomena which could not be explained by classical physics. Atomic oscillators emit or absorb electromagnetic radiations only in discrete amounts, called **quanta**. *The energy of each quantum is proportional to the frequency of radiation.* The constant of proportionality (**h**) is called **Planck's constant**.

$$E_{photon} = hf \quad \text{or} \quad E_{photon} = \frac{hc}{\lambda}$$

Where:
E = energy
f = frequency
h = Planck's constant
c = speed of light
λ = wavelength

1.2 Photoelectric Effect

The **photoelectric effect** is the emission of photoelectrons (electrons that are responsive to light) from an object when a certain electromagnetic radiation strikes it. Wave theory did not explain the observations of the photoelectric effect. According to the wave theory, the maximum kinetic energy should be related to the intensity of the radiation and any radiation should cause the emission of photoelectrons if sustained long enough.

In 1905, **Albert Einstein** proposed that **electromagnetic radiation is always quantized**. Using this, he explained the photoelectric effect as a particle phenomena.

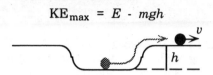

$$KE_{max} = E - mgh$$

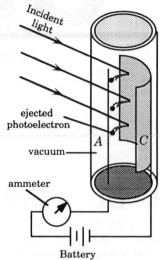

The illustration above is a mechanical analogy of the photoelectric effect. The example is of a ball in a ditch which when given kinetic energy E greater than mgh, will escape from the ditch. After the ball's escape, the maximum kinetic energy is $KE_{max} = E - mhg$.

The illustration at the right is a schematic view of an apparatus for observing the photoelectric effect. Electrons, collected at wire A, are ejected when light strikes plate C, causing a current.

The equation for the photoelectric effect is:

Energy out = Energy in - work function

(or)

$$KE_{max} = hf - w_o$$

Where:

    KE_{max} is the maximum kinetic energy of the emitted electron

    h is Planck's constant: 6.63×10^{-34} joule-sec

    f is the incident frequency

    w_o is the work function (minimum energy for an electron to escape)

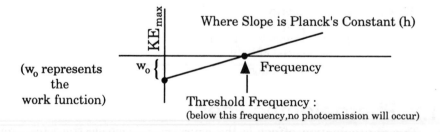

In the region of the spectrum where photoemission can occur, the **rate** of emission depends on the **intensity** of the incident light. Doubling the illumination or intensity, doubles the number of electrons emitted. The **maximum kinetic energy** depends only on the **frequency** of the incident radiation.

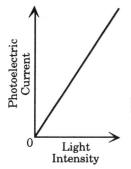

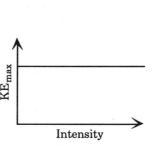

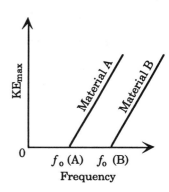

Photoelectric current plotted against the intensity of the incident light for a case in which photoelectrons are emitted.	A graph of KE_{max} of the emitted photoelectrons versus intensity of the light causing the emission. KE_{max} is constant for a given light source.	The maximum kinetic energy of photoelectrons as a function of the frequency of the incident light for two materials. Each material exhibits a threshold frequency f_o.

For each photoemissive material there is a minimum frequency below which no photoelectrons will be emitted. This is called the **threshold frequency** (f_o). The energy associated with the threshold frequency is the **work function** (w_o).

$$w_o = hf_o$$

1.3 Photon–Particle Collisions
(The Compton Effect)

In 1922, **Arthur Compton** used x-rays for photon-particle collisions. The **Compton effect** can be explained in terms of conservation of energy and momentum. The illustration at the right shows the result of a collision between an electron in an atom and a very high energy x-ray photon. Both an electron and a photon of lower energy are emitted. Energy and momentum are conserved.

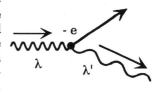

Using the equations $E = mc^2$, $E = hf$, and $c = f\lambda$, the momentum of photons is expressed as:

$$p = \frac{E}{c} = \frac{hf}{c} = \frac{h}{\lambda}$$

Where: **p**   is momentum
           c   is the speed of light
           h   is Planck's constant
           f   is frequency
           λ is wavelength
           E   is energy

The momentum of the photon is inversely proportional to its wavelength. In Compton's experiments the wavelength of the photon increased, indicating a momentum loss. Momentum is a particle property.

Although the photon carries momentum and can exert a force, it does not and cannot have, rest mass. In any frame of reference in space, the **photon moves with the speed of light** and cannot be at rest.

1.4 Matter Waves (De Broglie)

In 1924, **Louis De Broglie** made the proposal that moving particles have wave properties. It was based on his intuitive feeling that nature is symmetrical. The dual nature of light should be matched by a dual nature of matter. His **matter waves** would be represented by:

$$\lambda = \frac{h}{p} = \frac{h}{mv}$$

Where: p   is momentum
          h   is Planck's constant

The wavelength of a particle is inversely proportional to its momentum. *Note: No particle can move at the speed of light; hence, De Broglie simply changed c to v, for his prediction.*

Theoretically, all matter has wave characteristics. Under ordinary circumstances, the wave nature of the object is not significant. One does not notice the wavelength of a moving baseball or golf ball. The wavelength is very small since the momentum is relatively large. If the particle moving was an electron rather than a baseball, the wave movement would become significant, because of a much smaller mass. In 1927, diffraction patterns of electrons were observed by **Davisson** and **Germer**. The observed wavelength was equal to **h/p**.

Questions

Base your answers to questions 1 through 5 on the information below:

Photons with an energy of 3.0 eV, strike a metal surface and eject electrons with a maximum energy of 2.0 eV.

1   The work function of the metal is
    (1) 1.0 eV   (2) 2.0 eV   (3) 3.0 eV   (4) 5.0 eV
2   If the photons had a higher frequency, what would remain constant?
    (1) the energy of the photons   (3) the energy of the electrons
    (2) the speed of the photons   (4) the speed of the electrons
3   If the photon intensity were decreased, there would be
    (1) an increase in the energy of the photons
    (2) a decrease in the energy of the photon
    (3) an increase in the rate of electron emission
    (4) a decrease in the rate of electron emission

4   Compared to the frequency of 3.0 eV photons, the threshold frequency for the metal is
(1) lower   (2) higher   (3) the same

5   If a metal with a greater work function were used and the photon energy remained constant, the maximum energy of the ejected electrons would
(1) decrease   (2) increase   (3) remain the same

6   Which phenomenon can be explained only in terms of the particle model of light?
(1) reflection   (3) photoelectric effect
(2) refraction   (4) diffraction

7   Which formula may be used to compute the energy of a photon?
(1) $E = hf$ (3) $E = \frac{1}{2} mv^2$
(2) $E = mgh$ (4) $E = Fs$

8   All of the following particles are traveling at the same speed. Which has the greatest wavelength?
(1) proton   (3) neutron
(2) atom   (4) electron

Base your answers to questions 9 through 13 on the diagram at the right, which represents monochromatic light incident upon photoemissive surface A. Each photon has 8.0×10^{-19} joule of energy. B represents the particle emitted when a photon strikes surface A.

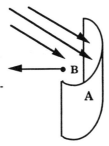

9   What is particle B?
(1) an atom   (3) a neutron
(2) an electron   (4) a proton

10 If the work function of metal A is 3.2×10^{-19} joule, the energy of particle B is
(1) 3.0×10^{-19} joule   (3) 8.0×10^{-19} joule
(2) 4.8×10^{-19} joule   (4) 11×10^{-19} joule

11 The frequency of the incident light is approximately
(1) 1.2×10^{15} hertz   (3) 3.7×10^{-15} hertz
(2) 5.3×10^{-15} hertz   (4) 8.3×10^{-16} hertz

12 If only the frequency of the incident light is increased, the rate of particle emission will
(1) decrease   (2) increase   (3) remain the same

13 If only the intensity of the incident light is increased, the energy of each emitted particle will
(1) decrease   (2) increase   (3) remain the same

14 The momentum of a photon whose wavelength is 5.0×10^{-7} m is
(1) 1.3×10^{-27} N-s   (3) 6.0×10^{-14} N-s
(2) 3.3×10^{-34} N-s   (4) 4.0×10^{-22} N-s

15 What is conserved in a photon - particle collision?
(1) mass, only   (3) momentum, only
(2) energy, only   (4) mass, energy, and momentum

16 The wave nature of atomic particles is demonstrated by showing that they
    (1) travel through a vacuum   (3) produce diffraction patterns
    (2) travel in straight lines   (4) engage in collisions with photons
17 Matter waves associated with relatively large masses such as an automobile are not observable because their
    (1) wavelengths are too small   (3) amplitudes are too small
    (2) wavelengths are too long   (4) energy is too small
18 The ratio of the photon's energy to its frequency is
    (1) its speed   (3) its wavelength
    (2) its amplitude   (4) Planck's constant
19 Which of the following are not quantized?
    (1) radiation   (3) number of people in a room
    (2) electric charge   (4) all are quantized
20 If a particle is to exhibit wave behavior, it must be
    (1) charged   (2) uncharged   (3) moving   (4) stationary

II. Models Of The Atom
2.1 The Rutherford Model

In 1909, **Rutherford** investigated the scattering of alpha particles by thin gold foil. The experimental arrangement is shown below.

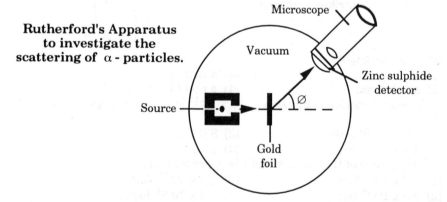

Rutherford's Apparatus to investigate the scattering of α - particles.

A narrow beam of alpha particles from a radon source inside a metal block was incident on a thin gold foil. A glass screen coated with zinc sulphide was used to detect the scattered alpha particles. Whenever a particle hit the screen, it produced a faint flash of light (a **scintillation**). The experiment was carried out in a dark room, and the scintillations were observed through a microscope. The apparatus was evacuated, because the range of alpha particles in air was about 5 cm. Without the evacuation, the particles would not have reached the screen.

The majority of the alpha particles were scattered through small angles, *but a few (about 1 in 8,000) were deviated by more than 90°.* The angle at which the particle is deflected is called the **scattering angle**.

In 1911, **Ernest Rutherford** (based on his experiment) proposed a model of the atom in which the positive charge and most of the mass of the atom are considered to be concentrated in a small dense core called the nucleus. Most of the atom is empty space.

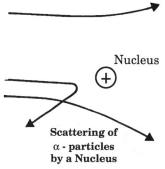

Nucleus

Scattering of
α - particles
by a Nucleus

Large angle scattering occurred whenever an alpha particle (*+2 charge*) was incident, almost head on, to a nucleus. *This unexpected and significant result* is explained by the **Coulomb Force**, between the nucleus and the approaching alpha particle. The distances between the path leading to a head on collision with the nucleus and the original path is called the impact parameter. As the impact parameter approaches zero, the scattering angle approaches 180°.

The probability of a head on collision is extremely small. Alpha particles, which come very close to the nucleus, are deflected into **hyperbolic paths**, due to the Coulomb force.

Very few particles were scattered through large angles. This indicates that the nucleus occupies only a small portion of the available space. (The nuclear radius is in the order of 10^{-15} m; the radius of the atom, as a whole, is about 10^{-10} m.)

2.2 The Bohr Model

The major limitation of Rutherford's Model was that it did not account for (1) the lack of emission of radiation, as electrons move about the nucleus, and (2) the unique spectrum of each element.

The electrons moving around the nucleus are accelerated and should radiate energy of changing frequency, thus the electrons should eventually collide with the nucleus. Since atoms emit only radiation of specific frequencies and do not collapse spontaneously, Rutherford's Model required modification.

Neil Bohr's model of the hydrogen atom consists of a positively charged nucleus and a single electron revolving in a circular orbit. In order to explain this model, he made assumptions which were contrary to classical theory.

First, Bohr's orbiting electron does not lose energy even though it has an acceleration (centripetal) towards the center. According to classical physics, the electron should lose energy by emitting electromagnetic radiation and spiral into the nucleus. Bohr assumed that all forms of energy were quantized.

Second, only a limited number of specified orbits of radius **r** is permitted. Each orbit represents a particular energy state. The permitted orbits are those for which the angular momentum of the electron is an integral multiple (**n**) of Planck's constant divided by **2π**:

$$\mathbf{mvr} \;=\; \frac{\mathbf{nh}}{\mathbf{2\pi}} \quad \text{(mvr, angular momentum is quantized)}$$

Third, when an electron changes from one energy state to another, a quantum of energy equal to the difference between the energies of the two states is emitted or absorbed. The change in energy is given by:

$$hf = E_1 - E_2$$

Where: E_1 and E_2 are the respective energies of the two states and f is the frequency of the photon emitted or absorbed: absorbed when electrons move to a high energy state, and emitted when they drop to a lower energy state.

In 1914, **J. Franck** and **G. Hertz** further strengthened the concepts of stationary states or fixed energy levels, by bombarding gas molecules with electrons. The gas molecules can only accept energy in discrete amounts. The process of raising the energy of atoms is called **excitation**. Excitation energies were different for different gases. Excited atoms subsequently released the energy as photons. Electrons with energies lower than the discrete excitation energies, collided elastically with gas molecules. The Franck-Hertz experiment demonstrated one way of exciting atoms. Other methods included thermal excitation, electrical discharge, and electromagnetic excitation.

In the energy level diagram, **n = 1** is the lowest possible energy level or the **ground state.** The **ionization potential** is the minimum energy necessary to remove an electron from the ground state to infinity; that is, to ionize the atom. For example, the ionization potential of hydrogen is 13.6 eV and mercury is 10.38eV.

The negative (–) signs of energy levels indicate that the electron is controlled by the nucleus with that energy.

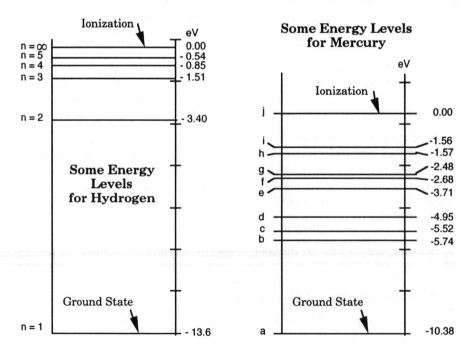

As the electron orbits the nucleus, the probability of finding the electron at a particular position can be described by a standing wave, which can exist at only certain distances from the nucleus. This standing wave will only occur when:

$$mvr = \frac{nh}{2\pi}$$

So: $$2\pi r = \frac{nh}{mv}$$

and $$n\lambda = 2\pi r$$

Therefore: $$n\lambda = \frac{nh}{mv}$$

Thus: $$\lambda = \frac{h}{p}$$

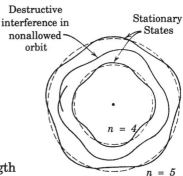

Destructive interference in nonallowed orbit

Stationary States

$n = 4$

$n = 5$

In a **Stationary State**, there is *no destructive interference*. The n's refer to the energy levels of the hydrogen (see Energy Levels for Hydrogen).

Where, **n** is an integer, λ is the wavelength of the standing wave, **h** is Planck's constant, and **p** is the momentum. An electron in a standing wave about the nucleus will not radiate energy. Bohr's model of the atom did not successfully predict other aspects of the hydrogen atom, nor did it explain the electron orbits of large atoms having many electrons.

Atomic Spectra

Each element has a characteristic **spectra**. Bohr's model helped to explain the spectra of elements. Atoms with electrons, excited to an energy level above the ground state, emit energy as photons, as their electrons fall to lower energy levels.

Absorption Spectra

Any atom can absorb those photons whose energies are equal to the energies of the photon it can emit when excited. Absorbing a photon will only occur if the incident energy is exactly the right energy to raise it to a particular energy state. If a hydrogen electron in the ground state, is hit by a photon of 10.5 eV or 10.0 eV, the photon will either pass through or scatter elastically. However, a 10.2 eV photon will be absorbed and the electron will jump to the $n = 2$ level. If any photon equal to or greater than 13.6 eV hits the hydrogen atom in the ground state, the atom will ionize.

Emission Spectra

The energies associated with the lines in an emission spectrum of an element may be determined by using an energy level diagram. The reference charts contain the energy level diagrams for both hydrogen and mercury. The emission spectra is emitted when electrons in higher energy levels fall to lower energy levels.

Example

Using the energy level diagram of hydrogen, calculate the number and energy of photons possible, as an $n = 3$ state electron returns to the ground state.

$$\mathbf{E_{photon}} = \mathbf{E_i} - \mathbf{E}_f$$

Solution: In returning to the ground state from the $n = 3$ level, the electron could jump directly to $n = 1$ or go to $n = 2$, and then $n = 1$. Each jump represents a different energy level photon.

$n = 3$ to $n = 1$	$n = 3$ to $n = 2$
$hf = E_3 - E_1$	$hf = E_3 - E_2$
$= -1.5\text{eV} - (-13.6 \text{ eV}) = 12.1 \text{ eV}$	$= (-1.5 \text{ eV}) - (-3.4 \text{ eV}) = 1.9 \text{ eV}$

$$n = 2 \text{ to } n = 1$$
$$hf = E_2 - E_1$$
$$= -3.4 \text{ eV} - (-13.6 \text{ eV}) = 10.2 \text{ eV}$$

In hydrogen, any jump to the $n = 1$ level represents high energy. The radiation would be in the ultraviolet range. This series is commonly called the **Lyman Series**. Any jump to the $n = 2$ level (from higher energy levels) represents lower energy radiation. Many of these lines are visible light. This is the **Balmer Series**. Other series have recently been discovered.

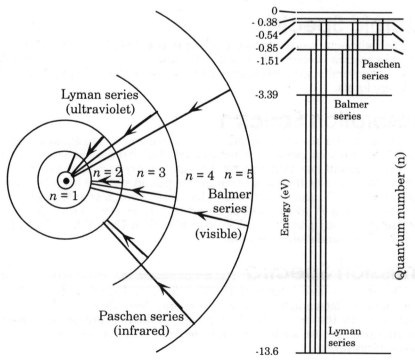

2.3 The Cloud Model

Schrödinger interpreted matter waves as probability waves which can only give the probable position of an electron at any given distance, not an exact position. The electrons are not confined to a particular orbit as in the Bohr model. The highest probability is that an electron will be at a distance from the nucleus which agrees with one of Bohr's radii.

In the **cloud model** of the atom, there is no specific orbit for an electron as it moves about the nucleus. Instead, there is a region of most probable electron location called a state. Each electron occupies a state. A state can hold no more than two electrons. The high probability volume for an electron is called the **electron cloud**, which is most dense in regions where the probability of finding an electron is the greatest.

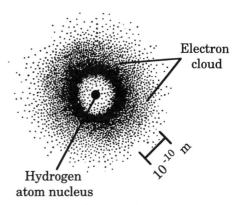

Electron cloud

Hydrogen atom nucleus

10^{-10} m

Each cloud corresponds to an electron orbital. Each orbital may have 0, 1, or 2 electrons. Two is the maximum number to fill an orbital. Each electron in the orbital has an opposite spin. At the $n = 1$ level, the cloud is a sphere with the nucleus at the center. At $n = 2$, the cloud resembles a pair of "dumbbells." This new model (derived from the Rutherford - Bohr ideas and modified by wave - mechanics) allows us to accurately construct models of electron arrangements for all elements.

III. Uncertainty Principle
(Heisenberg)

The **Uncertainty Principle** was first advanced by **Werner Heisenberg** in 1927. As a mathematical statement, it can be written as:

$$\Delta x \Delta p \geq h$$

Where: Δx represents the uncertainty in position

Δp represents the uncertainty in momentum

h represents Planck's constant

The uncertainty principle limits how well we can know simultaneous values of the position and the linear momentum of a particle or wave. As we make the observation, a photon of light would hit the object. The photon imparts momentum to the electron. You cannot locate the electron without changing its momentum.

Questions

Base your answers to questions 1 through 5 on Rutherford's experiments, in which alpha particles were allowed to pass into a thin gold foil. All alpha particles had the same speed.

1   The paths of the scattered alpha particles were
    (1) hyperbolic   (3) parabolic
    (2) circular   (4) elliptical

2   Some of the alpha particles were deflected. The explanation for this phenomena, is that
    (1) electrons have a small mass
    (2) electrons have a small charge
    (3) the gold leaf was only a few atoms thick
    (4) the nuclear charge and mass are concentrated in a small volume

3   The alpha particles were scattered because of
    (1) gravitational forces   (3) magnetic forces
    (2) Coulomb forces   (4) nuclear forces

4   As the distance between the nuclei of the gold atoms and the paths of the alpha particles increases, the angle of scattering of the alpha particles
    (1) decreases   (2) increases   (3) remains the same

5   If a foil were used whose nuclei had a greater atomic number, the angle of scattering of the alpha particles would
    (1) decrease   (2) increase   (3) remain the same

6   When an electron changes from a higher energy state to a lower energy state within an atom, a quantum of energy is
    (1) fissioned   (2) fused   (3) emitted   (4) absorbed

7   Electromagnetic radiation may be generated by
    (1) neutrons moving with constant velocity
    (2) electrons moving with constant velocity
    (3) accelerating neutrons
    (4) accelerating electrons

8   If an orbiting electron falls to a lower orbit, the total energy of that atom will
    (1) decrease   (2) increase   (3) remain the same

9   Which photon energy could be absorbed by a hydrogen atom that is in the $n=2$ state?
    (1) 1.5 eV   (2) 1.9 eV   (3) 2.1 eV   (4) 2.4 eV

10 A hydrogen atom is excited to the $n = 3$ state. In returning to the ground state, the atom could not emit a photon with an energy of
    (1) 1.9 eV   (2) 10.2 eV   (3) 12.1 eV   (4) 12.75 eV

11 How much energy is needed to raise a hydrogen atom from the $n = 2$ energy level to the $n = 4$ energy level?
    (1) 10.2 eV   (2) 2.55 eV   (3) 1.90 eV   (4) 0.65 eV

12 A photon having an energy of 15.5 electron volts is incident upon a hydrogen atom in the ground state. The photon may be absorbed by the atom and
    (1) ionize the atom   (3) excite the atom to $n = 3$
    (2) excite the atom to $n = 2$ (4) excite the atom to $n = 4$

13 In returning to the ground state from the $n = 2$ state, the maximum number of photons a hydrogen atom can emit is
    (1) 1   (2) 2   (3) 3   (4) 4

14 A hydrogen atom emits a photon with an energy of 1.63×10^{-18} joule as it changes to the ground state. The radiation emitted by the atom would be classified as
     (1) infrared   (2) ultraviolet   (3) blue light   (4) red light

15 Which phenomenon provides evidence that the hydrogen atom has discrete energy levels?
     (1) emission spectra   (3) alpha particle scattering
     (2) photoelectric effect   (4) natural radioactive decay

16 Compared to the energy of photons of blue light, the energy of the photons of red light is
     (1) less   (2) greater   (3) the same

17 The momentum of a photon of light whose wavelength is 5.0×10^{-7} m is
     (1) 1.3×10^{-27} N · s   (3) 6.0×10^{-14} N · s
     (2) 3.3×10^{-34} N · s   (4) 4.0×10^{-12} N · s

18 According to the uncertainty principle, it is not possible, when observing an atomic particle to measure exactly at the same time both of the following quantities
     (1) mass, velocity   (3) position, momentum
     (2) momentum, velocity   (4) position, mass

19 The wave nature of atomic particles can be demonstrated by showing that
     (1) they pass through a vacuum
     (2) they travel in straight lines
     (3) they produce diffraction patterns
     (4) they engage in collisions with photons

20 Threshold frequency is to work function as hertz is to
     (1) joules   (2) watts   (3) force   (4) coulombs

21 The diagram shows the characteristic spectral line patterns of four elements. Also shown are spectral lines produced by an unknown mixture. Which pair of elements is present in the unknown?

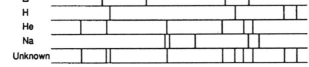

     (1) lithium and sodium   (3) lithium and helium
     (2) sodium and hydrogen   (4) helium and hydrogen

Thinking Physics

1 The "electric eye" makes use of the photoelectric effect. In addition to opening a door at the supermarket, where else have you observed use of the photoelectric effect?

2 Compare the optical and electron microscopes.

3 Spectra analysis is used in industry to determine components of a competitors brand. What would cause the spectrum of one incandescent light bulb to differ from another brand of incandescent light bulb?

4 Why will helium, rather than hydrogen, more readily leak through an inflated rubber balloon?

5 Suntanning results in cell damage to the skin. Why can ultraviolet light, even on a cloudy day, produce this damage but infrared radiation cannot?

Free Response Questions

1   The following data was collected from
    an experiment with two different
    photoemissive metals, A and B.

KE$_{max}$ x 10^{-20} J		Frequency x 10^{14} Hz
Metal A	Metal B	
0	0	1
6.6	0	2
13.2	0	3
19.8	6.6	4
26.4	13.2	5
33.0	19.8	6
	26.4	7
	33.0	8

   a   Sketch the graph with properly
       labeled axis.

   b   Determine the slope of each graph.

   c   Determine the work function of
       each metal.

   d   If monochromic light with a period
       of 2.0 x 10^{-15} seconds is incident on
       both metals, describe in full
       sentences how the energy of the photoelectrons emitted by A
       compared to the energy of those emitted by B.

2   Base your answers to parts a through c on
    the experiment described below and on your
    knowledge of physics.

    A photoemissive metal was illuminated suc-
    cessively by various frequencies of light.
    The maximum kinetic energies of the emit-
    ted photoelectrons were measured and re-
    corded on the table shown at the right.

Data Table

Frequency (x 10^{14} Hz)	Maximum Kinetic Energy (eV)
8.2	1.5
7.4	1.2
6.9	0.93
6.1	0.62
5.5	0.36
5.2	0.24

   a   Using the information and the data
       table, construct a graph plotting
       frequency versus KE$_{max}$.

   b   Based on the graph, what is the threshold frequency of the metal
       used in this experiment?

   c   If the data from a different photoemissive metal were graphed, which
       characteristic of the graph would remain the same?

Self–Help Questions

Base your answers to questions 1 through 3 on the
graph that represents the maximum kinetic
energy of the photoelectrons emitted by a metal
surface, upon exposure to a beam of light of
varying frequency.

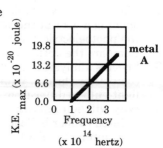

1   The work function for metal A is
    (1) 0.0 joule
    (2) 6.6 x 10^{-20} joule
    (3) 6.6 x 10^{-34} joule
    (4) 6.6 x 10^{-48} joule

2 The slope of the line for metal *A* is
(1) 6.6 joule-sec.   (3) 6.6 x 10^{-20} joule-sec.
(2) 6.6 x 10^{-6} joule-sec.   (4) 6.6 x 10^{-34} joule-sec.

3 The frequency of light incident upon metal *A* is increased from 1 x10^{14} hertz, to 3 x 10^{14} hertz. The kinetic energy of the photoelectrons will
(1) decrease   (2) increase   (3) remain the same

Base your answers to questions 4 through 8 on the graph at the right, which represents the maximum kinetic energy of photoelectrons, as a function of incident electromagnetic frequencies for two different photoemissive metals, *A* and *B*.

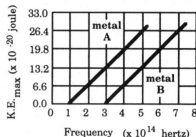

4 The slope of each line is known as
(1) Bohr's constant
(2) the photoelectric constant
(3) Compton's constant
(4) Planck's constant

5 The threshold frequency for metal *A* is
(1) 1.0 x 10^{14} hertz   (3) 3.0 x 10^{14} hertz
(2) 2.0 x 10^{14} hertz   (4) 0.0 hertz

6 The work function for metal *B* is closest to
(1) 0.0 joules   (3) 3.0 x 10^{-19} joules
(2) 2.0 x 10^{-19} joules   (4) 1.5 x 10^{-14} joules

7 Compared to the work function for metal *B*, the work function for metal *A* is
(1) less   (2) greater   (3) the same

8 Monochromatic light, with a period of 2.0 x 10^{-15} second, is incident on both of the metals. Compared to the energy of the photoelectrons emitted by metal *A*, the energy of the photoelectrons emitted by metal *B* is
(1) less   (2) greater   (3) the same

Base your answers to questions 9 through 13 on the following information:
Incident photons with an energy of 6.0 x 10^{-19} joule per photon, cause electrons to be ejected from a surface. The work function of the surface is 3.0 x 10^{-19} joule.

9 What is the frequency of the incident photons?
(1) 9.0 x 10^{29} hertz   (3) 1.1 x 10^{-15} hertz
(2) 9.0 x 10^{14} hertz   (4) 4 .0 x 10^{-37} hertz

10 Compared to the original incident photons, photons with higher energy would have a
(1) longer wavelength   (3) higher frequency
(2) higher intensity   (4) greater speed

11 What is the photoelectric threshold frequency of the surface?
(1) 2.2 x 10^{-15} hertz   (3) 2.2 x 10^{15} hertz
(2) 4.5 x 10^{-14} hertz   (4) 4. 5 x 10^{14} hertz

12 What is the maximum kinetic energy possible for an ejected electron?
(1) 3.0 x 10^{-38} joule   (3) 3.0 joules
(2) 3.0 x 10^{-19} joule   (4) 9.0 joules

13 If the surface were replaced by a surface with a higher threshold frequency, the maximum energy possible for an ejected electron would be
(1) less   (2) greater   (3) the same

14 Rutherford's model of the atom showed that most of the atom's volume is composed of
(1) protons   (2) electrons   (3) neutrons   (4) empty space

15 In the Rutherford scattering experiment, metal foils were bombarded with
(1) alpha particles   (3) protons
(2) beta particles   (4) neutrons

16 An atomic model in which the electrons can exist only in specified orbits was suggested by
(1) Bohr   (2) Planck   (3) Einstein   (4) Rutherford

17 When alpha particles are scattered by thin metal foils, which observation indicates a very high percentage of space in atoms?
(1) Thicker foils scatter more.
(2) The paths are hyperbolic.
(3) Most pass through with little or no deflection.
(4) The scattering angle is related to the atomic number.

18 Which is a characteristic of both the Bohr and Rutherford atomic models?
(1) The nucleus contains protons and neutrons.
(2) Only a limited number of specified orbits are permitted.
(3) The nucleus is concentrated in a small dense core.
(4) Electron energy level changes are in discrete amounts.

19 Which diagram best illustrates the path of an alpha particle, as it passes near the nucleus of an atom?

nucleus   (1)   nucleus   (2)   nucleus   (3)   (4) nucleus

20 If a hydrogen atom absorbs 1.9 eV of energy, it could be excited from energy level
(1) $n = 1$ to $n = 2$ (3) $n = 2$ to $n = 3$
(2) $n = 1$ to $n = 3$ (4) $n = 2$ to $n = 4$

21 An atom, changing from an energy state of -0.54 eV to an energy state of –0.85eV, will emit a photon whose energy is
(1) 0.31 eV   (2) 0.54 eV   (3) 0.85 eV   (4) 1.39 eV

22 Compared to the amount of energy required to excite an atom, the amount of energy released by the atom, when it returns to the ground state is
(1) less   (2) greater   (3) the same

23 Electromagnetic radiations can not be generated by accelerating
(1) electrons   (2) protons   (3) neutrons   (4) alpha particles

24 The direction taken by a photon of light that passes through a narrow slit and falls upon a screen
(1) is a continuation of its original direction
(2) is determined by its probability wave
(3) is most likely to be about 45° to its original direction
(4) is parallel to the plane of the slit

Optional
Unit

5.1 Nuclear Energy

Important Terms To Be Understood

proton
neutron
electron
nucleon
nuclear force
isotope
atomic number
mass number
atomic mass unit
nuclide
mass-energy
mass defect
binding energy
radioactivity
conservation of baryon number

alpha particle
beta particle
positron decay
gamma radiation
transmutation
half life
fission
chain reaction
thermal neutron
critical mass
fuel rods
moderator
control rods
coolant

shielding
radioactive waste
core
breeder reactor
fusion
particle accelerator
ionization
meson
muon
pion
antiparticles
leptons
hadrons
baryons
quarks

I. Structure Of The Nucleus: The Nucleons

Every atom has a central, positively charged nucleus. Nuclear diameters are approximately 10^{-15} m. Over 99.9% of the mass of an atom is in its nucleus. Atomic nuclei are not affected by chemical reactions.

Nuclei contain protons and neutrons which, because they are found in the nucleus are collectively referred to as **nucleons**. The properties of the proton, neutron, and electron are compared in the table below.

	Electron	Proton	Neutron
Mass	9.11×10^{-31} kg	1.67×10^{-27} kg	1.67×10^{-27} kg
Charge	-1.6×10^{-19} C	$+1.6 \times 10^{-19}$ C	none

The Nucleons Compared with the Electrons

The nucleons are held together by one of the four fundamental forces, the *strong interaction*. (The other three are weaker than this and in descending order of effectiveness are the electromagnetic force, the weak interaction, and the gravitational force.)

The strong force, also called the **nuclear force**, is a very strong short-range force, and it more than offsets the considerable electrostatic repulsion of positively charged protons. Nuclear forces are the strongest forces known. They exceed the electrostatic force by one to six orders of magnitude. They not only act between protons and protons, but also between neutrons and neutrons, as well as, between neutrons and protons. The nuclear force accounts for the stability of the nucleus. Nuclear forces operate when the distance between nucleons is less than 10^{-15} m.

II. Isotopes: Definitions

Two atoms which have the same number of protons but a different number of neutrons are said to be **isotopes** of each other. Since each atom contains the same number of electrons as each other isotope in the neutral state, their chemical properties are identical. Isotopes cannot be separated by chemical methods. Some isotopes like gold and cobalt have only one naturally occurring isotope; tin has ten isotopes.

The **atomic number, Z,** of an element is the number of protons in the nucleus of an atom of the element. The atomic number indicates the charge of the nucleus. The **mass number, A,** of an atom is the number of nucleons (protons and neutrons) in its nucleus. The various isotopes of an element whose chemical symbol is represented by **X** are distinguished by using a symbol of the following form:

$$\small{}^{A}_{Z}\normalsize X$$

Where: A is the mass number and **Z** the atomic number of the isotope. The most abundant isotope of lithium (lithium-7) has three protons and four neutrons, so that $Z = 3$ and $A = 3 + 4 = 7$. It is represented by $^{7}_{3}$Li. Lithium-6 has only 3 neutrons and is written as $^{6}_{3}$Li. The Z value is sometimes omitted because it gives the same information as the chemical symbol. For example, *all* lithium atoms have three protons.

Hydrogen has three isotopes which are given different names. The most abundant isotope has one proton and no neutron and is actually called protium ($^{1}_{1}$H) ordinary hydrogen. The other isotopes are deuterium ($^{2}_{1}$H) and tritium ($^{3}_{1}$H).

The atomic mass unit **u** is defined as $1/12$ of the mass of an atom of carbon-12.

$$\textbf{1 u} = \textbf{1.66} \times \textbf{10}^{-27} \textbf{ kg}$$

The term **nuclide** is used to specify an atom with a particular number of protons and a particular number of neutrons. Thus, $^{6}_{3}$Li, $^{7}_{3}$Li, $^{16}_{8}$O, and $^{18}_{8}$O are four different nuclides. Isotopes are nuclides with the same number of protons.

Example

How many neutrons are in $^{238}_{92}U$?

Given: Since there are 238 nucleons (protons and neutrons) and 92 protons, the difference between the two numbers will give the number of neutrons.

Solution: 238 - 92 = 146 neutrons

Questions

1 Elements which have the same number of protons but different number of neutrons are called
(1) molecules   (2) isotopes   (3) ions   (4) electrons

2 How many protons are in the nucleus of an atom of $^{17}_{9}F$?
(1) 8   (2) 9   (3) 17   (4) 26

3 How many electrons will a neutral atom of carbon have, if the carbon nucleus has 6 protons and 8 neutrons?
(1) 6   (2) 2   (3) 8   (4) 14

4 The total number of neutrons in the nucleus of any atom is equal to the
(1) number of the atom
(2) atomic number of the atom
(3) atomic number minus the mass number
(4) mass number minus the atomic number

5 Positively charged particles in the nucleus of an atom are called
(1) protons   (2) photons   (3) neutrons   (4) electrons

6 The nucleus of isotope *A* of an element has a larger mass than isotope *B* of the same element. Compared to the number of protons in the nucleus of isotope *A*, the number of protons in the nucleus of isotope *B* is
(1) less   (2) greater   (3) the same

7 An atom consists of 9 protons, 9 electrons and 10 neutrons. The number of nucleons in this atom is
(1) 0   (2) 9   (3) 19   (4) 18

8 What type of particle has a charge of 1.6 x 10⁻¹⁹ coulomb and a rest mass of 1.67 x 10⁻²⁷ kilogram?
(1) proton   (2) electron   (3) neutron   (4) alpha particle

9 A neutral atom could be composed of
(1) 4 electrons, 5 protons, 6 neutrons
(2) 5 electrons, 5 protons, 6 neutrons
(3) 6 electrons, 3 protons. 6 neutrons
(4) 0 electrons, 5 protons, 5 neutrons

10 Which is an isotope of $^{44}_{21}Se$?
(1) $^{44}_{20}Ca$ (2) $^{46}_{20}Ca$ (3) $^{46}_{21}Se$ (4) $^{44}_{22}Ti$

11 What is the force which holds the nucleus of an atom together?
(1) coulomb force   (3) atomic force
(2) magnetic force   (4) nuclear force

III. Mass–Energy Relationship

According to the special theory of relativity as developed by Einstein, mass and energy are different forms of the same thing and are equivalent. The energy equivalence of mass is proportional to the mass and the speed of light squared, according to the formula:

$$E = mc^2$$

Where: c is the speed of light (3.0×10^8 m/s)

   m is the mass in kilograms

Whenever a nuclear reaction occurs, there is a release of energy, and this is associated with a decrease in mass.

Example

How much energy in joules and in MeV is released when **1u** is completely changed into energy? (1 MeV $= 10^6$ eV)

Given: $1u = 1.66 \times 10^{-27}$ kg

Solution: $E = mc^2$
   $= (1.66 \times 10^{-27}$ kg$)(3.0 \times 10^8$ m/s$)^2$
   $= 1.5 \times 10^{-10}$ J
   $= 1.5 \times 10^{-10}$ J$/(1.6 \times 10^{-19}$ J/eV$)$
   $= 9.30 \times 10^8$ eV
   $= 930$ MeV (This constant appears in the Reference Tables)

3.1 Binding Energy

The mass of a nucleus is always less than the total mass of its constituent nucleons. The difference in mass is called the mass defect of the nucleus.

Mass Defect = Mass of Nucleons - Mass of Nucleus

The mass defect is equivalent to the energy given away to surroundings when the nucleons came together to form the nucleus. Binding energy is the energy equivalent of the mass defect. This amount of energy must be supplied to the nucleus in order to separate the nucleus into its nucleons.

Binding Energy =   Mass Defect
   (*"The mass defect is equivalent to the energy"*)
   $E = mc^2$ (*Einstein's equation*)
   $J = $ (kg) x (m/s)^2 (*units*)

Binding Energy =   930 MeV/u x Mass Defect
   (*conversion to electron volts*)

Example

A nitrogen isotope $^{15}_{7}\text{N}$ has a nucleus whose mass is 15.0001 u.

Calculate: a) the mass defect of the nucleus
                 b) binding energy of the nucleus
                 c) the binding energy per nucleon

$$\text{mass of the proton } (^1_1\text{H}) \quad = \quad 1.0078 \text{ u}$$

$$\text{mass of the neutron} \quad = \quad 1.0086 \text{ u}$$

Given: $^{15}_{7}\text{N}$ has 7 (^1_1H) and 8 neutrons

$$^1_1\text{H} = 1.0078 \text{ u}$$

$$^1_0\text{n} = 1.0087 \text{ u}$$

$$^{15}_{7}\text{N (nucleus)} = 15.0001 \text{ u}$$

Find: a) mass defect
            b) binding energy (total)
            c) binding energy (per nucleon)

Solution: a)

Mass of 7 (^1_1H)	$= 7(1.0078 \text{ u})$	$= \quad 7.0546 \text{ u}$
Mass of 8 neutrons	$= 8(1.0087 \text{ u})$	$= \quad \underline{8.0696 \text{ u}}$ (addition)
		15.1242 u
Mass of Nitrogen nucleus	$=$	$\underline{15.0001 \text{ u}}$ (subtraction)
Mass defect $=$		0.1241 u

Solution: b) Since 1 u = 930 MeV:

$$0.1241 \text{ u } \times \frac{930 \text{ MeV}}{\text{u}} = 115 \text{ MeV}$$

Solution: c) There are 15 nucleons in $^{15}_{7}\text{N}$

$$\frac{115 \text{ MeV}}{15 \text{ nucleon}} = 7.7 \text{ MeV/nucleon}$$

Binding energies are usually compared in terms of binding energy per nucleon. The highest binding energy per nucleon occurs for a mass number of about 60. For elements above carbon, the binding energy per nucleon is about 8 MeV.

The graph represents the relationship between binding energy per nucleon with mass number. The nuclides of intermediate mass number have the largest values of binding energy per nucleon. $^{56}_{26}$Fe has a value of 8.8 MeV and is one of the most stable nuclides. Three nuclides, $^{4}_{2}$He, $^{12}_{6}$C, and $^{16}_{8}$O lie significantly above the main curve. Note that $^{12}_{6}$C and $^{16}_{8}$O are combinations of three and four alpha–particles.

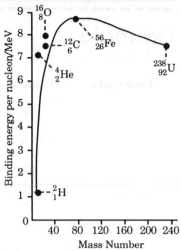

**Graph of the
Variation of Binding Energy
per Nucleon with Mass Number**

IV. Natural Radioactivity

Radioactivity is the disintegration of the nuclei of atoms. Practically all naturally occurring radioactive isotopes have atomic numbers greater than 82. In all nuclear reactions, the sum of the charges (atomic numbers) and the sum of the mass numbers on both sides of the equation are equal. This fact is an example of **conservation of baryon number**. As an element decays it may emit alpha particles, beta particles, and/or gamma radiation.

An **alpha particle** is the nucleus of a helium atom. It has a mass number of **4** and a charge of **+2**. It is symbolized by $^{4}_{2}$**He**.

A **beta particle** comes from the disintegration of a neutron. The neutron splits into a proton which remains in the nucleus and an electron which is ejected. Energy and other sub-atomic particles are released. A beta particle is a negative electron with a mass number of zero, a charge of **-1** and is symbolized by $^{0}_{-1}$e. A **positron**, positive electron, has a mass number of zero, a charge of **+1** and is symbolized by $^{0}_{+1}$**e**.

Generally, an antineutrino is also emitted during negative beta decay and a neutrino during positive beta decay.

Gamma radiation is made up of high energy photons which have no mass or charge. The symbol for gamma radiation is $^{0}_{0}\gamma$. Gamma radiation is emitted when a nucleus in an excited state changes to a more stable state. It is often given as evidence of nuclear energy levels.

Alpha decay causes a change in mass and charge number. Beta decay causes a change in charge number. Gamma radiation causes no change in atomic number or mass. The Uranium Disintegration Series diagrams the steps as Uranium-238 goes to lead. There are 8 alpha decays and 6 beta decays, illustrated in the series found on the Reference Tables.

4.1 The Radioactive Series

Most of the radioactive nuclides which occur naturally have atomic numbers that are greater than that of lead. Each of the nuclides can be arranged in one of the **radioactive series**. The series are known as the **Uranium Series**, the **Thorium Series**, and the **Actinium Series**.

URANIUM DISINTEGRATION SERIES

The first eight members of the Uranium Series are listed in the table at the bottom of the page. The final member in each of the above three series is a stable isotope of lead, with a different isotope for each series.

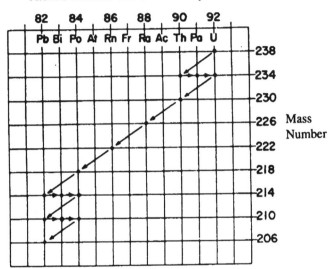

Atomic Number and Chemical Symbol

Of the elements whose atomic numbers are less than that of lead, indium and rhenium are the only elements whose abundant isotopes are radioactive. The half–life is very long: 6×10^{14} years for $^{145}_{49}$In and $>10^{10}$ years for $^{187}_{75}$Re.

First Eight Members of the Uranium Series

Element	Nuclide	Half–life	Radiation	Energy of α or β in MeV
Uranium	$^{238}_{92}$U	4.51×10^9 years	α　γ	4.2
Thorium	$^{234}_{90}$Th	24.1 days	β　γ	0.19
Protactinium	$^{234}_{91}$Pa	6.75 hours	β　γ	2.3
Uranium	$^{234}_{92}$U	2.47×10^5 years	α　γ	4.77
Thorium	$^{230}_{90}$Th	8.0×10^4 years	α　γ	4.68
Radium	$^{226}_{88}$Ra	1620 years	α	4.78
Radon	$^{222}_{86}$Rn	3.82 days	α	5.49
Polonium	$^{218}_{84}$Po	3.05 minutes	α	6.0

Examples

1. In the equation below, determine the unknown particle **X**.

$$^{226}_{88}\text{Ra} \rightarrow {}^{222}_{86}\text{Rn} + {}^{A}_{Z}\text{X}$$

Solution: The mass numbers at the top of each element (particle) in the reaction must be equal on both sides of the arrow. This is conservation of baryon number.

$$226 = 222 + A \qquad\qquad A = 4$$

Therefore, the number of nucleons in element **X** is 4. The atomic numbers at the bottom of the reaction must be equal on both sides of the arrow. This is conservation of charge.

$$88 = 86 + Z \qquad\qquad Z = 2$$

Therefore, the number of protons in element **X** is 2. Since the emitted particle has an atomic number of 2 and the mass number of 4, the emitted particle is an alpha particle.

2. In the equation below, determine the atomic number and the mass number of the daughter produced.

$$^{64}_{29}\text{Cu} \rightarrow {}^{A}_{Z}\text{X} + {}^{0}_{-1}\text{e}$$

Solution: The mass numbers on the top of the equation must be equal on both sides of the arrow.

$$64 = A + 0 \qquad\qquad A = 64$$

The number of nucleons is 64.

The atomic numbers on the bottom of the equation must be equal on both sides of the arrow.

$$29 = Z + (-1) \qquad\qquad Z = 30$$

There are 30 protons in the daughter element.

The unknown element is $^{64}_{30}\text{X}$

A **periodic table** is necessary to determine that the element with an atomic number of 30 is zinc.

4.2 Half–Life

Radioactive material disintegrates at a fixed rate. There is *nothing* that can speed up or slow down this decay rate. The half-life is the term generally used to discuss the decay rate of radioactive material. **Half-life** is the time for one half of the sample to decay. Although it is impossible to predict when any particular

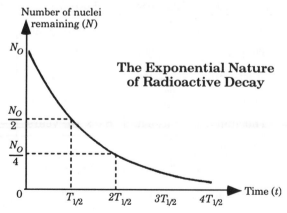

Number of nuclei remaining (N)

The Exponential Nature of Radioactive Decay

nucleus will disintegrate, it is possible to say what proportion of a very large number of nuclei will disintegrate during any given time.

The determination of half–life uses the following important formula:

$$m_f = \frac{1}{2^n} m_i$$

Where: m_f = mass remaining
m_i = initial mass
n = number of half-lives

Example

The half life of thorium 234 is 24 days. How much thorium will remain in a 100 gram sample after 48 days?

Solution: 48 days is two half-lives so n = 2
$$m_f = m_i (1/2)^n$$
$$= 100 \text{ grams} \times (1/2)^2$$
$$= 25 \text{ grams}$$

4.3 Artificial Transmutation

Transmutation is a change from one nuclide to another different nuclide because of a gain or loss of protons and/or neutrons by the nucleus. Radioactivity is an example of natural transmutation. In 1919, Rutherford bombarded nitrogen with alpha particles and produced oxygen.

$$^{14}_{7}N + ^{4}_{2}He \rightarrow ^{17}_{8}O + ^{1}_{1}H$$

This was the first artificial transmutation. Artificial transmutation may be produced by bombardment of nuclei with alpha particles, protons, neutrons, or other particles.

In 1932, the Joliot-Curies bombarded aluminum with alpha particles to produce the first example of induced radioactivity. This was the radioactive isotope of phosphorus that does not occur naturally in our environment.

$$^{27}_{13}Al + ^{4}_{2}He \rightarrow ^{30}_{15}P + ^{1}_{0}n$$

Another type of artificial transmutation may be produced by high energy photons. These photons may cause the ejection of a proton or a neutron from the nucleus, or the photons may cause an unstable nucleus to disintegrate and produce a new nuclide.

4.4 Positron Emission

A positron is produced when a nuclear proton changes into a neutron.

$$_{1}^{1}p \rightarrow _{0}^{1}n + _{+1}^{0}e$$

This may occur if a nucleus has too many protons relative to its number of neutrons. The process is called **positron emission** or **positive beta decay.**

Beta decay in artificial transmutation includes the emission of positive electrons (positrons) as well as negative electrons from the nucleus. A neutrino is also emitted during positive beta decay. The emission of a positive beta particle decreases the atomic number by one but does not change the mass number.

$$_{29}^{64}Cu + _{28}^{64}Ni \rightarrow _{+1}^{0}e$$

The emission of negative beta particle increases the atomic number by one.

$$_{11}^{24}Na + _{12}^{24}Mg \rightarrow _{-1}^{0}e$$

4.5 Electron Capture (K–capture)

Electron capture may be experienced if a nucleus has too many protons relative to its number of neutrons. Since the electron that is captured generally comes from the innermost electron level, or the K shell of the same atom, it is sometimes called K–capture. The electron captured by the proton can be converted into a neutron according to the equation:

$$_{1}^{1}p + _{-1}^{0}e \rightarrow _{0}^{1}n$$

Electron capture decreases the atomic number of the nucleus by 1, but does not change the mass number.

$$_{19}^{40}K + _{-1}^{0}e \rightarrow _{18}^{40}Ar$$

4.6 Neutron

Neutrons were first discovered by bombarding beryllium with alpha particles. Chadwick identified the particle in 1932, according to the equation:

$$_{4}^{9}Be + _{2}^{4}He \rightarrow _{6}^{12}C + _{0}^{1}n$$

Neutrons have no charge so they are not repelled by the positively charged nuclei. When the neutron is very close to the nucleus, the nuclear force actually attracts the neutron.

A neutron is unstable outside the nucleus. The decay of a neutron produces a proton and an electron.

$$_{0}^{1}n \rightarrow _{1}^{1}p + _{-1}^{0}e$$

Questions

Base your answers to questions 1 through 6 on the nuclear equations below.

(A) $^{230}_{90}\text{Th} \rightarrow ^{4}_{2}\text{He} + ^{226}_{88}\text{Ra}$

(B) $^{14}_{7}\text{N} + ^{4}_{2}\text{He} \rightarrow ^{17}_{8}\text{O} + ^{1}_{1}\text{H}$

(C) $^{3}_{1}\text{H} + ^{1}_{1}\text{H} \rightarrow ^{4}_{2}\text{He} + \text{energy}$

(D) $^{235}_{92}\text{U} + ^{1}_{0}\text{n} \rightarrow ^{90}_{38}\text{Sr} + ^{140}_{54}\text{Xe} + \text{X}^{1}_{0}\text{n}$

1 Which reaction best represents alpha decay?
   (1) A   (2) B   (3) C   (4) D
2 Which reaction is part of the Uranium Disintegration Series?
   (1) A   (2) B   (3) C   (4) D
3 In nuclear equation D, coefficient **X** is equal to
   (1) 1   (2) 6   (3) 8   (4) 4
4 In reaction C, the particle produced by the reaction is
   (1) a proton   (3) a beta particle
   (2) a neutron   (4) an alpha particle
5 What is the number of neutrons in an atom of $^{230}_{90}\text{Th}$?

   (1) 90   (2) 140   (3) 230   (4) 320
6 Which nuclear equation is an example of a natural transmutation?
   (1) A   (2) B   (3) C   (4) D

Base your answers to questions 7 through 11 on the information below:

When tellurium is bombarded with protons, the following reaction occurs:

$$^{1}_{1}\text{H} + ^{130}_{52}\text{Te} \rightarrow ^{130}_{53}\text{I} + \text{y}$$

The iodine produced is radioactive, has a half-life of 12.6 hours, and decays according to the reaction:

$$^{130}_{53}\text{I} \rightarrow ^{130}_{54}\text{Xe} + ^{0}_{-1}\text{e}$$

7 The number of neutrons in a nucleus of $^{130}_{52}\text{Te}$ is

   (1) 52   (2) 78   (3) 130   (4) 182
8 Particle y is
   (1) a proton   (2) a neutron   (3) a positron   (4) an electron
9 The first equation represents an example of
   (1) induced transmutation   (3) nuclear fission
   (2) natural radioactivity   (4) nuclear fusion

10 A 2.4 x 10^{-5} kilogram sample of $^{130}_{53}\text{I}$ is left to decay. After 37.8 hours,

   the amount of iodine present will be
   (1) 8.0 x 10^{-6} kg   (3) 3.0 x 10^{-6} kg
   (2) 2.0 x 10^{-5} kg   (4) 4.0 x 10^{-6} kg

11 When it decays, $^{130}_{53}\text{I}$ emits

   (1) a proton   (3) a beta particle
   (2) an alpha particle   (4) a neutron

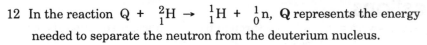

12 In the reaction $Q + {}^{2}_{1}H \rightarrow {}^{1}_{1}H + {}^{1}_{0}n$, **Q** represents the energy needed to separate the neutron from the deuterium nucleus.

Given: Mass of deuterium $({}^{2}_{1}H)$ = 2.0141 amu

                  Mass of hydrogen $({}^{1}_{1}H)$ = 1.0078 amu

                  Mass of neutron $({}^{1}_{0}n)$ = 1.0087 amu

What is the value of **Q**?
(1) 0.0009 amu   (3) 2.0165 amu
(2) 0.0024 amu   (4) 4.0306 amu

Base your answers to questions 13 through 16 on the nuclear equation below:

$$ {}^{30}_{15}P \rightarrow {}^{A}_{Z}Si + {}^{0}_{+1}X $$

13 In the equation, **X** represents
(1) a positron   (3) a proton
(2) an electron   (4) a gamma photon
14 What is the value of **A** in the equation?
(1) 28   (2) 29   (3) 30   (4) 31
15 What is the value of **Z** in the equation?
(1) 14   (2) 15   (3) 16   (4) 17
16 The nucleus of ${}^{30}_{15}P$ has

(1) 30 protons   (2) 30 neutrons   (3) 15 nucleons   (4) 15 neutrons

Base yours answers to questions 17 through 20 on the following nuclear reactions. Q represents heat energy.

         (1) ${}^{14}_{7}N + {}^{4}_{2}He \rightarrow {}^{17}_{8}O + {}^{1}_{1}H$

         (2) ${}^{24}_{11}Na \rightarrow {}^{24}_{12}Mg + {}^{0}_{-1}e$

         (3) ${}^{266}_{88}Ra \rightarrow {}^{222}_{86}Rn + {}^{4}_{2}He$

         (4) ${}^{235}_{92}U + {}^{1}_{0}n \rightarrow {}^{90}_{38}Sr + {}^{143}_{54}Xe + y{}^{1}_{0}n + Q$

         (5) ${}^{1}_{1}H + {}^{2}_{1}H \rightarrow {}^{3}_{2}He + Q$

17 Which reaction represents beta decay?
(1) 1   (2) 2   (3) 3   (4) 5
18 Which reaction represents a transmutation in the Uranium Disintegration Series?
(1) 1   (2) 2   (3) 3   (4) 4
19 In reaction 4, the number represented by **y** is
(1) 1   (2) 2   (3) 3   (4) 6
20 Which reaction best represents alpha decay?
(1) 5   (2) 2   (3) 3   (4) 4

V. Nuclear Stability
5.1 Fission And Fusion

Nuclear **fission** is the disintegration of a heavy nucleus into two lighter nuclei. Energy is released by this process because the average binding energy per nucleon of the fission products is greater than that of the parent.

As an example of a fission reaction, consider the bombardment of U^{235} by slow neutrons. This can result in the capture of a neutron and the formation of U^{236} which is unstable and undergoes fission. Many different pairs of nuclei can be produced by fission of U^{236}. One possible reaction is

$$^{235}_{92}U + ^{1}_{0}n \rightarrow ^{236}_{92}U \rightarrow ^{141}_{56}Ba + ^{92}_{36}Kr + 3\,^{1}_{0}n + Q$$

The energy (Q) released by fission of a single uranium atom is about 200 MeV and about 80% of this goes into the kinetic energy of the two **fission fragments**. These are often radioactive and subsequently decay. Nuclear reactors make use of controlled fission reactions. The atom bomb makes use of an uncontrolled fission reaction.

5.2 Nuclear Reactors

When a U^{235} nucleus captures a **thermal** (**slow moving**) neutron and undergoes fission at least 2 or 3 neutrons are released. (The actual number depends on which set of fission products is formed). The principle of the thermal reactor is to cause these neutrons to produce more fission by being captured by other U^{235} nuclei so a **chain reaction** occurs. A chain reaction is a self-sustaining reaction which, once started, steadily provides the energy and neutrons necessary to continue the reaction.

Thermal neutrons are neutrons with kinetic energies nearly equal to those of the molecules of a substance at ordinary temperature. Thermal neutrons with small kinetic energies are more likely to undergo absorption by a nucleus and are therefore more likely to cause fission reactions.

The **critical mass** is the minimum quantity of fissionable material necessary to sustain a fission reaction.

A Representative Fission Reactor

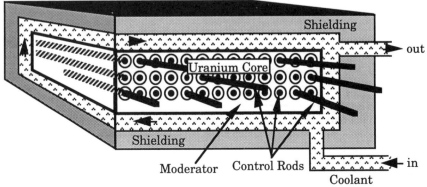

The main parts of a nuclear reactor include:

1. **Fuel rods** – contain pellets of fissionable material. Mined uranium (U^{238}) is 99.3% non-fissionable. Only 0.7% (U^{235}) is obtained from mined ore. The uranium is enriched to about 3%, then formed into pellets that are stacked into fuel rods. Examples: **Uranium-235, Plutonium-239**.

2. **Moderator** – materials that have the ability to slow down neutrons quickly with little tendency to absorb them. The neutrons are slowed down most effectively by a head-on collision with a particle of similar mass. Examples: water, heavy water, beryllium and graphite.

3. **Control rods** – a device used to absorb neutrons and controls the rate of the fission process. The control rods can be lowered into the reactor to control the rate of fission reaction and reduce the amount of heat generated. If the control rods are inserted to their maximum depth, the reactor will "shut down." Examples: boron, cadmium

4. **Coolant** – used to keep temperatures in the reactor at reasonable levels and to carry heat to heat exchangers and turbines. Examples: water, air, heavy water, molten sodium.

5. **Shielding** – internal shield protects the walls of the reactor from radiation damage. The external shield protects the personnel from radiation. Examples: heavy water, concrete or lead.

6. **Radioactive wastes** – fission products which must be stored for long periods of time. Solid and liquid wastes are stored in double walled containers and buried underground in isolated areas. Low level wastes are diluted and released into the environment.

7. **Core** – contains fissionable fuel, control rods and moderators.

8. **Thermal neutrons** – have kinetic energies approximately equal to those of molecules at ordinary temperatures. These slow moving neutrons are more likely to be absorbed than the fast moving neutrons that are the products of fission.

5.3 The Breeder Reactor

Breeder reactors produce more fuel than they consume. U^{233}, produced by neutron capture by thorium-232 and plutonium-239, produced from neutron capture by U^{238} are used as fuels in breeder reactors. U^{238} does not undergo fission.

$$^{238}_{92}U \; + \; ^{1}_{0}n \; \rightarrow \; ^{239}_{92}U \; \rightarrow \; ^{239}_{93}Np \; + \; ^{0}_{-1}e$$

$$^{239}_{93}Np \; \rightarrow \; ^{239}_{94}Pu \; + \; ^{0}_{-1}e$$

Since plutonium-239 is an isotope that can be made to undergo fission and produce energy, it is considered a fuel, not a waste.

5.4 Nuclear Fusion

Nuclear **fusion** is the combining of two lighter nuclei to produce a heavier nucleus. When two light nuclei fuse into a heavier nucleus, they form a more stable nucleus having a greater binding energy per nucleon. The mass of the heavier nucleus formed is less than the sum of the masses of the lighter nuclei, and the difference in mass is converted into energy.

The energy released per nucleon in a fusion reaction is much greater than the energy released per nucleon in a fission reaction. An example is the fusion of two deuterium nuclei to produce helium-3.

$$_1^2\text{H} + _1^2\text{H} \rightarrow _2^3\text{He} + _0^1\text{n} + Q$$

Where: $Q = 4.0 \, \text{MeV}$

Reactions of this type are the source of the Sun's energy. Temperatures in excess of 10^8 K are required to provide the nuclei, which are to fuse, with the kinetic energy needed to overcome their mutual electrostatic repulsion.

Deuterium and tritium may be used as fuels for the fusion process.

5.5 Methods Of Studying Atomic Nuclei

A. Particle Accelerators

A **particle accelerator** is a device used to project charged particles at high speeds into matter. Electric and magnetic fields are used to provide the force to accelerate and control these charged particles at speeds approaching the speed of light. When these charged particles bombard the nucleus of an atom, they alter its stability and new particles are produced. Particle accelerators include the Van de Graff generator, cyclotron, betatron, synchrotron and linear accelerators. Charged particles include electrons, protons, alpha particles, and deuterons.

B. Detection Devices

Almost all of the methods of detecting high energy particles depend on the ionization process. When a charged particle moves through matter, it removes electrons from atoms in its path leaving a trail of positive ions. Various devices which use ionization to detect subatomic particles include the Geiger counter, ionization chamber, bubble chamber, cloud chamber, and photographic film.

The Geiger–Müller tube can be used to detect the presence of X-rays, gamma rays, and beta particles. Tubes with very thin mica windows can also detect alpha particles. The tube is connected to a scalar which records the number of pulses registered. The tube should be operated within a certain voltage range to insure that all ionizations are recorded.

The **Wilson Cloud Chamber** shows up the paths of ionizing "particles" which pass through it. The chamber makes use of the fact that a supersaturated vapor condenses more readily on ions than on neutral atoms.

Effect Of Alpha Particles Passing Through A Cloud Chamber

alpha particle

(path of ejected electron appears as condensed vapor droplets)

The **bubble chamber** makes use of a superheated liquid. The path of the ionizing radiation shows up as a train of bubbles. Ionizing radiations passing through photographic emulsions shows up as a well defined track when the plate is developed. This method provides a permanent record of the "particle" tracks, but the tracks are typically short (1 mm for a beta particle), and microscopes have to be used to study them.

The scintillation counter does not depend on ionization. This device employs the fact that particles striking a phosphorescent screen produce a small flash of light. A photomultiplier tube converts the small flash of light to a burst of current that can operate a counter. The scintillation counter has a high sensitivity to all types of radiation.

5.6 Subatomic Particles

By 1935, six elementary particles had been discovered: protons, electrons, neutrons, positrons, neutrinos and photons. In the decades that followed, hundreds of other elementary particles were discovered.

Some of these elementary particles mediate the forces in nature. In the case of the electromagnetic force between two charged particles, it is the photons that are exchanged between the two particles that give rise to the force.

Type	Relative Strength	Field Particle
Strong Nuclear	1	Mesons
Electromagnetic	10^{-2}	Photons
Weak Nuclear	10^{-13}	W particle
Gravitational	10^{-40}	Graviton (?)

The Japanese physicist Hideki Yukawa predicted the existence of a particle that in some way would mediate the strong nuclear force. The new particle would have a mass intermediate between that of the electron and the proton. It was called the **meson** meaning in the middle.

In 1937, a particle was discovered close to the predicted mass. The new particle called a muon (or mu meson) did not strongly interact with matter and could not mediate the strong nuclear force.

Finally, in 1947 the particle predicted by Yukawa was found. It is called the "π" or **pi meson** or simply **pion**. A number of other mesons have since been discovered. These also mediate the strong nuclear force.

Theorists believed that the weak nuclear and the gravitational force are also mediated by particles. In 1983, Carla Rubbia and Simon Van der Meer identified the **W** and **Z** particles. They were awarded the 1984 Nobel Prize in Physics. Yet, despite extensive searching, the graviton has not been positively identified.

5.7 Antiparticles

Antiparticles are produced in nuclear reactions when sufficient energy is present. Antiparticles do not live very long in the presence of matter. For example, when an electron encounters its antiparticle, the positron, the two annihilate each other. The energy of their vanished mass and any kinetic energy they had is converted to gamma rays or other particles.

5.8 Particle Classification

The important way to understand elementary particles is to arrange them in categories according to their properties. One such classification involves their interactions. The photon is in a class by itself. It only takes part in the electromagnetic force.

The **leptons** are those particles that do not interact by way of the strong force but do interact by means of the weak nuclear force as well as the much weaker gravitational force. Those that carry an electrical charge also interact via the electromagnetic force. The four known leptons are the electron, the muon and two types of neutrino: the electron neutrino and the muon neutrino.

The third category is the hadron. **Hadrons** are those particles that can interact via the strong nuclear force. They also interact via the other forces but the strong force predominates at short distances. The hadrons include nucleons, pions and a large number of other particles. They are divided into two subgroups: **baryons**, which are particles that have a baryon number of **+1** (*or -1 in the case of their antiparticle*); and **mesons**, which have a baryon number = 0. (The baryon number is the same as the nucleon number - both numbers are conserved in reactions).

Recent theories propose that certain nuclear particles may be composed of constituent particles called **quarks**. At the present time, no unit of charge smaller than $\pm e$ has been detected. However, if the existence of quarks is confirmed, they would have charges of $\pm e/3$ and $\pm 2e/3$.

Questions

1   If 100 MeV of energy is released by an atom during fission, the amount of mass that is converted to energy is approximately
   (1) 1 amu   (2) 0.1 amu   (3) 9 amu   (4) 9×10^4 amu
2   When a neutron is emitted from a nucleus, the mass number of the nuclide
   (1) decreases   (2) increases   (3) remains the same
3   Which device is used to detect nuclear radiation?
   (1) synchrotron   (3) linear accelerator
   (2) cloud chamber   (4) cyclotron
4   Which device is used to accelerate a charged particle?
   (1) a photographic plate   (3) a cyclotron
   (2) an electroscope   (4) a cloud chamber
5   As a sample of uranium disintegrates, the half-life of the remaining uranium
   (1) decreases   (2) increases   (3) remains the same

6   Which device is used to produce a stream of high - velocity charged particles?
   (1) Geiger counter   (3) cyclotron
   (2) electroscope   (4) photographic plates

7   The nuclear binding energy of an atom is proportional to the
   (1) mass of the atom   (3) nuclear protons of the atom
   (2) mass defect of the atom (4) nuclear neutrons of the atom

8   Which statement most accurately describes the interaction which binds a nucleus together?
   (1) long-range and weak   (3) short-range and weak
   (2) long-range and strong   (4) short-range and strong

Base your answers to questions 9 through 11 on the information below.

The equations represent a two - stage nuclear reaction.

$$^{27}_{13}\text{Al} + ^{4}_{2}\text{He} \rightarrow ^{30}_{15}\text{P} + \text{X}$$

$$^{30}_{15}\text{P} \rightarrow ^{30}_{14}\text{Si} + \text{Y} + \text{energy}$$

9   Which nucleus in the two equations has the greatest number of neutrons?

   (1) $^{27}_{13}\text{Al}$ (2) $^{4}_{2}\text{He}$ (3) $^{30}_{15}\text{P}$ (4) $^{30}_{14}\text{Si}$

10 What is particle X?
   (1) a positron   (2) an electron   (3) a proton   (4) a neutron

11 Particle Y represents

   (1) $^{1}_{0}\text{n}$ (2) $^{1}_{1}\text{H}$ (3) $^{0}_{+1}\text{e}$ (4) $^{0}_{-1}\text{e}$

12 When a nucleus captures an electron, the atomic number of the nucleus
   (1) decreases   (2) increases   (3) remains the same

13 In a nuclear reactor, control rods are used to
   (1) slow down neutrons   (3) absorb neutrons
   (2) speed up neutrons   (4) produce neutrons

14 During nuclear fusion, energy is released as a result of the
   (1) splitting of heavy nuclei   (3) combining of light nuclei
   (2) combining of heavy nuclei   (4) splitting of light nuclei

15 Moderators are used to slow neutrons down in a nuclear reactor in order to
   (1) reduce nuclear reactions
   (2) remove radioactive impurities in the core
   (3) improve the probability of fission
   (4) achieve critical mass

Thinking Physics

1   Diagnostic medicine is a large user of radioactive traces. What are the half lives of these materials?

2   Coal contains trace quantities of radioactive materials. Environmental radiation surrounding a coal–fired power plant is greater than a fission power plant. How can this be explained?

3 A sample of radioactive material is always a little warmer than its surroundings. Why is the center of the Earth so hot?

4 Film badges are worn by people that work around radioactivity. What kind of exposure do these badges monitor?

5 If a uranium nucleus was split into three pieces of approximately the same size, instead of being split into two pieces, would more or less energy be released? Explain your reasoning.

Self–Help Questions

1 Which is an isotope of $^{237}_{93}$Np?
 (1) $^{237}_{92}$U (2) $^{237}_{94}$Np (3) $^{235}_{92}$U (4) $^{235}_{93}$Np

2 The total number of nucleons in an atom of $^{8}_{5}$B is

 (1) 5 (2) 8 (3) 3 (4) 13

3 Which nucleus has the greatest nuclear charge?
 (1) $^{2}_{1}W$ (2) $^{8}_{5}X$ (3) $^{7}_{3}Y$ (4) $^{4}_{2}Z$

4 What is the mass number of an atom with 9 protons, 11 neutrons, and 9 electrons?
 (1) 9 (2) 18 (3) 20 (4) 29

5 What is the energy equivalent of a mass of 1 kilogram?
 (1) 9×10^{16} J (2) 9×10^{13} J (3) 9×10^{10} J (4) 9×10^{7} J

6 If the mass of one proton is totally converted into energy, it will yield a total energy of
 (1) 5.1×10^{-19} J (2) 1.5×10^{-10} J (3) 9.3×10^{8} J (4) 9.0×10^{16} J

7 How much energy is released when 1×10^{-3} kilogram of matter is converted to energy?
 (1) 3×10^{5} J (2) 3×10^{8} J (3) 9×10^{13} J (4) 9×10^{16} J

8 Which force between the protons in a helium atom will have the greatest magnitude?
 (1) gravitational force (3) nuclear force
 (2) electrostatic force (4) magnetic force

9 In the reaction $^{24}_{11}$Na $\rightarrow$ $^{24}_{12}$Mg + x, what does x represent?

 (1) an alpha particle (3) a neutron
 (2) a beta particle (4) a positron

10 The diagram at the right represents an inverted test tube over a sample of a radioactive material. Helium has collected in the test tube. The presence of helium indicates that the sample is most probably undergoing the process of
 (1) alpha decay
 (2) beta decay
 (3) neutron decay
 (4) gamma emission

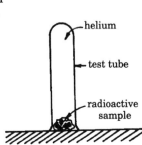

Base your answers to questions 11 through 15 on the following information:

A certain radioactive isotope with a half-life of 5.0 minutes decays to a stable (non-radioactive) nucleus by emitting one alpha particle.

11 If the mass of the original sample was 1.00 kilogram, how long would it take to form 0.75 kilogram of the stable material, leaving 0.25 kilogram of the original sample?
(1) 2.5 minutes   (3) 10.0 minutes
(2) 5.0 minutes   (4) 15.0 minutes

12 The difference between the mass number of the original nucleus and the mass number of the new stable nucleus is
(1) 1   (3) 3
(2) 2   (4) 4

13 The difference between the atomic number of the original nucleus and the atomic number of the new stable nucleus is
(1) 1   (3) 3
(2) 2   (4) 4

14 Compared to the binding energy per nucleon of the original radioactive nucleus, the binding energy per nucleon of the new stable nucleus is
(1) less   (2) greater   (3) the same

15 As the absolute temperature of the radioactive isotope increases, its half-life
(1) decreases   (2) increases   (3) remains the same

Base your answers to questions 16 and 17 on the information below.

Mass of a proton = 1.007277 amu
Mass of a neutron = 1.008665 amu
Mass of an electron = 0.0005486 amu

Mass of a $_2^4$He nucleus = 4.001509 amu

16 How many nucleons are in a $_2^4$He nucleus?

(1) 8   (2) 2   (3) 6   (4) 4

17 What is the mass defect of a $_2^4$He nucleus?

(1) 1.985567 amu   (3) 0.030375 amu
(2) 1.985018 amu   (4) 0.029278 amu

Base your answers to questions 18 through 20 on the following information:

$_{53}^{131}$I initially decays by emission of beta particles.

18 Beta particles are
(1) protons   (3) neutrons
(2) electrons   (4) electromagnetic waves

19 When $_{53}^{131}$I decays by beta emission, it becomes
(1) $_{53}^{130}$I   (2) $_{51}^{129}$Sb   (3) $_{54}^{131}$Xe   (4) $_{54}^{135}$Xe

20 The half-life of $_{53}^{131}$I is 8 days. After 24 days, how much of a 100.-gram sample would remain?
(1) 0 g   (2) 12.5 g   (3) 25.0 g   (4) 50.0 g

Base your answers to questions 21 through 25 on the graph which represents the disintegration of a sample of a radioactive element. At time = 0 the sample has a mass of 4.0 kilograms.

21 What mass of the material remains at 4.0min.?
   (1) 1 kg   (3) 0 kg
   (2) 2 kg   (4) 4 kg

22 What is the half-life of the isotope?
   (1) 1.0 min.   (3) 3.0 min.
   (2) 2.0 min.   (4) 4.0 min.

23 How many half-lives of the isotope occurred during 8.0 minutes?
   (1) 1   (2) 5   (3) 8   (4) 4

24 How long did it take for the mass of the sample to reach 0.25 kilogram?
   (1) 1 min.   (2) 5 min.   (3) 3 min.   (4) 8 min.

25 If the mass of this material had been 8.0 kilograms at time t = 0, its half-life would have been
   (1) less   (2) greater   (3) the same

26 In the Uranium Disintegration Series, when an atom of $^{238}_{92}$ U decays to $^{206}_{82}$ Pb, the total number of beta particles emitted is

   (1) 6   (2) 2   (3) 8   (4) 14

27 According to the Uranium Disintegration Series in the reference tables, what is the total number of alpha particles emitted when a single U^{238} nucleus completely decays to Pb^{206}?
   (1) 8   (2) 6   (3) 3   (4) 4

Base your answers to questions 28 through 30 on the information below.
   An atom of $^{238}_{92}$ U absorbs a neutron as indicated in the equation

$$^{238}_{92} U + ^{1}_{0} n \rightarrow Y.$$

28 The atomic number of element Y is
   (1) 92   (2) 91   (3) 90   (4) 89

29 The mass number of Y is
   (1) 240   (2) 239   (3) 238   (4) 237

30 How many neutrons does $^{238}_{92}$ U have in its nucleus?

   (1) 146   (2) 147   (3) 237   (4) 238

31 The half–life of $^{223}_{88}$ Ra is 11.4 days. If M kilograms of this radium isotope are present initially, how much remains at the end of 57 days?
   (1) ½ M   (2) ¼ M   (3) ⅕ M   (4) 1/32 M

32 A 100-kilogram sample of a substance having a half–life of 300 years decays. How much time will it take to have only 25 kilograms of the original left?
   (1) 75 years   (2) 300 years (3) 600 years (4) 1,200 years

33 The half–life of an isotope is 14 days. How many days will it take 8 grams of this isotope to decay to 1 gram?
   (1) 14   (2) 21   (3) 28   (4) 42

34 Which isotope is used in defining the atomic mass unit?

(1) $_1^1 H$ (2) $_{92}^{238} U$ (3) $_8^{16} O$ (4) $_6^{12} C$

35 Which statement best describes the fission products from nuclear reactors?
   (1) They are nonradioactive and may be safely discarded.
   (2) They are nonradioactive and must be treated and/or stored.
   (3) They are intensely radioactive and may be safely discarded.
   (4) They are intensely radioactive and must be treated and/or stored.

36 Which part of a nuclear reactor would most likely contain plutonium?
   (1) control rod   (2) fuel rod   (3) moderator   (4) shielding

37 In the nuclear reaction $_{84}^{218} Po \rightarrow \ _{82}^{214} Pb + X$, the X represents

(1) $_2^4 He$ (2) $_{-1}^0 e$ (3) $_{+1}^0 e$ (4) $_0^1 n$

38 Which reaction is an example of nuclear fusion?

(1) $_{88}^{226} Ra \ \rightarrow \ _{86}^{222} Rn + _2^4 He + Q$

(2) $_{83}^{214} Bi \ \rightarrow \ _{84}^{214} Rn + _{-1}^0 He + Q$

(3) $_{92}^{235} U + _0^1 n \ \rightarrow \ _{36}^{92} Kr + _{56}^{141} Ba + 3_0^1 n + Q$

(4) $\quad _1^3 H + _1^1 H \ \rightarrow \ _2^4 He + Q$

39 The uranium isotope $_{92}^{238} U$ is used to produce

(1) shielding   (3) control rods
(2) fissionable plutonium   (4) heavy water

40 Neutrons are used in some nuclear reactions as bombarding particles because they are
   (1) positively charged and are repelled by the nucleus
   (2) uncharged and are not repelled by the nucleus
   (3) negatively charged and are attracted by the nucleus
   (4) uncharged and have negligible mass

41 What is the purpose of a cloud chamber?
   (1) to accelerate particles   (3) to fuse particles
   (2) to split particles   (4) to detect particles

42 Which device can be used to separate isotopes of an element?
   (1) a mass spectrometer   (3) an induction coil
   (2) an electroscope   (4) two closely spaced double slits

43 What kind of nuclear reaction is represented by the equation below?

$$_7^{14} N + _2^4 He \rightarrow \ _8^{17} O + _1^1 H$$

   (1) alpha decay   (3) artificial transmutation
   (2) beta decay   (4) nuclear fission

44 The fission process in a nuclear reactor is controlled by regulating the number of available
   (1) electrons   (2) neutrons   (3) protons   (4) positrons

45 The splitting apart of a heavier nucleus to form lighter nuclei is called
   (1) fusion   (3) positron emission
   (2) fission   (4) beta emission

Optional Unit 5.2 — Solid State Physics

Important Terms To Be Understood

conductor
insulator
semiconductor
conductivity
resistivity
electron–sea model
band model
valence electrons
ions
doping
N–type semiconductor

donor material
P–type semiconductor
acceptor material
diode
P–N junction
forward biasing
reverse biasing
transistors
emitter
base
collector

PNP transistor
NPN transistor
collector current
integrated circuit
anode
cathode
LED
avalanche voltage
triode
analog
digital

I. Conduction In Solids

A solid may be classified as a conductor, an insulator or a semiconductor based on its ability to carry an electric current. One property of a substance that determines the amount of current that will flow through it is called **resistivity**. Numerical magnitudes of resistivity are determined experimentally for a specific temperature:

$$\rho = \frac{RA}{L}$$

Where:
ρ = resistivity in ohm·meters
R = resistance in ohms
A = cross–sectional area in meters squared
L = length in meters

The resistivity of a material is constant for any given temperature.

Note: Devices which measure resistance as a function of temperature are called **thermistors**. The term is short for **resistance thermometer**. Thermistors measure temperature by a change in resistance. As the temperature of a thermistor increases, the resistance decreases.

The unit of conductivity is the (ohm·meter)$^{-1}$. Therefore, if a material has a resistivity of 20ohm·meter, then the conductivity is

$$\frac{1}{\rho} = \frac{1}{20 \text{ ohm·m}} = .05 \text{ (ohm·meter)}^{-1}$$

Conduction is measured by **conductivity**, the reciprocal of resistivity. **Conductors** are typically metallic solids such as aluminum and copper. **Insulators** are typically nonmetallic solids such as phosphorus, iodine, glass, and rubber.

Semiconductors are typically metalloids whose electrical conductivities are higher than those of insulators but less than those of conductors. We normally think of semiconductors as being solids, but some liquids are semiconducting. Commonly used semiconducting materials include silicon, germanium, and gallium arsenide.

In semiconductors, the electrical resistivity decreases with increasing temperature. Insulators behave in a similar manner and the distinction between semiconductors and insulators is only one of degree – all insulators are semiconductors at high temperatures and all semiconductors are insulators at low temperatures. Classification is based on their room temperature resistivities.

The extent to which a semiconductor conducts electricity is affected by the presence of impurities. Very pure semiconductors and those which have their conductivity increased by temperature are called intrinsic semiconductors. Those to which impurities have been added are called extrinsic semiconductors.

1.1 Theories Of Solid Conductors

Two models which explain electrical conduction in solids are the Electron Sea Model and the Energy Band Model.

A. Electron Sea Model

The Electron Sea Model relates conduction to the number of valence electrons which can be dislodged and moved freely through the solid. Valence electrons are those which occupy the outermost principal energy level of an atom. Since the valence electrons are loosely bound to the nucleus, these electrons are available to form an **"electron sea."** *All metals are therefore good conductors.* Insulators which tightly bind their valence electrons do not allow formation of an "electron sea" for charge transportation. Although semiconductors have some free electrons, they have fewer electrons than conductors have. This model does not explain the conductive properties of semiconductors.

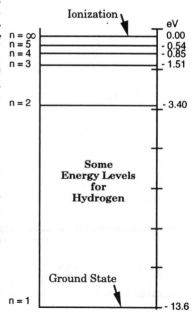

B. Energy Band Model

The Energy Band Model, based on quantum mechanics, is better in explaining conduction in solids. The model shows that only a limited number of energy levels are available to the electrons in the atom. In an isolated hydrogen atom, the permitted levels appear on the chart below.

An electron *cannot* exist at any intermediate state between energy levels. These other energies are called **forbidden states.**

Solids are made up of a large number of atoms, each with many electrons. **Quantum theory** predicts the interaction and merging of atom's energy levels. According to the **exclusion principle,** only two electrons of opposite spin can occupy the same energy level. When two atoms interact, the energy levels of both atoms become almost, but not exactly, the same energy. As more and more atoms interact, each energy level becomes an **energy band** of many closely spaced energy levels. The **valence band** is an energy band occupied by all the valence electrons of the solid.

All electrons occupy the lowest energy levels that are available. For an electron to leave its position and move through the solid, it must be first raised to a higher energy level. If there is an unoccupied energy level available, only a small amount of energy will raise the valence electron to the next available level in the band. If all of the levels are occupied within the valence band, an electron may absorb enough energy to leave the valence band and enter a higher energy band called the **conduction band.** Electrons in the conduction band can move freely throughout the solid and transport charge.

In many solids, there is an "energy gap" between the highest level of the valence band and the lowest level of the conduction band. This gap is called the **forbidden band.** *An electron may not exist in the forbidden band.* To enter the conduction band, an electron must acquire an amount of energy at least equal to the energy gap. The resistivity of any material depends on the structure of the valence band and the width of the forbidden band.

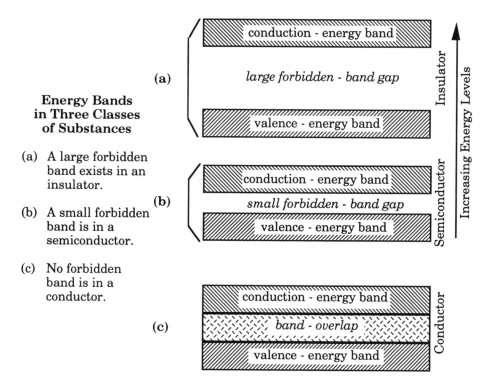

**Energy Bands
in Three Classes
of Substances**

(a) A large forbidden band exists in an insulator.

(b) A small forbidden band is in a semiconductor.

(c) No forbidden band is in a conductor.

1.2 Electron Bands In Conductors

The valence level of every atom can hold a maximum of eight electrons. Metals have one, two, or three valence electrons. There are a large number of empty energy levels in the valence band.

Valence electrons can move to an unoccupied level with very little energy input. **Thermal energy** or the energy from a photon of radiation may be sufficient to make the transition. Since the valence band of a metal touches or overlaps its conduction band, there is no forbidden band. When a small potential difference is applied across the metal, large numbers of electrons will begin to move. This movement is in response to the electrostatic force exerted on the charged particles in the electric field.

1.3 Resistance In Conductors

Resistance in a conductor is due mainly to the **frictional effect** of collisions between moving electrons and atoms. The temperature of the conductor increases when the friction is converted into heat. This increases the rate of collisions and the energy loss. As the temperature of the conductor increases, a greater potential difference is needed to keep the electrons moving at the same rate. The resistance increases because the resistivity of the metallic conductor increases when the temperature increases.

1.4 Band Patterns In Insulators

In insulators, the atoms are held together by pairs of shared electrons that form what is called a **covalent bond**. The two electrons in a covalent bond in an insulator occupy the *same* energy level in *both* atoms. The covalent bond in an insulator has the effect of filling up the valence band of the material. It is almost impossible for an electron to acquire the energy necessary to leave the valence band and enter the conduction band because of two reasons. First, there are no vacant levels in the valence band. Second, there is a wide forbidden band. The result is that resistivity is very high.

1.5 Band Pattern In A Semiconductor

Elements with an intermediate valence of four, such as silicon, carbon, and germanium have unusual conduction properties. In the physical state, pure silicon exists as a crystal. The four silicon atoms in the crystal structure share their valence electrons to complete the orbit. In a silicon crystal, the stability of the lattice atoms is influenced by the external temperature.

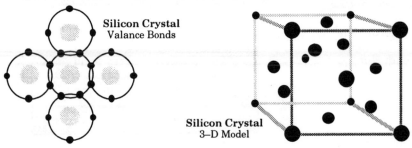

Silicon Crystal
Valance Bonds

Silicon Crystal
3–D Model

At absolute zero (0 Kelvin), atoms have minimum energy (slow vibration) in their lattice. Silicon is a perfect electrical insulator (poor conductor). At room temperature (293 K), the valence electrons are agitated because the crystal lattice absorbs heat and the atoms vibrate. Silicon, carbon, and germanium are semiconductors at room temperature. At 1700°C, the crystal lattice is destroyed, free electrons are liberated, and silicon becomes a good conductor.

The movement or liberation of valence electrons is responsible for conduction in materials. In silicon, a thermally dislodged electron can pass from one atom to another under the electric field established by the external voltage. In the figure below, the isolated atom of silicon is suspended between the plates of a capacitor. Assume that the voltage is adjusted to zero volts; hence, no field exists. If the voltage is increased, the electrostatic field increases. Addition of 1.1 eV is sufficient to cause a valence shell electron to break away from the parent atom. *The liberated electron would be attracted to the positive plate of the capacitor.*

In order to liberate electrons from various levels, energy must be added to the atom. The orbiting electrons already possess energy because of their position. The valence electrons, because of their position, are high energy, while electrons in the first shell are lower energy. The valence electrons have greater **potential energy** and are capable of doing more work.

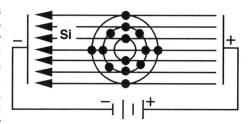

Silicon Atom in an Electrostatic Field

The liberated electron in silicon is excited to a **conduction band**. As the temperature increases, more electrons with conduction band energy will be produced and resistivity will decrease.

1.6 Electron–Hole Concept

The conduction electrons flow in the conduction band. The electron flows to the positive electrode. The electrons are current carriers.

It was discovered in semiconductors that current can also flow in the opposite direction. Suppose that atom A in the figure below has a deficiency of one valence electron, or a **hole**, and that an electric field **E** is applied.

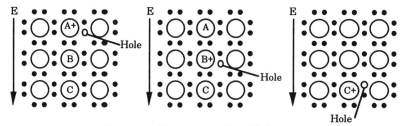

Current Transport by Holes

A valence electron on atom **B** can move to atom **A** and fill the deficiency with the result that the hole is now associated with atom **B**. Similarly, a valence electron can move from atom **C** to atom **B**. The net result of these processes is the passage of valence electrons from atoms **C** to **A**, and the motion of the **hole** in the *opposite* direction. Although holes flow with ease, they experience more opposition than electron flow and do not flow with the same velocity. An electron flow is twice that of a hole flow for each volt/cm.

When an electron is energetically ejected from a valence band, a covalent bond is broken and a hole is left behind. This hole can travel to an adjacent atom by acquiring an electron from that atom, breaking an existing covalent bond and establishing a covalent bond by filling the hole. *The number of holes equals the number of freed electrons.*

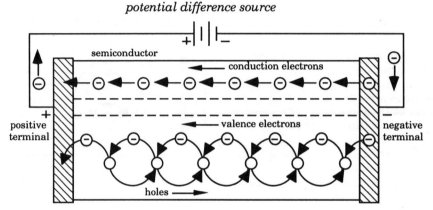

Current in a circuit containing a semiconductor

When the semiconductor is put into a circuit, the electrons from the valence band enter the positive terminal leaving a hole behind. Holes arrive at the negative terminal and are filled by electrons entering the material. In addition, electrons are flowing in the conduction band. In the external circuit, only the electrons carry current.

1.7 Extrinsic Semiconductors

Small amounts (~ 1 part in 10^6) of certain elements can be added to germanium (and silicon) without producing any distortion of the basic crystal structure. The process is known as **doping** and produces a considerable increase in conductivity. The resulting material is called an **extrinsic semiconductor**.

The atoms which are added are often referred to as **impurities**, but this is not meant to imply that there is anything accidental or haphazard about their inclusion. This process is very carefully controlled in growing the semiconductor material from a melt or in depositing the impurity on the surface of the crystal and heating to diffuse the impurity inward. The atoms added have a valence of 5 or 3.

A glance at the right hand side of the periodic table shows elements with one less valence electron, such as boron (**B**), or one more valence electron, such as phosphorous (**P**), than silicon (**Si**).

A. Acceptors

When a pure germanium crystal is doped with atoms of valence (3) such as indium, the indium acts as an acceptor of germanium valence electrons.

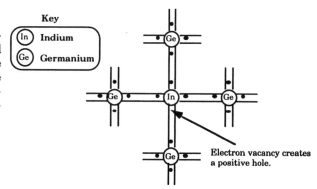

Key

(In) Indium
(Ge) Germanium

Electron vacancy creates a positive hole.

The indium atom is still lacking a desired electron in its valence shell. The space that remains is a hole. The hole exists within the confines of the crystal and is positively charged. It requires very little energy for an electron in a nearby germanium–germanium bond to move across and fill the vacancy in the Indium. At room temperature, lattice vibrations readily provide this energy, thus creating a hole in one of the germanium atoms. Germanium crystals with an excess of **positive holes** are called *P*–type.

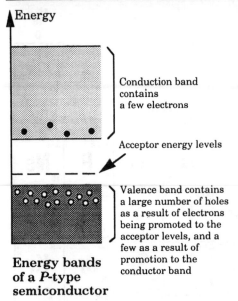

Conduction band
contains
a few electrons

Acceptor energy levels

Valence band contains
a large number of holes
as a result of electrons
being promoted to the
acceptor levels, and a
few as a result of
promotion to the
conductor band

**Energy bands
of a P-type
semiconductor**

In **P–type** crystals, thermal agitation produces equal pairs of electron–hole pairs. It is important to stress that the addition of acceptor impurities leaves the crystal electrically neutral. The excess positive charge associated with the charge carriers is balanced by the excess negative charge at the nuclei sites.

The addition of an acceptor impurity to an intrinsic semiconductor creates extra energy levels just above the top of the valence band. At room temperature these levels are occupied by electrons which have been thermally excited from the valence band. This leaves a large number of holes in the valence band and increases the conductivity.

Because the majority of electric current in **P–type** semiconductors is composed of holes, the holes are majority current carriers, and the electrons are minority current carriers.

B. Donors

Atoms with five valence electrons, such as phosphorous and antimony, act as donor dopants. When antimony is substituted into the germanium crystal lattice, the extra electron is not held in the valence bond.

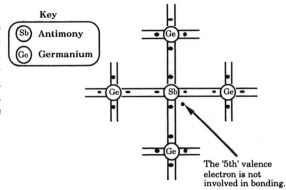

Key

Sb Antimony

Ge Germanium

The '5th' valence
electron is not
involved in bonding.

The extra electrons of negative charge produce an **N–type** (negative) semi–conductor.

The addition of a donor impurity to an intrinsic semiconductor creates extra energy levels just below the bottom of the conduction band. At room temperatures, thermal agitation is easily capable of providing the small amount of energy necessary to raise the electrons in these levels to the conduction band. Once in the conduction band, the electrons can conduct an electric current.

In *N–type* material, the electrons are the majority carriers, and the holes are the minority carriers. The *N–type* crystal is neutrally charged throughout. Due to thermal agitation, equal numbers of electron–hole pairs are produced. The doping material adds no net charge. For each free electron, a positive donor ion forms.

The resistivity of the germanium crystal is directly proportional to the number of free electrons or holes. Pure germanium has few current carriers and a high resistivity.

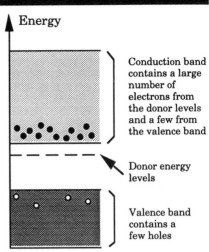

Energy bands of a
N-type semiconductor

1.7 Positive Charge Carriers

About 100 years ago, **E. H. Hall** performed an experiment proving that positive charge carriers *do* exist. When a magnetic field is applied at right angles to the current carrying strip of metal, the electrons are forced to one side of the strip. It is easy to determine the resulting potential difference and the internal electric field.

When the strip is replaced with a **P–type** semiconductor, the direction of the electric field and the potential difference reverse. Hall concluded that positive charge carriers must exist. It was not until a century later that modern physics identified this charge carrier as a "**hole.**"

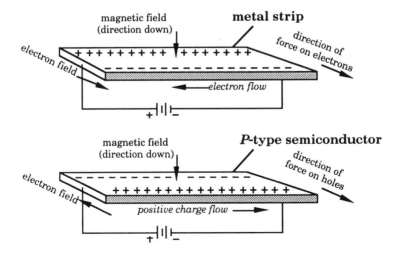

Questions

1   Which of the atoms represented below can be classified as a semiconductor?

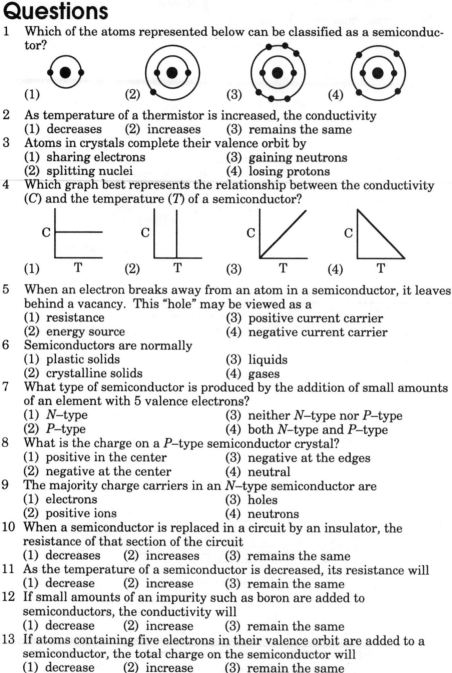

(1)   (2)   (3)   (4)

2   As temperature of a thermistor is increased, the conductivity
    (1) decreases   (2) increases   (3) remains the same

3   Atoms in crystals complete their valence orbit by
    (1) sharing electrons   (3) gaining neutrons
    (2) splitting nuclei   (4) losing protons

4   Which graph best represents the relationship between the conductivity
    (C) and the temperature (T) of a semiconductor?

    (1)   T   (2)   T   (3)   T   (4)   T

5   When an electron breaks away from an atom in a semiconductor, it leaves
    behind a vacancy. This "hole" may be viewed as a
    (1) resistance   (3) positive current carrier
    (2) energy source   (4) negative current carrier

6   Semiconductors are normally
    (1) plastic solids   (3) liquids
    (2) crystalline solids   (4) gases

7   What type of semiconductor is produced by the addition of small amounts
    of an element with 5 valence electrons?
    (1) N–type   (3) neither N–type nor P–type
    (2) P–type   (4) both N–type and P–type

8   What is the charge on a P–type semiconductor crystal?
    (1) positive in the center   (3) negative at the edges
    (2) negative at the center   (4) neutral

9   The majority charge carriers in an N–type semiconductor are
    (1) electrons   (3) holes
    (2) positive ions   (4) neutrons

10 When a semiconductor is replaced in a circuit by an insulator, the
    resistance of that section of the circuit
    (1) decreases   (2) increases   (3) remains the same

11 As the temperature of a semiconductor is decreased, its resistance will
    (1) decrease   (2) increase   (3) remain the same

12 If small amounts of an impurity such as boron are added to
    semiconductors, the conductivity will
    (1) decrease   (2) increase   (3) remain the same

13 If atoms containing five electrons in their valence orbit are added to a
    semiconductor, the total charge on the semiconductor will
    (1) decrease   (2) increase   (3) remain the same

14 Which element could be used as a semiconducting material?
    (1) antimony   (3) copper
    (2) arsenic   (4) germanium

15 As the amount of dopant in a semiconductor increases, the resistance of the semiconductor
   (1) decreases   (2) increases   (3) remains the same
16 To produce a *P*–type semiconductor from a crystal of silicon, the number of electrons in the valence level of the dopant element should be
   (1) 0   (2) 5   (3) 3   (4) 4
17 The width of the forbidden band or energy gap is
   (1) the same for all semiconductors
   (2) is greatest in metals
   (3) is greatest for semiconductors
   (4) is the greatest for insulators

II. Semiconductor Devices

The **diode** is the simplest and most fundamental semiconductor device used in electronics. It acts as a one way valve letting charge flow through it in one direction while blocking charge flow in the other. The figure below shows the schematic symbol for a diode.

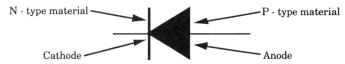

N - type material ——→ ←—— P - type material

Cathode ——→ ←—— Anode

The presence of an "arrow" in any semiconductor device always points from *"P"* to *"N."*

The *P*–**type** material makes up the **anode** of the diode. The word "anode" is used in electrical and electronic devices to identify the terminal which attracts electrons. The *N*–**type** material makes up the **cathode**. The word "cathode" refers to the terminal that gives off, or emits, electrons. Note that the forward electron current moves from the cathode to the anode.

2.1 *P–N* Junction Diode

When a silicon crystal is doped with donors and acceptors, respectively, *N*–**type** and *P*–**type** crystals are formed. The boundary at which the doped areas meet is called a *P–N* **Junction**.

Holes and electrons are evenly distributed through both *N* and *P*–**type** regions. Majority current carriers (electrons) in the *N*–**type** material are just as likely to cross the boundary due to thermal agitation into the *P*–**region** as to move further into the *N*–**region**. When this occurs, the migrating electrons and holes will combine to form a region at the junction which leaves behind positively (**+**) charged atoms. Similarly holes near the junction which are filled with electrons leave behind negatively charged (**–**) atoms.

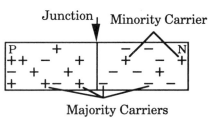

Junction | Minority Carrier

Majority Carriers

The dual region near the junction is called the **depletion region**. There is a depletion or lack of holes and electrons in this area. The region is devoid of free electrons in the *N*–**material** and holes in the *P*–**material**.

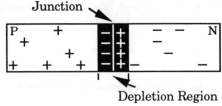

Depletion Region

2.2 Barrier Potential

The electric field in the depletion region is created by charged atoms (ions). The field acts as a potential energy barrier.

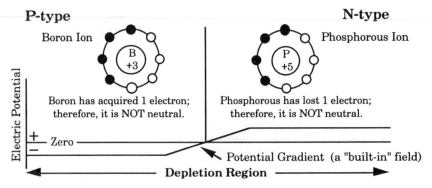

P-type   **N-type**

Boron Ion   Phosphorous Ion

Boron has acquired 1 electron; therefore, it is NOT neutral.

Phosphorous has lost 1 electron; therefore, it is NOT neutral.

Electric Potential

Zero

Potential Gradient (a "built-in" field)

Depletion Region

How heavily the Germanium is doped, determines the Potential Gradient height.
The Potential Gradient is established by the lattice ions, near the junction.

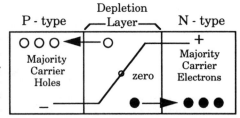

Any additional electrons that attempt to diffuse through the depletion junction (barrier) from the *N*–**region** into the *P*–**region** are repelled by negatively charged atoms (ions). Any holes attempting to diffuse from the *P*–**region** to the *N*–**region** are repelled by the positively charged atoms of the barrier.

Depletion
P - type   —Layer—   N - type

Majority Carrier Holes

zero

Majority Carrier Electrons

The Potential Gradient sweeps the Majority and Minority Carriers out of the junction, leaving a Depletion Layer.

2.3 Bias

When a battery is connected to a *P–N* **junction**, an electrical external potential is applied. The battery supplies extra energy to overcome the electric potential at the *P–N* **junction**.

In the figure top of opposite page, the negative terminal of the battery is connected to the *N*–**region** end, and the positive battery terminal is connected to the *P*–**region** of the *P–N* junction diode.

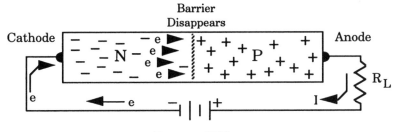

Forward Bias

The (+) positive terminal of the battery repels the holes in the **P–region** toward the junction, and the negative terminal (–) repels the electrons on the **N–region** towards the barrier. The forward push of the many electrons from the battery helps the free electrons to overcome the barrier, and the combination causes conduction through the barrier.

When this happens, electrons enter the **N–type** region via the external wire under the influence of the battery. Likewise, a hole is created in the **P–type** material by an electron breaking a covalent bond and entering the positive terminal of the battery. Under the influence of the external battery, electrons flow in the circuit.

The above condition is known as **forward biasing** the junction. A **P–N** junction that is forward biased conducts a current through the **P–N** junction. A diode is forward biased when the anode of the diode is connected to the positive battery terminal and the cathode is connected to the negative battery terminal.

If the battery connections are reversed (positive battery terminal connected to **N–region** and negative terminal to **P—region**), the holes in the **P–region** are attracted to the negative battery terminal away from the junction. Electrons in the **N–region** are attracted to the positive terminal away from the junction.

Both types of majority carriers (holes and electrons) move away from the barrier, leaving behind more charged atoms to add to the junction barrier. This process continues until the barrier equals the potential of the external barrier; then, the current stops. This condition is called **reverse** or **back bias** since it offers maximum resistance to the external flow of majority carriers.

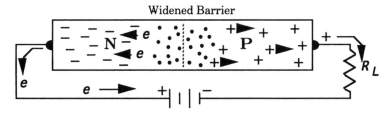

Reverse Bias

2.4 Electrical Devices: *P–N* Junction

A. Diodes

In its simplest form, a **diode** consists of a silicon crystal doped in one region with **P–type** and in the other region with **N–type**. The semiconductor diode has largely replaced the vacuum tube diode. They are more reliable and consume less power. In the forward bias, direct current will flow from a DC source. In the reverse bias, hardly any current will flow.

A **zener diode** is used as a voltage regulator. As long as it is operated over its normal range, the voltage across the diode will equal the rated voltage plus or minus a small error voltage. It is operated with reverse bias.

Diodes may also be used as limiters to change the shape of a signal or remove noise from a signal. Another important type of diode is the **LED** (Light Emitting Diode). These have a long life and can be switched rapidly. They have wide applications in numeric displays like calculators and low-power indicator lamps..

B. Rectifier

Silicon rectifiers are junción type diodes connected in series to an alternating (**AC**) current source. An **AC** potential applied to the circuit results in a pulsating direct current. An **AC** rectifier supplies alternating forward and then reverse bias. The rectifying properties of a *P–N* junction make it ideal for detection of modulated radio signals.

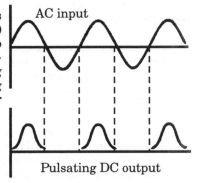

Pulsating DC output

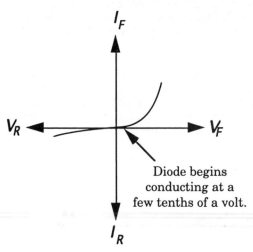

Diode begins conducting at a few tenths of a volt.

**A Voltage – Current
for a diode characteristic curve**

The diode does not obey Ohm's Law. The diode characteristic curve shows that the diode is not a linear device. The nonlinearity can be an advantage or a disadvantage depending on the exact circuit application.

V_F represents voltage applied when forward biased. With zero volts across the diode, the diode will not conduct . The diode will not conduct until a few tenths of a volt are applied across it. This voltage is necessary to collapse the depletion region and the barrier potential. It requires about

0.2V to turn on a germanium diode and about 0.6V to turn on a silicon diode.

V_R represents increasing reverse bias applied to the diode. The graph above shows some reverse current I_R which is due to the minority carriers. Leakage current does not become significant until there is a large reverse bias across the diode.

The graph at the right shows how silicon and germanium compare under conditions of reverse bias.

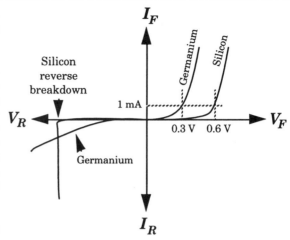

Comparison of germanium and silicon diodes

At voltages below a certain critical voltage, the leakage current of the silicon diode is very low compared to the germanium diode. Once the critical value of V_R is reached, the silicon diode shows a rapid increase in reverse current. This point is referred to as the **avalanche voltage** at approximately 200–300 V.

Questions

1 The boundary between P–type and N–type semiconductors is called a
   (1) thermistor   (3) majority carrier
   (2) junction   (4) minority carrier
2 When a diode type semiconductor has its forward bias changed to reverse bias, the current will
   (1) decrease   (2) increase   (3) remain the same

Base your answers to 3 and 4 on the diagram at the right which represents a silicon semiconductor.

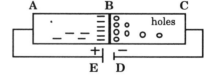

3 In the diagram, C represents
   (1) N–type silicon   (3) cathode
   (2) P–type silicon   (4) diode
4 The P–N junction in the diagram is biased
   (1) reverse   (2) forward   (3) C to D (4) A to E
5 Which is produced when a negative potential is applied to the P–type semiconductor and a positive potential applied to the N–type semiconductor of a P–N junction?
   (1) a forward bias   (3) pulsating direct current
   (2) a reverse bias   (4) alternating current

6   Pulsating direct current results when a *P–N* junction is connected to
    (1) a battery   (3) a source of direct current
    (2) an oscilloscope   (4) an alternating potential
7   Which one of the following diagrams represents reverse bias?

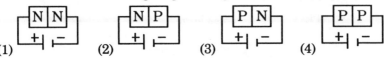

    (1)   (2)   (3)   (4)

8   When *P*–type and *N*–type semiconductor materials are joined to form a
    P–N junction, this results in an electric
    (1) barrier   (2) conductor   (3) current   (4) circuit

Base your answers to questions 9 through 11 on the diagram
at the right of a circuit containing a semiconductor device.

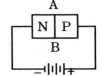

9   Which type of semiconductor device is shown?
    (1) emitter   (3) diode
    (2) resistor   (4) transistor
10 In the diagram, line *AB* identifies the
    (1) emitter   (2) base   (3) collector   (4) junction
11 As drawn, this device is
    (1) forward biased   (3) grounded
    (2) reversed biased   (4) open

III. Transistors

A circuit device which contains a thin layer of one type of semiconductor
between two layers of the other type semiconductor is called a **transistor**. A
transistor is a triode capable of amplifying a change in current.

A layer of **P-type** between two layers of **N-type** is called a *NPN* transistor. A layer of **N-type** between two layers of **P-type** is called a *PNP* transistor.

The middle layer of the transistor is called the **base**. One of the outer
layers is more heavily doped with charge carriers than the other. This layer
is called the **emitter**. In the **N-type** semiconductor, the emitter contains conduction electrons. In the **P-type** semiconductor, the emitter contains holes.
The other layer is less heavily doped with the same charge carrier.

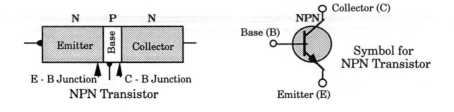

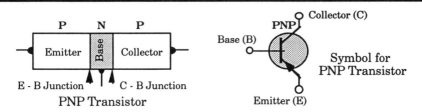

The only difference between the two types is the direction in which the current flows in the emitter.

3.1 Conduction In Transistors

In the figure below, when the voltage V_{BE} is applied, the electrons move out into the emitter lead and enter the emitter material. These electrons, together with the free electrons of the emitter material, are pushed towards the base region. The greater the forward bias, the greater the number of electrons that enter the base region. The number of holes in the base region is very small because it is so very thin and it is so slightly doped. A few electrons will find holes and combine to form a small base current. The rest of the electrons (about 98%) will not find a hole and will be pushed across the junction by the electrons newly arriving from the emitter. Once they get past the collector – base junction, they are attracted by the large collector voltage V_{CB} and become part of the collector current. They return to the battery, join up with I_B, enter V_{BE}, and start around again. As you can see, the emitter current is the sum of the base current and the collector current.

An increase in the forward bias will cause an increase in the emitter current and a corresponding increase in the collector current and the base current.

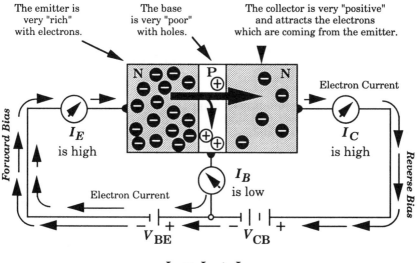

$$I_E = I_B + I_C$$

You cannot replace an *NPN* transistor with a *PNP* transistor because of polarity. In an *NPN* transistor, the collector will have to be **positive** with respect to the base. In a *PNP* transistor the collector will have to be **negative** with respect to the base. In both the base emitter must be forward biased.

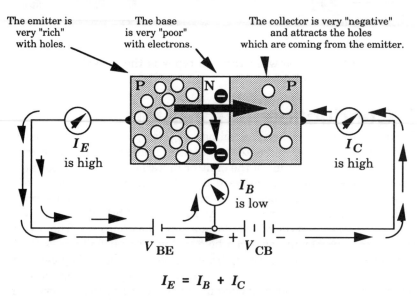

The emitter is very "rich" with holes.

The base is very "poor" with electrons.

The collector is very "negative" and attracts the holes which are coming from the emitter.

I_E is high

I_C is high

I_B is low

V_{BE} V_{CB}

$$I_E = I_B + I_C$$

A forward biased semiconductor junction has a low resistance. A reversed biased junction has a high resistance. It is this large difference in junction resistances that makes the transistor capable of power gain. Transistor amplification is based on the difference in the resistance of the input circuit to the resistance of the output circuit. This is known as the **resistance ratio.**

3.2 Amplification By Transistors

If a small varying potential difference is applied across the emitter–base circuit, an amplified copy of the wave appears at the collector output. This amplification appears because the ratio of collector current to base current is large and constant. The transistor functions this way because of doping its extremely thin structure and properly applied bias voltages. The small varying base currents produce large varying collector currents as energy is drawn from the collector battery or voltage source. *(Note the illustration at the top of the opposite page of a transistor and the corresponding "input" and "output" graphs.)*

A single transistor can amplify the input several hundred times. If there is a resistance in the collector circuit, the potential difference across the resistor can become the input to the base of a second resistor. This results in additional amplification. By using a *series* of resistors, a small input can be amplified millions of times. Transistor radios amplify the tiny amount of power from the broadcast signal to produce a much more powerful output signal at the speaker.

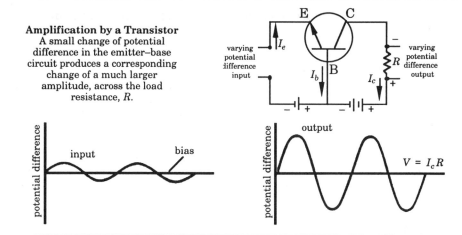

Amplification by a Transistor
A small change of potential difference in the emitter–base circuit produces a corresponding change of a much larger amplitude, across the load resistance, R.

3.3 Integrated Circuits

An **integrated circuit** is a miniaturized semiconductor circuit etched on a small silicon chip. The circuit can contain thousands of electronic parts like capacitors, diodes, resistors and transistors. An integrated circuit can do the same job as a conventional circuit. Because the conductors are so tiny, the components of the integrated circuit can only handle small amounts of current. Since they require little space, are easy to mass produce, use little energy, and produce little heat, integrated circuits have replaced vacuum tubes and changed our standard of living.

The *NPN* and the *PNP* transistors today are used in computers, calculators, entertainment, communications, and process control systems. The *NPN* transistors are favored for circuits where a fast use time is needed. Hence it turns on quickly since electrons are more mobile than holes. The *PNP* transistors are used in circuits where fast fall time is required. Hence it turns off quickly.

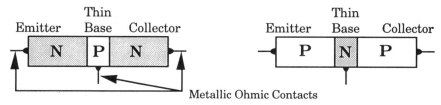

Metallic Ohmic Contacts

The major advantages of semiconductor circuits compared to vacuum tubes are as follows:

		Semi	Vacuum
1.	Energy Requirements	low	high
2.	Efficiency	25%	10%
3.	Reliability	good	fragile
4.	Miniaturization	small	larger
5.	Cost	low	high

Today, electronics is such a big field that it is often necessary to divide it into smaller subfields. One way that it can be divided is into **analog** and **digital**.

A digital device or circuit will respond to or produce an output of several (usually two) states. Most digital circuits will respond to only two conditions: high or low voltage. An analog device or circuit will recognize or produce an output of an infinite number of states. In theory, there is an infinite number of voltages possible.

Analog systems have been in use for a long time. Your height and weight are analog quantities. Computer circuits are based on digital electronics. These circuits had to be able to make highly reliable logical decisions.

Questions

1   Which is not an advantage of a semiconductor?
    (1) It is more reliable.   (3) It can be made smaller.
    (2) It uses no electricity.   (4) It uses less energy.
2   What kind of bias does a collector P–N junction have
    (1) reverse   (3) negative
    (2) forward   (4) positive
3   The bias applied across the junction of two semiconductors may be measured in
    (1) amperes   (3) ohms
    (2) volts   (4) mhos

Base your answers to questions 4 and 5 on the diagram at the right which represents a transistor circuit.

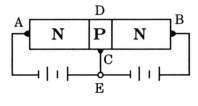

4   The current in the circuit will flow from
    (1) A to B (3) E to C
    (2) A to E (4) C to B
5   At which point is the emitter located?
    (1) A (2) B (3) C (4) E
6   In the *NPN* transistor shown in the diagram, which is the collector?
    (1) 1
    (2) 2
    (3) 3
    (4) 4
7   The maximum number of P–N junctions in a n–p–n transistor is
    (1) 1   (2) 2   (3) 3   (4) 4
8   For amplification to occur in transistors, the emitter and collector must have differences in
    (1) resistance   (3) current
    (2) temperature   (4) bias
9   The P–N junction of a transistor with reverse bias is called the
    (1) emitter   (3) donor
    (2) base   (4) collector

10 Which will occur if the battery connection to semiconductor A shown in the diagram is reversed?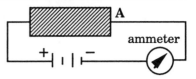
   (1) The current would decrease only if *A* is *N*- type.
   (2) The current would decrease only if *A* is *P*-type.
   (3) The current would decrease if *A* is either *N*–type or *P*–type.
   (4) The current will remain the same if *A* is either *N*–type or *P*–type.
11 The section of a transistor with forward bias is the
   (1) base   (2) emitter   (3) collector   (4) junction
12 What does the circuit diagram represent?
   (1) forward – biased diode
   (2) reverse – biased diode
   (3) forward – biased transistor
   (4) reverse – biased transistor

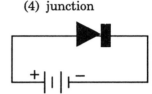

Thinking Physics

1 Light emitting diodes (LEDs) consume little energy and are inexpensive. LED segments can be used to display numbers. What devices incorporate LEDs?

2 Liquid crystal displays (LCDs) do not emit light. Under the influence of a varying potential, they operate by changing their reflectivity. What devices incorporate LCDs?

3 Alternating current can enter a diode but only direct current leaves the diode. How can this pulsating direct current be smoothed to pure DC?

4 Integrated circuits are now being produced with a half million electronic parts on an extremely thin layer of silicon. What are the limits of this technology?

Self–Help Questions

1 As the temperature of a semiconductor increases, its conductivity
   (1) decreases   (2) increases   (3) remains the same
2 Compared to the number of free electrons in an insulator of a given size, the number of free electrons in a conductor of the same size is
   (1) less   (2) greater   (3) the same
3 As the temperature of a semiconductor increases, the resistance of the semiconductor
   (1) decreases   (2) increases   (3) remains the same
4 According to accepted atomic models, metals are good conductors because their atoms have
   (1) more electrons than protons
   (2) unstable nuclei that emit electrons
   (3) negative electrons that are attracted to positive protons
   (4) a number of valence electrons that can move easily

5   Doping material that contains fewer valence electrons per atom than the
    original semiconductor is classified as
    (1) a donor   (3) an insulator
    (2) an acceptor   (4) an emitter

6   A doping agent that adds electrons to a semiconducting material is called
    (1) an N–type semiconductor   (3) a donor
    (2) a P–type semiconductor   (4) an acceptor

7   A section of P–type semiconductor has a potential difference applied
    across it, as shown in the diagram at the right. Which statement best
    describes the flow of charge through the semiconductor?
    (1) Holes flow toward the positive terminal.
    (2) Holes flow toward the negative terminal.
    (3) Protons flow toward the positive terminal.
    (4) Protons flow toward the negative terminal.

8   A hole in the atoms of a semiconductor crystalline lattice structure can be
    defined as a
    (1) free electron
    (2) negative atom
    (3) region in which an electron is located
    (4) region from which an electron has vacated

9   If silicon has a small amount of arsenic as an impurity in the crystal
    structure, the current in this N–type semiconductor is carried by
    (1) electrons   (2) neutrons   (3) protons   (4) ions

10 In an N–type semiconductor there is an excess of
    (1) ions   (2) protons   (3) atoms   (4) free electrons

11 Current carriers in a semiconductor are
    (1) electrons, only   (3) electrons and positive holes
    (2) protons, only   (4) protons and negative holes

12 What type of semiconductor is represented by
    the diagram at the right?
    (1) P–type
    (2) N–type
    (3) N–P–type
    (4) P–N–type

13 A potential difference is applied to a P–type
    semiconductor as shown in the diagram at the
    right. Current is conducted through this
    semiconductor by
    (1) positive holes moving from end A to B
    (2) positive holes moving from end B to A
    (3) conducting protons moving from end A to B
    (4) conducting protons moving from end B to A

14 A minute quantity of gallium, which has 3 valence electrons, is added to
    silicon during the process of crystallization. Which type of semiconductor
    is formed by this process?
    (1) N   (2) P   (3) N–P   (4) P–N

15 The very narrow region where P–type and N–type semiconductors are joined is called a
   (1) bias   (3) collector
   (2) valence region   (4) junction
16 The diode in the circuit diagram is considered to be
   (1) open biased   (3) forward biased
   (2) closed biased   (4) reverse biased
17 An incandescent light bulb is connected in series with a diode and a source of direct current. If the diode is forward biased, the light bulb will
   (1) light and stay lighted
   (2) flicker with a period dependent on the DC current
   (3) not light at all
   (4) light at first, then dim out
18 The diagram at the right shows the alternating current input signal to a diode. What will the output signal look like?

(1)   (2)   (4)

Base your answers to questions 19 through 23 on the diagram at the right which represents a germanium device.

19 In the circuit diagram, the diode is forward biased when electron current flows from
   (1) N to B   (3) P to N
   (2) A to P   (4) N to P
20 In the circuit diagram, the electric field barrier is represented by the region between
   (1) A and P   (2) P and N   (3) N and B   (4) B and A
21 In this type of circuit, holes migrate from
   (1) P to N   (2) A to B   (3) N to P   (4) B to A
22 According to the circuit diagram, which symbol best represents the orientation of the device shown?

   (1)   (2)   (3)   (4)

23 If an alternating potential difference is substituted for the battery, the result will be
   (1) an alternating current   (3) a steady direct current
   (2) a pulsating direct current   (4) no current flow
24 A three–element device made up of semiconductor materials is called a
   (1) collector   (3) thermistor
   (2) crystal diode   (4) transistor
25 As the emitter–base current in a transistor increases, the base–collector current
   (1) decreases   (2) increases   (3) remains the same
26 Which device is used for amplifying the flow of electrons in a solid state circuit?
   (1) diode   (2) transistor   (3) resistor   (4) switch

27 The semiconductor material located between the emitter and the collector of a transistor is called
   (1) base   (2) junction   (3) bias   (4) rectifier

28 An *N–P–N* transistor is connected as shown in the diagram at the right. Within the transistor, a forward–biased current will flow from the
   (1) collector to the base
   (2) base to the collector
   (3) base to the emitter
   (4) emitter to the base

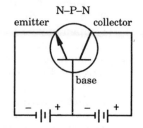

29 Which is the base of the *P–N–P* semiconductor device shown at the right?
   (1) left *P*   (3) *N*
   (2) right *P*   (4) *P–N* junction

30 Which device could be used to amplify the input signal in a microphone–speaker circuit?
   (1) diode   (2) resistor   (3) transistor   (4) solenoid

31 Which diagram below correctly represents the basic operating circuit of an *N–P–N* transistor?

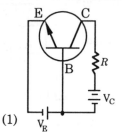

(1)

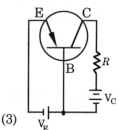

(3)

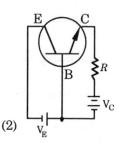

(2)

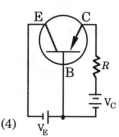

(4)

Appendices

I.   **Math Review** 282

II.   **Physics Reference Tables** 287

III. **The College Board** - Physics Achievement Test© 295

IV. **Self—Help Question Answers** 305

V.   **Glossary - Index** 307

      Sample Exams 316

Appendix 1 — Math Review

Physics describes physical phenomena and relationships in the world around us. Since mathematics is the most precise way of expressing these relationships, it is used extensively in the course. Students should have a knowledge of algebra, plane geometry, and trigonometry of right triangles.

A physics equation differs from a pure mathematical equation, because it is based on measurement. No measurement can be "exactly" correct. Each measurement consists of three parts:

First,   there is a *number* reading, read off the measuring device.
Second,   there is a *unit*.
Third,   there is a *statement* of the measurements accuracy. This accuracy is determined by the number of significant digits.

A measurement taken with a meter stick might be read as, **5.17 cm**. The last digit is an estimate to the nearest tenth of a scale division and is considered significant. This measurement has three significant digits. This reading is the same as, **0.0517 meters**. The zero (0) simply locates the decimal point. The reading still has the same accuracy and three significant digits.

When measurements combine arithmetically, the accuracy of the answer depends on both the accuracy of the measurements and the way in which they are combined.

I. Rules For Significant Figures

Students should apply the following rules for determining the number of significant digits.

1) **Initial zeros** are never significant. Since they are only used to locate a decimal point, the number 0.0**203** has three significant figures.
2) **Final zeros** are ambiguous. The student cannot tell if they are significant, except in the context of the measurement process. Thus, numbers should be written in scientific notation to eliminate ambiguity. The number **2500** could have *two, three,* or *four* significant figures.
3) **Final zeros**, after the decimal point, are only written if they are significant. Therefore, **2.50** x 10^2 has three significant digits.
4) In **adding** or **subtracting** measured quantities, first round off each quantity to the number of decimal places in the measurement having the least number of decimal places. Now, add or subtract the rounded off quantities.
5) In **multiplying** or **dividing** two measured numbers, the product or quotient should have the same number of significant digits as the least significant quantity.

II. Scientific Notation

To simplify working with very large and very small numbers, scientists usually write numbers in scientific notation.

a) The general procedure for writing any number in scientific notation is listed below:

    1) Move the decimal point of the given number to the right or left until there is only one digit to the left of the decimal point.

    2) Count the number of places you moved the decimal point. Raise ten to the power of this number. Moving the decimal point to the left is positive, while moving the decimal point to the right is negative.

b) To **add** or **subtract** numbers in scientific notation:

    1) Make sure the **powers** of ten are equal

$$2.34 \times 10^3 + 1.33 \times 10^3$$

    2) Add or subtract the numbers **before** the power of ten

$$\mathbf{2.34} \times 10^3 + \mathbf{1.33} \times 10^3$$

    3) Retain the **same power** of ten in the answer

$$3.67 \times 10^3$$

c) To **multiply** numbers in scientific notation:

    1) Multiply the two numbers **preceding** the power of ten

$$\mathbf{2.3} \times 10^4 \cdot \mathbf{1.2} \times 10^3$$

    2) Multiply the **powers** of ten by adding exponents

$$2.3 \times 10^4 \cdot 1.2 \times 10^3$$

    3) The answer includes both steps (1) and (2)

$$\mathbf{2.76 \times 10^7} = \mathbf{2.8 \times 10^7}$$

(expressed to the correct number of significant digits)

d) To **divide** numbers expressed in scientific notation:

    1) Divide the number **preceding** the power of ten in the numerator, by the number **preceding** the power of ten in the denominator

$$\mathbf{4.9} \times 10^5 / \mathbf{2.3} \times 10^2$$

    2) Subtract the **exponent** of the power of ten in the numerator. This gives the exponent of the power of ten in the answer

$$4.9 \times 10^5 / 2.3 \times 10^2$$

    3) The answer includes both steps (1) and (2)

$$\mathbf{2.13 \times 10^3} = \mathbf{2.1 \times 10^3}$$

(expressed to the correct number of significant digits)

III. Order Of Magnitude

Orders of magnitude are useful when it is desired to make a quick esti-mate of the measurements of some quantity. The power of ten that is the nearest approximation of a measurement is defined as the order of magnitude of that measurement.

Examples Of Magnitude Estimates

To obtain an order of magnitude estimate, the input data only needs to have one significant figure. A scientist or an engineer may wish to design a particular instrument. By making an order of magnitude calculation based on the sensitivity of the instrument, the properties of the material, the size of the phenomenon itself, and the feasibility of the project can be estimated.

Sometimes these questions are called Fermi questions. Examples include:

1. How many hairs does a person have on his head?
2. How many red headed female physics teachers are there in the world?
3. How many kilograms of food are consumed by each person in a lifetime?
4. How many grains of uncooked rice are there in a cup?
5. How many times does your heart beat in one average lifetime?
6. What is the volume of the Earth?
7. What thickness of rubber is worn off a car's tire during each revolution? The size of an atom is about 10^{-10} m. How many atoms does this correspond to?

IV. Dimensional Analysis

Measurements in physics always have units associated with them. Addi-tion and subtraction are only possible, if the quantities have the same units. It is possible to multiply or divide quantities with different units and come out with still another kind of unit. In these cases, the units should be treated as algebraic quantities.

In addition, mathematical expressions can be simplified by manipulating units. Units can be canceled out if they appear in both the numerator and de-nominator, or combined into other units.

Examples Of Dimensional Analysis

1. Students derived the following equation in which x refers to distance, **v** the speed, **a** the acceleration (m/s^2), **t** the time, and the subscript $(_0)$ means the quantity at t = 0. Which equation(s) could be correct?

    a) $x = vt^2 + 2at$
    b) $x = v_0t + \frac{1}{2}at^2$
    c) $x = v_0t + 2at^2$
    d) $x = va + \frac{1}{2}t^2$
    e) $x = x_0 + v_0t$

2. The speed **v** of a body is given by the equation $\mathbf{v = At^2 - Bt}$, where **t** refers to the time. What are the dimensions of **A** and **B**?

3. Using the equations below, calculate the dimensions of the constant if **F** is given in newtons, **m** is mass in kg, **r** refers to distance, **q** is the change in coulombs, and **E** is the electric field in newtons/m.

$$\mathbf{F} = \frac{\mathbf{m_1 m_2}}{\mathbf{r^2}} \qquad\qquad \mathbf{F} = \frac{\mathbf{q_1 q_2}}{\mathbf{r^2}} \qquad\qquad \mathbf{E} = \frac{\mathbf{kQ}}{\mathbf{r^2}}$$

4. If **P** and **Q** have different dimensions, which of the following operations are possible?

    a) $P + Q$ c) $P - \sqrt{Q}$ e) $P^2 Q - \sqrt{P/Q}$

    b) PQ d) $1 - P/Q$

5. A student is asked to determine a formula for the period of a pendulum. The period (**T**) of the pendulum is the time it takes to complete one swing. (Assume that **T** is proportional to **m**, *l*, and **g**.) How does **T** depend on the mass **m** of the bob, the length *l*, and the acceleration due to gravity **g**?

$$\mathbf{T} \propto \mathbf{C m^x l^y g^z}$$

Where: C is a dimensionless constant and x, y, and z are to be determined.

Since: T must be in seconds, what combination of x, y, and z will provide equality on both sides of the equation?

Note: Students are reminded to always change derived units back to fundamental units before canceling.

V. Graphing Functions

The laws of physics are often expressed in the form of a mathematical function. This function is simply a statement of how one quantity corresponds to another quantity.

Functions may be direct or inverse relationships. In direct relationships, as one quantity increases (or decreases), the other quantity does the same.

In the graph (Fig.B-1), as the time increases, the velocity increases at a constant rate. The equation of this graph is $\mathbf{y = mx}$, where **m** is the slope of the line. This is the graph of a **direct relationship**.

Fig.B-1

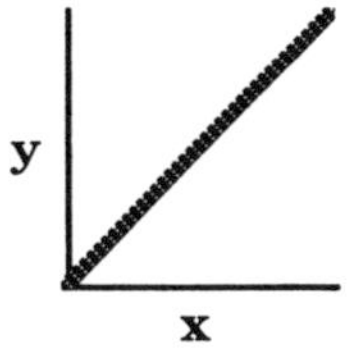

In the graph (Fig.B-2), **y** increases as the square of **x**. This function is called a **direct square relationship**. The equation is **y = kx²**. Where **k** is a constant.

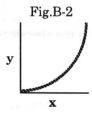

Fig.B-2

There are two different kinds of functions having an inverse relationship. In these functions, as the **x** quantity increases, the **y** quantity decreases.

The equation (Fig.B-3), is **k = xy** or **y = k/x**. This graph represents an **inverse proportionality**.

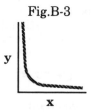

Fig.B-3

In the graph (Fig.B-4), an inverse square relationship is shown. The equation for the relationship is **y = k/x²**.

When **x** doubles, **y** becomes ¼.

When **x** triples, **y** becomes ⅑.

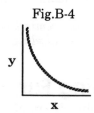

Fig.B-4

This inverse square relationship is used to express many laws in physics.

Appendix 2

Reference Tables

List of Physical Constants

Name	Symbol	Value(s)
Gravitational constant	G	6.7×10^{-11} N·m²/kg²
Acceleration due to gravity (*up to 16 km altitude*)	g	9.8 m/s²
Speed of light in a vacuum	c	3.0×10^8 m/s
Speed of sound at STP		3.3×10^2 m/s
Mass - energy relationship		1 u (amu) $= 9.3 \times 10^2$ MeV
Mass of the Earth		6.0×10^{24} kg
Mass of the Moon		7.4×10^{22} kg
Mean radius of the Earth		6.4×10^6 m
Mean radius of the Moon		1.7×10^6 m
Mean distance from Earth to Moon		3.8×10^8 m
Electrostatic constant	k	9.0×10^9 N·m²/C²
Charge of the electron (*1 elementary charge*)		1.6×10^{-19} C
One coulomb	C	6.3×10^{18} elementary charges
Electronvolt	eV	1.6×10^{-19} J
Planck's constant	h	6.6×10^{-34} J·s
Rest mass of the electron	m_e	9.1×10^{-31} kg
Rest mass of the proton	m_p	1.7×10^{-27} kg
Rest mass of the neutron	m_n	1.7×10^{-27} kg

Absolute Indices of Refraction

($\lambda = 5.9 \times 10^{-7}$ m)

Air	1.00
Alcohol	1.36
Canada Balsam	1.53
Corn Oil	1.47
Diamond	2.42
Glass, Crown	1.52
Glass, Flint	1.61
Glycerol	1.47
Lucite	1.50
Quartz, Fused	1.46
Water	1.33

Wavelengths of Light in a Vacuum

Violet	$4.0 - 4.2 \times 10^{-7}$ m
Blue	$4.2 - 4.9 \times 10^{-7}$ m
Green	$4.9 - 5.7 \times 10^{-7}$ m
Yellow	$5.7 - 5.9 \times 10^{-7}$ m
Orange	$5.9 - 6.5 \times 10^{-7}$ m
Red	$6.5 - 7.0 \times 10^{-7}$ m

Values of Trigonometric Functions

Angle	Sine	Cosine	Angle	Sine	Cosine
1°	.0175	.9998	46°	.7193	.6947
2°	.0349	.9994	47°	.7314	.6820
3°	.0523	.9986	48°	.7431	.6691
4°	.0698	.9976	49°	.7547	.6561
5°	.0872	.9962	50°	.7660	.6428
6°	.1045	.9945	51°	.7771	.6293
7°	.1219	.9925	52°	.7880	.6157
8°	.1392	.9903	53°	.7986	.6018
9°	.1564	.9877	54°	.8090	.5878
10°	.1736	.9848	55°	.8192	.5736
11°	.1908	.9816	56°	.8290	.5592
12°	.2079	.9781	57°	.8387	.5446
13°	.2250	.9744	58°	.8480	.5299
14°	.2419	.9703	59°	.8572	.5150
15°	.2588	.9659	60°	.8660	.5000
16°	.2756	.9613	61°	.8746	.4848
17°	.2924	.9563	62°	.8829	.4695
18°	.3090	.9511	63°	.8910	.4540
19°	.3256	.9455	64°	.8988	.4384
20°	.3420	.9397	65°	.9063	.4226
21°	.3584	.9336	66°	.9135	.4067
22°	.3746	.9272	67°	.9205	.3907
23°	.3907	.9205	68°	.9272	.3746
24°	.4067	.9135	69°	.9336	.3584
25°	.4226	.9063	70°	.9397	.3420
26°	.4384	.8988	71°	.9455	.3256
27°	.4540	.8910	72°	.9511	.3090
28°	.4695	.8829	73°	.9563	.2924
29°	.4848	.8746	74°	.9613	.2756
30°	.5000	.8660	75°	.9659	.2588
31°	.5150	.8572	76°	.9703	.2419
32°	.5299	.8480	77°	.9744	.2250
33°	.5446	.8387	78°	.9781	.2079
34°	.5592	.8290	79°	.9816	.1908
35°	.5736	.8192	80°	.9848	.1736
36°	.5878	.8090	81°	.9877	.1564
37°	.6018	.7986	82°	.9903	.1392
38°	.6157	.7880	83°	.9925	.1219
39°	.6293	.7771	84°	.9945	.1045
40°	.6428	.7660	85°	.9962	.0872
41°	.6561	.7547	86°	.9976	.0698
42°	.6691	.7431	87°	.9986	.0523
43°	.6820	.7314	88°	.9994	.0349
44°	.6947	.7193	89°	.9998	.0175
45°	.7071	.7071	90°	1.000	.0000

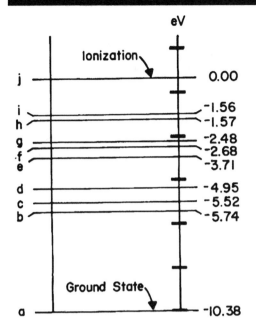

A few energy levels for the mercury atom

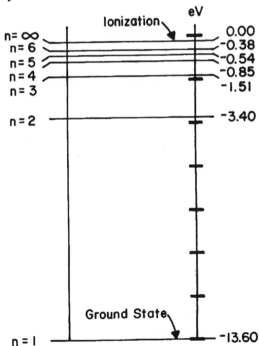

Energy levels for the hydrogen atom

Heat Constants

	Specific Heat (average) (kJ / kg·°C)	Melting Point (°C)	Boiling Point (°C)	Heat of Fusion (kJ / kg)	Heat of Vaporization (kJ/ kg)
Alcohol (ethyl)	2.43 (liq)	-117	79	109	855
Aluminum	0.90 (sol)	660	2467	396	10500
Ammonia	4.71 (liq)	-78	-33	332	1370
Copper	0.39 (sol)	1083	2567	205	4790
Iron	0.45 (sol)	1535	2750	267	6290
Lead	0.13 (sol)	328	1740	25	866
Mercury	0.14 (liq)	-39	357	11	295
Platinum	0.13 (sol)	1772	3827	101	229
Silver	0.24 (sol)	962	2212	105	2370
Tungsten	0.13 (sol)	3410	5660	192	4350
Water - Ice	2.05 (sol)	0	---	334	---
Water - Water	4.19 (liq)	---	100	---	2260
Water - Steam	2.01 (gas)	---	---	---	---
Zinc	0.39 (sol)	420	907	113	1770

URANIUM DISINTEGRATION SERIES

Atomic Number and Chemical Symbol

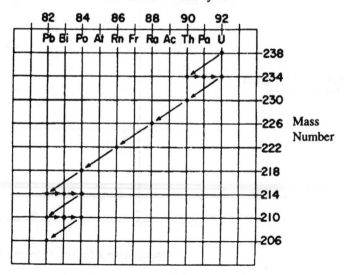

$$\bar{v} = \frac{\Delta s}{\Delta t}$$

$$\bar{v} = \frac{v_f + v_i}{2}$$

$$\bar{a} = \frac{\Delta v}{\Delta t}$$

$$\Delta s = v_i \Delta t + \frac{1}{2} a (\Delta t)^2$$

$$v_f^{\,2} = v_i^{\,2} + 2a \, \Delta s$$

$$F = ma$$

$$w = mg$$

$$F = \frac{G m_1 m_2}{r^2}$$

$$p = mv$$

$$J = F \Delta t$$

$$F \Delta t = m \, \Delta v$$

Mechanics

a	=	acceleration
r	=	distance between centers
F	=	force
g	=	acceleration due to gravity
G	=	universal gravitation constant
J	=	impulse
m	=	mass
p	=	momentum
Δs	=	displacement
t	=	time
v	=	velocity
w	=	weight

Energy

$$W = F \Delta s$$

$$P = \frac{W}{\Delta t} = \frac{F \Delta s}{\Delta t} = F \bar{v}$$

$$\Delta PE = mg \, \Delta h$$

$$KE = \frac{1}{2} mv^2$$

$$F = kx$$

$$PE_s = \frac{1}{2} kx^2$$

F	=	force
g	=	acceleration due to gravity
h	=	height
k	=	spring constant
KE	=	kinetic energy
m	=	mass
P	=	power
PE	=	potential energy
PE_s	=	potential energy stored in a spring
Δs	=	displacement
t	=	time
v	=	velocity
W	=	work
x	=	change in spring length from the equilibrium position

Geometric Optics

$$\frac{1}{d_o} + \frac{1}{d_i} = \frac{1}{f}$$

$$\frac{S_o}{S_i} = \frac{d_o}{d_i}$$

d_i	=	image distance
d_o	=	object distance
f	=	focal length
S_i	=	image size
S_o	=	object size

Electricity and Magnetism

$$F = \frac{kq_1 q_2}{r^2}$$

$$E = \frac{F}{q}$$

$$V = \frac{W}{q}$$

$$E = \frac{V}{d}$$

$$I = \frac{\Delta q}{\Delta t}$$

$$R = \frac{V}{I}$$

$$P = VI = I^2 R = \frac{V^2}{R}$$

$$W = Pt = VIt = I^2 Rt$$

d	=	separation of parallel plates
r	=	distance between centers
E	=	electric field intensity
F	=	force
I	=	current
k	=	electrostatic constant
P	=	power
q	=	charge
R	=	resistance
t	=	time
V	=	electric potential difference
W	=	energy

Series Circuits:

$$I_t = I_1 = I_2 = I_3 = \ldots$$

$$V_t = V_1 + V_2 + V_3 + \ldots$$

$$R_t = R_1 + R_2 + R_3 + \ldots$$

Parallel Circuits:

$$I_t = I_1 + I_2 + I_3 + \ldots$$

$$V_t = V_1 = V_2 = V_3 = \ldots$$

$$\frac{1}{R_t} = \frac{1}{R_1} + \frac{1}{R_2} + \frac{1}{R_3} + \ldots$$

$$T = \frac{1}{f}$$

$$v = f\lambda$$

$$n = \frac{c}{v}$$

$$\sin \theta_c = \frac{1}{n}$$

$$n_1 \sin \theta_1 = n_2 \sin \theta_2$$

$$n_1 v_1 = n_2 v_2$$

$$\frac{\lambda}{d} = \frac{x}{L}$$

Wave Phenomena

c	=	speed of light in a vacuum
d	=	distance between slits
f	=	frequency
L	=	distance from slit to screen
n	=	index of absolute refraction
T	=	period
v	=	speed
x	=	distance from central maximum
		to first order maxium
λ	=	wavelength
θ	=	angle
θ_c	=	critical angle of incidence
		relative to air

$$E = mc^2$$

$$m_f = \frac{m_i}{2^n}$$

Nuclear Energy

c	=	speed of light in a vacuum
E	=	energy
m	=	mass
n	=	number of half-lives

$$W_o = hf_o$$

$$E_{photon} = hf$$

$$KE_{max} = hf - W_o$$

$$p = \frac{h}{\lambda}$$

$$E_{photon} = E_i - E_f$$

Modern Physics

c	=	speed of light in a vacuum
E	=	energy
f	=	frequency
f_o	=	threshold frequency
h	=	Planck's constant
KE	=	kinetic energy
p	=	momentum
W_o	=	work function
λ	=	wavelength

$$v_{iy} = v_i \sin \theta$$

$$v_{ix} = v_i \cos \theta$$

$$a_c = \frac{v^2}{r}$$

$$F_c = \frac{mv^2}{r}$$

Motion in a Plane

a_c = centripetal acceleration

F_c = centripetal force

m = mass

r = radius

v = velocity

θ = angle

$$Q = mc\, \Delta T_c$$

$$Q_f = mH_f$$

$$Q_v = mH_v$$

Internal Energy

c =   specific heat

H_f =   heat of fusion

H_v =   heat of vaporization

m =   mass

Q =   amount of heat

T_c =   Celsius temperature

$$F = qvB$$

$$\frac{N_p}{N_s} = \frac{V_p}{V_s}$$

$$V_p I_p = V_s I_s$$
(ideal)

$$\% \text{ Efficiency} = \frac{V_s I_s}{V_p I_p} \times 100$$

$$V = Bl v$$

Electromagnetic Applications

B =   flux density

$F \cdot$ =   force

I_p =   current in primary coil

I_s =   current in secondary coil

N_p =   number of turns of primary coil

N_s =   number of turns of secondary coil

q =   charge

v =   velocity

V_p =   voltage of primary coil

V_s =   voltage of secondary coil

ℓ =   length of conductor

V =   electric potential difference

Appendix 3 The College Boards

The Physics Achievement Test©

(Excerpts from: *The College Board Achievement Tests in Science©*)

The Physics Achievement Tests consists of 75 multiple-choice questions. The test assumes that you have had a one-year introductory course in physics and that the course content was at a level suitable for college preparation. Questions appearing in the test have been tried out on college students who are taking an introductory physics course. The questions have also been approved by a committee of high school and college physics teachers appointed by the College Board.

The approximate percentages of the questions on each major topic in the Physics Achievement Test are listed in the chart below. The test emphasizes the topics that are covered in most high school courses. However, because high school courses differ, both in the percentage of time devoted to each major topic and in the specific subtopics covered, you may encounter questions on topics with which you are not familiar.

The test is not based on any one textbook or instructional approach, but concentrates on the common core of material found in most texts. You should be able to recall and understand the major concepts of physics and to apply physical principles to solve specific problems. You also should be able to organize and interpret results obtained by observation and experimentation and to draw conclusions or make inferences from experimental data.

Laboratory experience is a significant factor in developing reasoning and problem-solving skills. Although laboratory skills can be tested only in a limited way in a standardized test, there are occasional questions that ask you to interpret laboratory data. Reasonable laboratory experiences is an asset in helping you prepare for the Physics Achievement Test.

The Physics Achievement Test assumes that you understand simple algebraic, trigonometric, and graphical relationships and the concept of ratio and proportion, and that you can apply these concepts to word problems.

You will not be allowed to use an electronic calculator during the test.

Numerical calculations are not emphasized and are limited to simple arithmetic. In this test, metric units are used.

Content Of The Test

Topics Covered   *Approximate Percentage of Test*

I. Mechanics   **40 %**
   A. Kinematics (such as velocity, acceleration, motion in one dimension, motion of projectiles, and circular motion)
   B. Dynamics (such as force, torque, Newton's laws, centripetal force, and statics)
   C. Energy and Momentum (such as potential and kinetic energy, work, power, impulse, angular momentum, and conservation laws)
   D. Other (such as gravity and orbits of planets and satellites, simple harmonic motion, and pressure)

II. Electricity and Magnetism   **20 %**
   A. Electrostatics (such as Coulomb's law, and electric field & potential)
   B. Circuits (such as Ohm's law, Joule's law, and direct-current circuits with resistors and capacitors)
   C. Electromagnetism (such as production and effects of magnetic fields, and electromagnetic induction)

III. Optics and Waves   **20 %**
   A. General wave properties and sound (such as wave speed, frequency, wavelength, and Doppler effect)
   B. Geometrical Optics (such as reflection, refraction, and image formation in mirrors and lenses)
   C. Physical Optics (such as interference, diffraction, and polarization)

IV. Heat, Kinetic Theory, and Thermodynamics   **10 %**
   A. Thermal Properties (such as mechanical equivalent of heat, temperature, specific & latent heats, thermal expansion, & heat transfer)
   B. Kinetic Theory (such as ideal gas law from molecular properties)
   C. Thermodynamics (such as first and second laws, internal energy, and heat engine efficiency)

V. Modern Physics   **10 %**
   A. Atomic (such as the Rutherford and Bohr models, atomic energy levels, and atomic spectra)
   B. Nuclear (such as radioactivity and nuclear reactions)
   C. Relativity and other (such as photons, photoelectric effect, mass-energy equivalence, and limiting velocity)

Some questions are of a general nature and overlap several topics, illustrating the unified structure of physics.

Skills Specifications   *Approximate Percentage of Test*

Recall   **20-33 %**
   (generally involves only remembering the desired information)

Single-concept problem   **40-53 %**
   (recall and use of a single physical relationship)

Multiple-concept problem   **20-33 %**
   (recall and use of 2 or more physical relationships that must be combined)

75 Questions: Time - 60 minutes

Questions Used In The Test

A proven way to do well in standardized tests is the "three - reading approach."

(1) On the first reading of the exam questions, answer those questions which are simple recall or one-step concept questions. Remember that the difficulty of questions is irregular, and many easy questions are at the end of the exam. At this time, identify those questions which you can answer but will take some time and those questions which are unfamiliar.

(2) On the second reading, concentrate on those questions that you identified as questions that you know how to solve.

(3) On the third reading, tackle the most difficult questions.

Your raw score is determined by taking the number of correct answers minus one fourth of the number wrong answers. That raw score is then scaled for a particular test with scores ranging from 200 to 800.

There is no penalty for unanswered questions. If you can eliminate one or more answers to a question, it is worth guessing. If you have no knowledge of the answer, do not guess.

There are two types of questions that are commonly used in the exam:

(1) The **classification question** begins with five choices followed by several questions in a set. The five choices can be used for each question. Any choice may be reused.

(2) The **five choice completion question** can be either an incomplete statement or a question. It is often used to allow the possibility for more than one correct answer. Read all of the choices carefully because although one choice may be correct, another choice may be better since it may include additional correct answers.

Physics Achievement Additional Material

The main body of this review text covers all of the topics tested on the College Board Achievement Test except for: **Capacitance, Relativity, Air Columns,** and the **Laws of Strings.** The following section will review and summarize these additional topics.

Capacitance

A capacitor consists of two conductors separated by an insulator or dielectric.

$$\text{C of a capacitor} = \frac{\textbf{charge q on either plate}}{\textbf{potential difference V between plates}}$$

$$\text{C (farads)} = \frac{\text{q (coulombs)}}{\text{V (volts)}}$$

A capacitor has capacitance of 1 farad (f) if it requires 1 coulomb of charge per volt of potential difference between its conductors. Since 1 farad is a very large amount of capacitance, one microfarad ($1\mu\,f$ or $1 \times 10^{-6}\,f$) is a convenient submultiple.

Parallel Plate Capacitor

The capacitance of a parallel plate capacitor with two large plates, each having an area A square meter and a distance between the plates d meters, is

$$C = K e_o \frac{A}{d}$$

Where: K is dielectric constant *(unitless, for a vacuum $K = 1$)*

$$e_o = 8.85 \times 10^{-12} \; Fm^{-1}$$

As different materials are placed between the plates, the dielectric constant in the equation changes. The field between the plates is uniform.

Capacitors In Parallel and Series

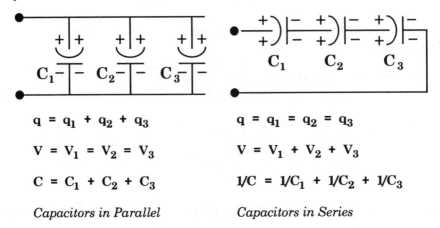

$$q = q_1 + q_2 + q_3 \qquad\qquad q = q_1 = q_2 = q_3$$

$$V = V_1 = V_2 = V_3 \qquad\qquad V = V_1 + V_2 + V_3$$

$$C = C_1 + C_2 + C_3 \qquad\qquad 1/C = 1/C_1 + 1/C_2 + 1/C_3$$

Capacitors in Parallel   *Capacitors in Series*

Energy Stored In A Capacitor

In a capacitor, potential difference is proportional to charge ($V = q/c$). While a capacitor is being charged, the charge builds up from an initial value of zero (0) to a final value of q. The potential builds up from an initial value of zero (0) to a final value of V with the average potential difference $\frac{1}{2}V$. The work required to transfer a total charge q through an average potential difference $\frac{1}{2}V$ is

$$W = q \, (\tfrac{1}{2}V)$$

The electrical energy stored in the charged capacitor is:

$$W = \tfrac{1}{2}qV = \tfrac{1}{2}C V^2 = \tfrac{1}{2}q^2/C$$

Using $q = CV$

Where W is in joules if q is in coulombs and C is in farads.

Vibrating Air Columns

Air columns can be made to resonate with vibrating objects. There are two types of columns, **open** and **closed**. Both columns are able to produce resonance.

In a closed tube, the sound wave resonates when a node appears at the closed end, and an antinode at the open end. The sound wave is drawn as a transverse wave for illustration purposes. Sound is a longitudinal wave. The wave is reflected from the fixed end and a 180° phase change occurs. There is no phase change at the open end. During resonance, a standing wave is set up. The minimum distance from a node to an antinode is $\frac{1}{4}$ wavelength. The wavelength of the sound is equal to four times the length of the closed air column that resonates.

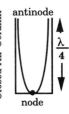

Example: Calculate the wavelength of the sound wave produced by a closed tube 0.30m long.

    Given: closed tube length is 0.30 m

    Find: λ

    Solution: $(\frac{1}{4})(\lambda)$ = 0.30 m

               λ = 1.2 m

In an open tube, the sound wave resonates when an antinode appears at each open end. The sound wave is drawn as a transverse wave for illustration purposes. Sound is a longitudinal wave. There is no phase change at the end of the open tube. During resonance a standing wave is set up. The minimum distance from one antinode to another antinode is $\frac{1}{2}$ a wavelength. Therefore, the wavelength of the sound produced is twice the length of the open air column.

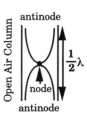

Example: Calculate the wavelength of the sound wave produced by a 0.5 m air column open at both ends.

    Given: open air column, length = 0.5 m
    Find: wavelength
    Solution: $\frac{1}{2}(\lambda)$ = 0.5 m

               λ = 1 m

Vibrating String Instruments

Strings can be made to vibrate at any length. As the length of a string is decreased, (usually by placing a finger on the string) the wavelength decreases, but the pitch or frequency increases. The frequency of a vibrating string is inversely proportional to its length. As the tension on the string is increased, the frequency or pitch of the string will increase. As the string is replaced with a thicker (more massive) string, the frequency is lowered.

Summary of the Law of Strings
Length: frequency inversely proportional
Mass: frequency inversely proportional
Tension: frequency directly proportional as square root.

Relativity

1. **The special theory of relativity** is concerned with bodies which are in *uniform* motion relative to each other. Einstein's postulates are:
    a) All motion is relative. It is therefore impossible to determine absolute motion.
    b) The speed of light in free space is constant, independent of the motion of the source or the motion of the observer.

2. **Mass variation** is predicted by the special theory of relativity. The mass of a moving body is larger than its mass when at rest, according to the formula:

$$m = \frac{m_o}{\sqrt{1 - \frac{v^2}{c^2}}}$$

Where: m = mass moving
        v = speed of body relative to observer
        m_o = mass at rest
        c = speed of light

3. **Mass—Energy.** In general, mass and energy are interchangeable according to the equation:

$$E = mc^2$$

If 1 kg of mass is annihilated, it would be replace by $(3 \times 10^8)^2$ joules of energy.

At ordinary speeds where v is small, the ratio v^2/c^2 is practically zero, and the measurements of mass, length, and time are approximately the same values (as Newtonian physics predicts). We intuitively add velocity vectors and get the correct answer.

For example, 10m/s North and 10m/s North will yield a resultant of 20m/s North. On the other hand, 0.9c North (or $9/10$ the speed of light) and 0.9c North will have a resultant less than c when added relativistically.

Physics Achievement Test Questions

The following questions have been selected (with permission from the Educational Testing Service) from a published Physics Achievement Test.

These questions have been chosen because we feel that the questions are
(1) *"typical" of the style and format of the actual achievement test questions that you will be expected to answer correctly.*
(2) *"typical" in respect to their difficulty and usage of graphs, illustrations, and diagrams.*
(3) *"typical" as to the broad range of material that will be covered on the actual achievement test, including areas not included on the New York State Regents Physics Examination.*
(4) *"typical" of the most often missed questions (generally, over 50%).*

Achievement Test Questions For Practice

Questions 1-2 refer to the following three diagrams each showing a light ray from an object at O that passes through a lens having focal points F. Assume that the thin lens approximation is used in drawing the diagrams.

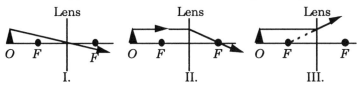

I.   II.   III.

(A) I only   (C) I and II only   (E) II and III only
(B) III only   (D) I and III only

1   Which of the diagrams above can be correct for a converging lens?
2   Which of the diagrams above can be correct for a diverging lens?

3   Which of the following statements is (are) true in the region of a positive point charge?
   I.   The electric field is directed toward the point charge.
   II.   A negatively charged body experiences a force directed toward the point charge.
   III.   The force on a second point charge is inversely proportional to the square of its distance from the first point charge.

(A) I only   (C) I and II only   (E) I, II, and III
(B) III only   (D) II and III only

4   A block of mass M slides down a plane inclined at an angle of 30° to the horizontal as shown at the right. Which of the following diagrams shows the sources and directions of the three forces acting on the block in any real situation?

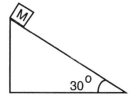

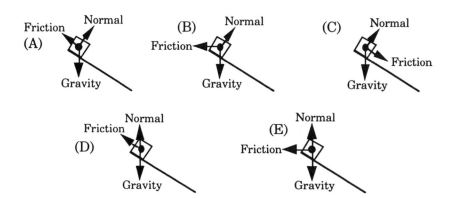

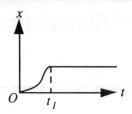

5　The graph above shows position x　as a function of time t　for a particle moving along the x -axis. Which of the following graphs of velocity v　as a function of time t　describes the motion of this particle?

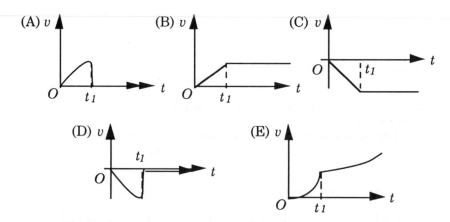

6　A 15-kilogram wagon is pulled to the right with a force of 45 newtons. The wagon accelerates at 2 meters per second squared. What is the net (unbalanced) force accelerating the wagon?
(A) 3 N
(C) 22 ½ N
(E) 45 N
(B) 15 N
(D) 30 N

7　A 75-kilogram astronaut sits in a 1,000-kilogram satellite as it orbits the Earth. In an inertial frame of reference, the astronaut and the satellite have the same
(A) acceleration
(B) gravitational force exerted on them
(C) kinetic energy
(D) momentum
(E) potential energy (assume zero at the Earth's surface)

8　A curved wave front could result from all of the following EXCEPT
(A) a disturbance emanating from a point source
(B) the passage of a plane wave through a very small opening
(C) the reflection of a plane wave from a concave surface
(D) the reflection of a plane wave from a convex surface
(E) the reflection from a parabolic surface of a circular wave originating at the focus of the parabola

Questions 9-10
   Two isolated point particles of masses M_1 and M_2
have charges of magnitude Q_1 and Q_2, respectively.
The gravitational force between the particles is bal-
anced by the electrostatic force between them.

Q_1 Q_2

•   •

M_1 M_2

9   Which of the following is true about the signs of the charges?
   (A) Q_1 must be positive and Q_2 must be negative.
   (B) Q_1 must be negative and Q_2 must be positive.
   (C) Both Q_1 and Q_2 must be positive.
   (D) Both Q_1 and Q_2 must be negative.
   (E) Q_1 and Q_2 must have the same sign but it does not matter whether
      they are both positive or both negative.

10 If G is the universal gravitational constant and k is the Coulomb's law
   constant, which of the following must be true about the magnitudes of the
   charges?

   (A) $Q_1 = Q_2$ (B) $\dfrac{Q_1}{Q_2} = \dfrac{M_1}{M_1}$ (C) $\dfrac{Q_1}{Q_2} = \dfrac{M_2}{M_2}$

     (D) $Q_1 Q_2 = \dfrac{G}{k} M_1 M_2$ (E) $Q_1 Q_2 = \dfrac{k}{G} M_1 M_2$

11 The illustration at the right represents a photograph
   of the spiral path of an electron in a bubble chamber.
   Which of the following is a possible direction of the
   magnetic field?
   (A) Counterclockwise in the plane of the paper.
   (B) Clockwise in the plane of the paper.
   (C) From left to right in the plane of the paper.
   (D) From right to left in the plane of the paper.
   (E) Perpendicular to the plane of the paper.

12 A source emits sound of frequency f_o in air. A person moving toward this
   source at a uniform speed would measure a frequency that is
   (A) constant and lower than f_o (D) continually decreasing
   (B) constant and the same as f_o (E) continually increasing
   (C) constant and higher than f_o

Questions 13-14
   An object with mass m and speed v_o directed to the right strikes a wall
   and rebounds with speed v_o directed to the left.

13 The change in the object's kinetic energy is
   (A) $-mv_o^2$ (C) zero   (E) mv_o^2
   (B) $-\tfrac{1}{2} mvo^2$ (D) $\tfrac{1}{2} mv_o^2$

14 The change in the object's momentum is
    (A) $2mv_o$ directed to the left
    (B) mv_o directed to the left
    (C) zero
    (D) mv_o directed to the right
    (E) $2mv_o$ directed to the right

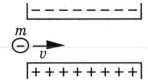

15 A negatively charged particle of mass m and speed v is projected into an
    electric field between the plates of a charged capacitor as shown above.
    Which of the following vectors best represents the direction of the acceler-
    ation of the charged particle?

Physics Achievement Test Answers

Answers		Location of Explanation			
1.	C	Unit 4	Section III	Pages	207-210
2.	D	Unit 4	Section III	Pages	207-210
3.	D	Unit 3	Section II	Pages	115-116
4.	A	Unit 1	Section III	Pages	25, 33-34
5.	A	Unit 1	Section II	Pages	16-20
6.	D	Unit 1	Section III	Pages	25-26
7.	A	Key word is same.			
		With the exception of choice (A), all other quantities involve mass.			
8.	E	Unit 4	Section III	Pages	175-178
9.	E	Unit 3	Section I	Page	112
10.	D	Unit 1	Section III	Pages	30, 112
11.	E	Unit 5.1	Section II	Page	158
12.	C	Unit 4	Section II	Page	177
13.	C	Unit 2	Section III	Page	78
14.	A	Unit 1	Section IV	Pages	36-40
15.	D	Unit 3	Section II	Pages	115, 118

Appendix 4 — Self-Help Answers

Col. 1 - **Question Number;**   Col. 2 - **Answer;**   Col. 3 - **Page Reference**

Unit 1
pg. 45

#	Ans	Pg
1.	3	10
2.	4	10
3.	3	25
4.	2	9
5.	4	9
6.	1	11
7.	3	12
8.	3	11
9.	1	12
10.	2	12
11.	3	11
12.	3	17
13.	2	16
14.	1	19
15.	3	16
16.	3	19
17.	2	19
18.	3	19
19.	4	18
20.	4	26
21.	4	18
22.	1	19
23.	2	18
24.	2	19
25.	1	17
26.	3	17
27.	4	18
28.	2	20
29.	4	26
30.	3	26
31.	1	26
32.	1	19
33.	4	33
34.	1	33
35.	4	33
36.	1	33
37.	4	36
38.	4	26
39.	3	30
40.	3	30
41.	2	30
42.	4	36
43.	1	30
44.	4	31
45.	2	25
46.	3	37
47.	3	39
48.	1	36
49.	1	39
50.	3	36

Optional Unit 1
pg. 67

#	Ans	Pg
1.	1	51
2.	4	51
3.	4	52
4.	2	55
5.	1	54
6.	1	56
7.	1	57
8.	4	57
9.	2	57
10.	4	57
11.	3	57
12.	3	57
13.	4	63
14.	3	63
15.	3	63
16.	3	63
17.	2	63
18.	1	64
19.	3	63
20.	1	57
21.	2	63
22.	3	63
23.	3	62
24.	4	62
25.	3	62

Unit 2
pg. 86

#	Ans	Pg
1.	3	83
2.	4	71
3.	2	72
4.	3	72
5.	3	72
6.	2	75
7.	2	75
8.	3	78
9.	4	78
10.	3	78
11.	3	82
12.	3	78
13.	3	78
14.	4	82
15.	3	78
16.	4	79
17.	1	79
18.	1	79
19.	2	17
20.	2	19
21.	4	78
22.	3	72
23.	3	78
24.	2	83
25.	3	26

Optional Unit 2
pg. 105

#	Ans	Pg
1.	4	90
2.	3	94
3.	2	89
4.	4	89
5.	1	89
6.	3	90
7.	3	90
8.	2	10
9.	1	103
10.	4	98
11.	3	98
12.	3	98
13.	1	96
14.	3	94
15.	3	95
16.	2	95
17.	2	92
18.	1	93
19.	2	93
20.	1	93
21.	2	93
22.	2	103
23.	3	94
24.	4	94
25.	3	100
26.	1	99
27.	3	100
28.	4	103
29.	1	100
30.	2	103

Unit 3
pg. 146

#	Ans	Pg
1.	4	109
2.	3	111
3.	4	110
4.	3	109
5.	4	112
6.	2	112
7.	4	110
8.	3	111
9.	3	110
10.	2	112
11.	2	287
12.	2	115
13.	3	116
14.	3	117
15.	1	117
16.	3	116
17.	3	116
18.	3	117
19.	2	117
20.	4	116
21.	2	118
22.	2	116
23.	3	116
24.	3	116
25.	1	118
26.	3	116
27.	2	118
28.	3	118
29.	2	115
30.	2	111
31.	2	115
32.	3	287
33.	3	123
34.	1	123
35.	4	121
36.	4	122
37.	2	121
38.	4	129
39.	2	129
40.	1	134
41.	3	126
42.	4	122
43.	3	122
44.	3	126
45.	3	131
46.	1	126
47.	2	126
48.	2	131
49.	1	134
50.	4	136
51.	1	134
52.	1	134
53.	4	134
54.	2	141
55.	2	135
56.	1	135

Col. 1 - **Question Number;**   Col. 2 - **Answer;**   Col. 3 - **Page Reference**

Optional Unit 3
pg. 170

Q	Ans	Pg
1.	4	153
2.	3	154
3.	2	154
4.	4	154
5.	2	154
6.	1	154
7.	3	154
8.	4	141
9.	3	158
10.	3	160
11.	4	157
12.	4	158
13.	4	158
14.	2	157
15.	1	159
16.	4	158
17.	2	157
18.	3	158
19.	3	158
20.	2	158
21.	2	158
22.	4	158
23.	3	158
24.	2	158
25.	2	158
26.	1	158
27.	1	163
28.	1	155
29.	3	155
30.	2	163
31.	3	163
32.	3	166
33.	4	166
34.	1	166
35.	2	166
36.	3	166
37.	1	166
38.	4	166
39.	2	166

Unit 4
pg. 196

Q	Ans	Pg
1.	2	175
2.	1	176
3.	4	86
4.	4	177
5.	2	186
6.	2	176
7.	2	177
8.	3	177
9.	2	185
10.	4	185
11.	4	185
12.	2	181
13.	4	176
14.	1	176
15.	1	181
16.	1	177
17.	1	177
18.	2	177
19.	2	177
20.	3	176
21.	4	190
22.	2	190
23.	1	176
24.	3	181
25.	2	186
26.	1	177
27.	2	186
28.	1	10
29.	4	186
30.	3	186
31.	4	177
32.	3	184
33.	4	176
34.	1	184
35.	1	187
36.	3	187
37.	3	185
38.	2	186
39.	4	186
40.	1	186
41.	3	190
42.	2	101
43.	2	191
44.	3	191
45.	4	191
46.	1	191
47.	1	191

Optional Unit 4
pg. 215

Q	Ans	Pg
1.	3	202
2.	1	202
3.	4	202
4.	2	202
5.	3	204
6.	2	203
7.	4	202
8.	2	203
9.	1	204
10.	1	203
11.	2	203
12.	3	204
13.	4	208
14.	3	209
15.	2	208
16.	1	208
17.	4	208
18.	1	209
19.	2	209
20.	2	209
21.	1	209
22.	1	207
23.	4	209
24.	3	204
25.	2	209
26.	3	210
27.	4	208

Unit 5
pg. 232

Q	Ans	Pg
1.	2	220
2.	4	220
3.	2	220
4.	4	220
5.	1	221
6.	2	221
7.	1	221
8.	1	221
9.	2	221
10.	3	220
11.	4	221
12.	2	221
13.	1	220
14.	4	225
15.	1	224
16.	1	224
17.	3	225
18.	3	225
19.	3	225
20.	3	289
21.	1	226
22.	3	226
23.	3	142
24.	2	229

Optional Unit 5.1
pg. 253

Q	Ans	Pg
1.	4	235
2.	2	235
3.	2	235
4.	3	235
5.	1	238
6.	2	238
7.	3	238
8.	3	250
9.	2	242
10.	1	240
11.	3	242
12.	4	236
13.	2	236
14.	2	238
15.	3	242
16.	4	239
17.	3	239
18.	2	240
19.	3	242
20.	2	243
21.	1	243
22.	2	243
23.	4	243
24.	4	243
25.	3	243
26.	1	241
27.	1	241
28.	1	242
29.	2	242
30.	1	242
31.	4	243
32.	3	243
33.	4	243
34.	4	236
35.	4	248
36.	2	248
37.	1	242
38.	4	249
39.	2	248
40.	2	248
41.	4	249
42.	1	249
43.	3	243
44.	2	248
45.	2	247

Optional Unit 5.2
pg. 277

Q	Ans	Pg
1.	2	257
2.	2	260
3.	1	258
4.	4	260
5.	2	262
6.	3	264
7.	2	262
8.	4	262
9.	1	264
10.	4	264
11.	3	261
12.	2	265
13.	2	265
14.	2	263
15.	4	267
16.	3	269
17.	1	269
18.	2	270
19.	4	269
20.	2	269
21.	1	269
22.	3	267
23.	2	260
24.	4	262
25.	3	273
26.	2	272
27.	1	272
28.	3	274
29.	3	273
30.	3	272
31.	1	274

Appendix 5
Glossary And Index

Absolute zero (89): The theoretical lowest limit of temperature at which the kinetic energy of the molecule is at a minimum; 0°K or -273°C.

Absolute index of refraction (186): The ratio of the speed of light in a vacuum to the speed of light in an optical substance.

Absorption spectrum (193): A continuous spectrum interrupted by dark bands or lines characteristic of the cool, low temperature medium through which the radiation passes. The colors removed are the same as those which the medium would emit as a line spectrum.

Acceleration (17): A vector quantity which represents the time rate of change of velocity.

Accelerator (249): Device used to increase the speed of charged particles.

Alpha particles (225): Helium nucleus, consisting of two protons and two neutrons.

Alpha scattering: see Rutherford's Model.

Alternating current (164): Electric current which reverses its direction in a regular, periodic manner.

Ammeter (121, 154): A low resistance meter used to measure current; always installed in series.

Ampere (121, 137): The SI Unit of current; one coulomb per second.

Amplitude (176): The magnitude of the maximum displacement of a vibrating particle from its equilibrium position.

Amu (236): A unit of mass in atomic physics described as $1/12$ the mass of the carbon-12 nucleus. One amu equals 930 MEV.

Angle of incidence (185): The angle between the incoming ray and a line drawn perpendicular to the surface which the ray strikes.

Angle of reflection (184): The angle between the reflected ray and the line drawn perpendicular to the surface from which the ray is reflected.

Antinode (181): Positions on a standing wave which vibrates at maximum amplitude.

Antiparticles (251): Particles of equal masses but opposite charge of their counterparts (such as a positron and an electron).

Artificial transmutatuion (243): A change from one nuclide to another due to bombardment by sub-atomic particles.

Back emf (153, 156): A potential difference induced in the armature of an operating motor that opposes the applied potential difference.

Balmer series (228): A series of related lines in the visible part of the hydrogen spectrum corresponding to energy jumps from higher energy levels to the n = 2 quantum level.

Baryon (251): Particles with one nucleon which interact with the strong force.

Beta decay (240): The emission of an electron or a positron from an atom.

Binding energy (238): Energy required to remove a particle from a system.

Bohr atom (225): A model of the atoms proposed by Niels Bohr in which electrons revolved around a nucleus in certain allowable orbits.

Breakdown voltage (271): A reverse potential difference that causes a large reverse current in a diode.

Bright line spectrum (193): A spectrum produced by gases at low pressures with characteristic bright lines for each element.

Boiling point (95): The temperature at which the vapor pressure of a liquid equals the atmospheric pressure.

Breeder reaction (211): Nuclear reactor capable of producing fissionable material.
Bubble chamber (250): Instrument to make the paths of ionizing particles visible.
Brushes (154): Used as connectors in motors and generators.

Capacitor (298): see Achievement Section. Device that consists of conducting
  plates separated by layers of dielectric that is used to store electric charge.
Cathode ray tube (157): A device in which a beam of electrons is accelerated towards
  a screen, then deflected by electric and/or magnetic fields. Used in oscilloscopes
  and television sets.
Celsius temperature (89): A temperature scale defined such that the ice point is 0°
  and the boiling point is 100° with 100 equal divisions between them.
Center of curvature (202): The center of the sphere of which the lens or mirror
  surface forms a part.
Center of mass: The point at which the entire mass of an object can be considered
  concentrated.
Centripetal acceleration (57): A vector quantity representing v/t and always
  directed towards the center of the circle.
Centripetal force (57): The force which causes an object to move in a circular path.
  This force is always directed towards the center of the circle.
Chain reaction (247): A series of reactions in which the material or energy that
  starts the reaction is also one of the products and the reaction is self-sustaining.
Chromatic aberration (211): The non-focusing of light of different colors.
Circuit (125): The path that charged particles follow around a closed loop.
Cloud Chamber (219): A device used to make the path of ionizing radiation visible.
Cloud model (229): No specific electron orbit. Electron is located in orbital.
Coefficient of friction (34): The ratio of the force necessary to overcome sliding
  friction to the normal force pressing the surfaces together.
Coherence (167, 190): The property of two waves with identical wavelengths and a
  constant phase relationship; laser light.
Collision, elastic (36): A collision in which the objects rebound from each other with
  no loss of kinetic energy.
Commutator (154): A split ring in a D.C. generator or a motor connected to the
  armature loop.
Component (12): One of several vectors that can replace a vector.
Compton effect (221): An increase in wavelength that takes place in a high energy
  photon when it collides with an electron. The effect illustrates that waves have
  the "particle" property of momentum.
Concave lens (207): A lens thinner in the center than at the edges; diverges parallel
  light.
Concurrent forces (11): Forces which act on the same point.
Conductor (121): A substance in which electrical charge flows easily.
Conservative force (83): A force for which the work done on an object is
  independent of path.
Conservation of charge (111): A principle that states electric charge can neither be
  created nor destroyed.
Conservation of energy (81, 92): A principle that states that energy can neither be
  created nor destroyed.
Constructive interference (181): The effect that occurs when two waves come
  together in such a way that a crest meets a crest or a trough meets a trough (in
  phase).
Contact (111): A method of charging an electroscope. The charge on the electroscope
  is the same as the charging body.
Continuous spectrum (193): Spectrum produced by gas under high pressure;
  contains all frequencies of electromagnetic radiation.
Control rod (248): A rod composed of neutron-absorbing material that can be moved
  into the core to slow down the reaction.

Converging lens (207): A lens that is thicker in the center than at the edges and focuses parallel rays of light.
Convex lens (207): A lens that converges parallel light.
Coolant (248): Used in nuclear reactors to prevent overheating of core.
Core of a reactor (248): The part of a nuclear reactor containing the fuel rods, moderator, coolant and control rods.
Coulomb (109): The quantity of charge on 6.25×10^{18} electrons.
Coulomb's Law (112): The force between two point charges is directly proportional to the product of the charges and inversely proportional to the square of the distance between them.
Critical angle (186): That angle of incidence in an optically denser material that results in an angle of refraction of 90°.
Critical mass (247): The minimum mass of fissionable material which will produce a chain reaction.
Cyclotron (249): A device for accelerating charged atomic particles using D-shaped electrodes.

De Broglie (222): Probability wave associated with moving particles, moving matter.
Derived unit (9): A new name for the unit made up of fundamental units. For example, a joule is a newton - meter.
Destructive interference (181): Effect that occurs when two waves come together in such a way that a crest meets a trough ($\frac{1}{2}$ λ out of phase).
Diamagnetic (133): Materials which are weakly repelled by a strong magnet; decrease of flux density.
Diffraction (190): The spreading of a wave disturbance into a region behind an obstruction; occurs when the obstruction or opening is narrow compared to the wavelength.
Diffraction grating (191): An optical surface with a large number of equally spaced lines grooved on it which cause interference patterns on a screen.
Diffuse reflection (184): The reflection that occurs when light strikes a rough surface. The light is scattered in all directions.
Dispersion (183, 187): The process of separating polychromatic light into its component wavelengths because of speed.
Displacement (10): A vector quantity that represents the length and direction of a straight line path from one point to another point between which the motion of an object has taken place; total displacement is a vector sum; distance a vibrating particle is from the midpoint of its vibration.
Distance (10): A scalar quantity that represents the length of a path from one point to another point.
Diverge (207): To move apart.
Diverging lens (207): A lens that is thicker at the edges than at the center and bends parallel rays so they appear to come from a common point.
Domain (133): A microscopic magnetic region consisting of atoms whose magnetic field are aligned in a common direction.
Doppler effect (193): The observed change in frequency of a moving object.
Double slit (190): Two closely spaced openings used in the study of diffraction and interference.
Dynamic equilibrium (25): The object is moving with constant velocity. The sum of all net forces and torques is zero.

Elastic collisions (36): Collision in which the sum of the kinetic energies before a collision equals the sum of the kinetic energies after the collision.
Elastic potential energy (79): Type of potential energy stored in a spring.
Electric field (115): Region where electrical force acts on a charged particle.
Electric field intensity (115): The force per unit positive charge at a given point in an electric field measured in N/C or V/m.

Electric potential (116): At any point is the work required to bring a charge from infinity to that point.

Electromagnet (135): Magnet whose field is produced by an electric current.

Electromagnetic spectrum (192): The entire range of electromagnetic radiation from low frequency radio waves to high frequency gamma rays.

Electromagnetic waves (142): Waves composed of electric and magnetic fields vibrating at right angles to each other.

Electromotive force (emf) (163): Energy per unit charge supplied by the source.

Electron (235): A negatively charged subatomic particle.

Electron volt (117): A unit of energy equal to the amount of work done moving an electron through a potential difference of one volt.

Electroscope (111): A device used to observe electrostatic charges.

Emission spectrum (227): A spectrum formed from an incandescent solid, liquid or gas.

Energy (78): A physical quantity that has the capacity to do work.

Equilibrant (11): The force that produces equilibrium. It is equal in magnitude to the resultant but in the opposite direction.

Escape velocity (64): The minimum velocity that is necessary to escape from the Earth's gravitational field.

Excitation (226): The process by which an electron jumps from the ground state of an atom to an excited state.

Excited state (226): Any allowed orbit except the one of lowest energy in the Bohr model of the atom.

Ferromagnetic (133): A material which is strongly attracted by a magnet.

Fiber optics (187): Light passes through small transparent fibers. Due to total internal reflection, it does not escape.

Fission (247): The splitting of a heavy nucleus into nuclei of medium mass accompanied by the release of neutrons and excess energy.

Flux density (136): A measurement of the field intensity per unit area whose units are N/amp·m, weber/m^2, or tesla.

Focal length (203): The distance between the principal focus of a lens or mirror and its optical center.

Focus (202): A point where light rays meet or from which rays of light diverge.

Force (25): A push or a pull that changes or tends to change the state of motion of an object.

Freefall (17): The motion of an object when the only force acting on the object is that of gravitational attraction.

Frequency (176): The number of cycles per unit time made by a vibrating object.

Friction (33): A force that resists the relative motion of objects that are in contact with each other.

Fuel rods (248): In a nuclear reactor, contain fissionable material.

Fundamental unit (9): SI system of units including meter, kilogram, second, ampere, and Kelvin.

Fusion (248): The change of phase from a solid to a liquid; melting. Also, a reaction in which light nuclei combine to form a more dense nucleus.

Galvanometer (137, 153): An electrical meter used to measure minute amounts of electric current.

Gamma radiation (240): High energy electromagnetic radiation emitted from the nucleus.

Geiger tube (249): An instrument used to detect ionizing radiation.

Geosynchronous orbit (64): The period of a satellite is the same as the period of the Earth.

Gravitational field (30): The region in which one massive body exerts a force of attraction on another massive body.

Gravitational mass (31): Mass measured by gravitational attraction.
Gravitational potential energy (78): Energy a body because of its position.
Gravity, force of (30): The force of attraction on an object by the earth.
Ground (111): An infinite source of electrons or sink for electrons.
Ground state (226): The lowest energy orbit available to an electron in the Bohr model of the atom.

Haldrons (251): Subatomic particles which interact by means of the strong force.
Half-life (242): Ave. time for half of a large number of radioactive nuclides to decay.
Heat (89): Thermal energy in the process of being added to, or removed from, a substance.
Heat of fusion (95): The energy required to melt a solid at its melting point.
Heat of vaporization (95): The energy necessary to change a liquid to a gas at its boiling point.
Hertz (176): The unit for measuring frequency (1/second).
Huygen's principle (189): Each point on a wave front acts as a source of new wavelets which in turn produce a pattern for the next wave front.

Image (201): Optical counterpart of an object produced by a lens or a mirror.
Impulse (39): The product of the force and the time interval in which it acts.
Index of refraction (186): Property of an optical substance measured by the ratio of the speed of light in a vacuum to the speed of light in the substance.
Induced current (163): Current made by changing magnetic field near a conductor.
Induced emf (163): Potential difference produced by a changing magnetic field.
Induced transmutation (243): Conversion of the nucleus of one element to another by means of a change in the number of protons brought about by particle bombardment.
Induction (111, 155): A process of charging an electroscope in which the charge induced is opposite that of the charging body.
Inertia (26): The property that opposes any change in its state of motion; proportional to mass.
Insulator (121): A substance which does not easily conduct electricity.
Interference (181): The effect produced when two or more waves overlap.
Internal energy (92): The sum of the internal potential and kinetic energies of a system due to motion and position of particles.
Ion (122): An atom or group of atoms having an electrical charge.
Ionization energy (226): The energy required to remove an electron from an atom in the ground state.
Isotopes (236): Atoms whose nuclei contain the same number of protons but a different number of neutrons.

Joule (72): Amount of work done when a force acts through a distance of one meter.

Kelvin scale (89): A temperature scale having a single fixed point, the triple point of water.
Kepler's Laws (61): 1. The planets move about theSun in ellipses. 2. The radius vector joining each planet with the Sun describes equal areas in equal times. 3. The ratio of the square of the planet's year to the cube of the planet's mean distance from the Sun is the same for all planets.
Kinetic energy (78): Energy possessed by a body because of its motion.
Kinetic friction (33): The force which must be overcome to keep an object moving.
Kinetic molecular theory (94): Molecules are in a constant state of motion.
Kirchhoff's laws (129): (1) In an electric circuit, the sum of the current entering a junction is equal to the sum of the currents leaving a junction. (2)The sum of the IR drop around a loop of a circuit equals the applied Emf.

Laser (165, 167): Light Amplification by Stimulated Emission of Radiation. Source of very intense, highly directional, monochromatic, and coherent beam of light.

Lens (207): A transparent object with one or two curved surfaces used to direct light rays by refraction.

Lenz's law (141, 156): The direction of an induced current is such that the magnetic field induced produces magnetic forces which oppose the original forces.

Lepton (251): Category of particles which interacts by means of weak nuclear force and the gravitational force.

Line spectrum (193): A spectrum produced by gases at low pressures with characteristic bright lines for each element.

Lines of force (134): Imaginary lines drawn in a field such that the tangent drawn at any point indicates the direction of a force on a test unit in that field.

Linear accelerator (249): A device for accelerating particles in a straight line through many stages of small potential differences.

Longitudinal wave (176): A wave in which the particles vibrate back and forth in the direction in which the wave travels.

Magnetic field (134): A region in which magnetic forces can be detected. The direction of the field is the direction of force on the north pole placed in that field.

Magnetic flux (134): Lines of flux through a region of a magnetic field.

Magnetic induction (137): Magnetic field strength in newtons per amp-meter or weber per square meter (or tesla).

Magnitude (9): The size of a quantity (for example, vector).

Mass (9): A specification of the inertia of an object.

Mass defect (238): The difference between the mass of the nucleus and the larger mass of its uncombined constitute parts.

Mass number (236): The total number of protons and neutrons (the number of nucleons) in a nucleus.

Matter waves (222): Another name for De Broglie waves.

Meter (9): The SI Unit of length based on the wavelength of krypton.

Millikan oil drop experiment (118): Determined that the smallest charge possible was the charge on one electron, 1.6×10^{-19} coulombs and all charges were whole number multiples of this fundamental value.

Moderator (248): In a nuclear reactor, material which slows down fast moving neutrons such that they become thermal neutrons.

Momentum (36): The product of the mass of an object and its velocity.

Muon (251): A lepton which interacts by means of the weak nuclear force.

Natural radioactivity (240): Radioactivity which occurs naturally in nature.

Neutrino (250): A subatomic particle of zero mass and zero charge.

Newton (9): The derived unit for force. One newton is equal to 1 kg.m/s^2

Node (181): The points on a standing wave at which no motion occurs.

Non-conservative force (83): One for which the work done on an object is dependent on the path taken. Friction is a non-conservative force.

Non-dispersive medium (187): A medium in which waves of different frequencies have the same speed. Air is a non-dispersive medium for light.

Normal (186): A line drawn perpendicular to a surface or line.

N—type semiconductor (264): A semiconductor doped so that it acquires free electrons.

Nucleon (235): A proton or neutron in the nucleus of an atom.

Nuclide (235): An atom of a particular mass and of a particular element.

Ohm (123): The SI Unit of electrical resistance.

Ohmic conductor (122, 131): Conductor which obeys Ohm's Law.

Ohm's law (122): At a constant temperature, the ratio of the potential difference across a resistor to the current flowing in the resistor is a constant.

Optical path (202): The path taken by a ray of light as it travels through a medium.

Orbital (or state) (229): Region of most probable location of the electron.

Order of magnitude (284): see Math Review. A numerical approximation to the nearest power of ten.

Parallel circuit (127): An electric circuit connected at the same two points so there are two or more paths for the current to follow.

Paramagnetic (133): Material which is weakly attracted by a strong magnet.

Pendulum (11, 40): Body suspended so it can swing about an axis.

Period (176): The time for one complete cycle or oscillation. The time required for a single wavelength to pass a given point.

Permeability (133): The property of a material by which it changes the flux density in a magnetic field from the value in air.

Phase (177): A condition of matter. In periodic motion, the stage within each oscillation.

Photoelectric effect (219): The emission of electrons by a surface when exposed to electromagnetic radiation.

Photoelectron (219): Electrons emitted from a light sensitive material when it is illuminated.

Photon (221): A quantum of light energy.

Planck's constant (219): A fundamental constant in nature that determines what values are allowed in quantum mechanics.

Plasma (122): A fourth stage of matter in which electrons are stripped from atoms leaving positive ions and free electrons.

Pion (251): A hadron which interacts by means of the strong nuclear force.

P–N diode (270): A rectifier consisting of a P–type semiconductor and an N–type semiconductor bonded together.

P–type semiconductor 267): A semiconductor doped so that it acquires many holes whose movement is equivalent to that of positive charge.

Polarization (190): Vibrations of the waves are aligned into one plane.

Positron (244): A particle with the same mass as an electron, but a positive charge.

Potential difference (117): The work done per unit charge as a charge is moved between two points in an electric field.

Potential energy (78): Energy that results from position or state of an object.

Power (75): The time rate of doing work. The SI Unit is the watt.

Principal axis (202): A line drawn through the center of curvature and the optical center of a lens.

Prism (187): An optical device used for the dispersion of light.

Projectile motion (51): The motion of an object characterized by constant speed in the horizontal direction and constant acceleration (gravity) in the vertical direction.

Pulse (175): A single non-repeated disturbance.

Quantum (219): An elemental unit of energy; a photon or energy hf.

Radiation (192): Any form of energy which moves along as a wave.

Radioactivity (240): Spontaneous nuclear disintegration by the emission of radiation.

Ray (203): A line drawn in the direction in which a wave is traveling.

Real image (201): The type of image formed when rays of light actually come together (can be caste on a screen).

Reflection (184): Light bounced off an optical surface.

Refraction (185): The bending or change of direction of a beam of light which occurs when the ray passes obliquely from one medium to another due to a change in speed.

Resistance (123): The opposition to flow of electrons.

Resistivity (123): A proportionality constant used to determine the resistance of a wire.

Resolving power of a lens (192): A measure of the ability of the lens to produce clearly distinguishable images of two very close points.

Resolution (12): Process of dividing a force into two or more components.

Resonance (182): The effect produced when a vibrating object is forced to vibrate at one of its natural frequencies.

Resultant (11): A vector representing the sum of several components.

Rutherford's Model (224): Model of atom with nucleus at center and mostly open space.

Satellite (63): A smaller body which revolves around a larger one.

Scalar (9): A physical quantity that is completely specified by magnitude.

Scintillation counter (224): A device that counts impacts of charged particles on a fluorescent screen using a photomultiplier tube.

Semiconductor (121): A material whose electrical resistance lies between that of insulators and conductors.

Series circuit (126): An electrical circuit in which there is only one path.

SI units (9): International System of Units.

Significant figures (282): see Math Review. Those digits in a number that are known for certain plus the first digit that is uncertain.

Simple harmonic motion (21): Repeating motion in which the acceleration is proportional to the displacement from the equilibrium position.

Snell's law (186): For a ray passing from one medium to another, the product of the index of refraction of the first medium and the sine of the angle of incidence is equal to the product of the index of refraction of the second medium and the sine of the angle of refraction.

Solenoid (134): A coil of wire frequently wound around a core.

Specific heat (137): The amount of heat necessary to raise the temperature of a unit mass one degree Celsius.

Special relativity (300): see Achievement Section. A theory based on the two postulates (1) all laws of physics are the same in all frames of reference moving with constant velocity relative to an inertial frame; and (2) the velocity of light is the same in all such frames of reference.

Spectrum (227): A range of electromagnetic frequencies.

Speed (17): A scalar quantity which represents the magnitude of the velocity.

Spherical aberration (212): The failure of parallel rays to meet at a single point after reflection or refraction from a spherical surface.

Standing wave (181): Interference effect which occurs when two identical waves travel through the same region of space but in opposite direction.

Static electricity (109): Electricity at "rest".

Static equilibrium (25): Body must be at rest and all forces and torques acting on the body must be zero.

Static friction (33): The force which must be overcome to start an object moving.

Strong nuclear interaction (250): The force of attraction between two nucleons in which mesons serve as the carrier.

Superposition (181): Combining the displacement of two or more waves to produce a resultant displacement.

Synchrotron (249): A type of cyclotron which synchronizes energizing pulses with changes in mass as the velocity of the accelerated particles increases.

Temperature (89): A measurement of the average kinetic energy of the molecules.

Tesla (136): SI Unit of magnetic flux density.

Terminal speed (20): The speed reached by a falling object at the instant a frictional resistance force equals the weight.

Thermionic emission (157): The emission of electrons from metals at high temperature.

Threshold frequency (221): The frequency below which electromagnetic radiation will not eject electrons from the surface of a given metal.

Torque (153): The product of the force perpendicular and the length of the torque arm.

Total internal reflection (186): When light traveling in a substance strikes the boundary of a surface in which its speed is greater, none passes into the "higher speed" substance if the angle of incidence is greater than the critical angle. All light is reflected (none absorbed).

Transmutation (243): The conversion of an atomic nucleus of one element into that of another element.

Transverse wave (175): A wave in which the vibrations are at right angles to the direction of propagation of the wave.

Uniform circular motion (57): Constant speed in a circular path.

Uniform motion (57): Constant speed in a straight line.

Van de Graaff generator (249): A particle accelerator that transfers charge from the electron source to an insulated sphere by a moving belt.

Vector (9): A quantity that is completely specified by both magnitude and direction.

Velocity (9, 177): A vector quantity which represents the time rate of change in displacement.

Virtual image (201): Type of image formed when rays of light appear to come from the image. This image can be seen but not focused on a screen.

Visible spectrum (192): The portion of the electromagnetic spectrum whose frequencies produce the sensations of color.

Volt (122): The SI Unit of potential difference; one volt is equal to one joule per coulomb.

Voltmeter (122, 154): An electrical device to measure voltage drop across a resistor. The meter is always attached in parallel. A galvanometer with a high resistance in series.

Watt (13, 75): The SI Unit of power; one watt is equal to one joule per second.

Wave (175): A series of pulses.

Wavelength (177): The distance between two consecutive points of a wave moving in the same direction with the same displacement.

Weak nuclear interaction (250): Nuclear reactions in which leptons (collective name for electrons, muons, and neutrinos) are emitted.

Weber (134, 157): The SI Unit for magnetic flux.

Weight (26): The force of gravitational attraction exerted on an object by the earth.

Work (31): The product of a force and the distance over which this force acts.

Work function (221): The minimum amount of energy required to remove an electron from the surface of a material.

W–particle (250): Carrier of the weak nuclear interaction.

X ray (192): A range of deeply penetrating high frequency electromagnetic radiations.

Physics Practice Exams

Part I Credits for the following Practice Examinations in Physics

Note:

Practice Exam 1 and Practice Exam 2 have sixty (60) questions in the Part I sections which are worth a total of seventy (70) credit points on each exam.

Practice Exam 3 has fifty five (55) questions in the Part I section which are worth a total of sixty five (65) credit points on the exam.

The following two Credit Tables should be used with these Practice Exams to determine the correct value of the Part I sections.

Part 1 Credits for 60 Questions (Total Credits = 70)

No. Right .... Credits		
60......... 70	41......... 55	19......... 37
59......... 69	40......... 54	18......... 36
58......... 68	39......... 53	17......... 36
57......... 68	38......... 52	16......... 35
56......... 67	37......... 52	15......... 34
55......... 66	36......... 51	14......... 32
54......... 65	35......... 50	13......... 29
53......... 64	34......... 49	12......... 27
52......... 64	33......... 48	11......... 25
51......... 63	32......... 48	10......... 23
50......... 62	31......... 47	9.......... 20
49......... 61	30......... 46	8.......... 18
48......... 60	29......... 45	7.......... 16
47......... 60	28......... 44	6.......... 14
46......... 59	27......... 44	5.......... 11
45......... 58	26......... 43	4.......... 9
44......... 57	25......... 42	3.......... 7
43......... 56	24......... 41	2.......... 5
42......... 56	23......... 40	1.......... 2
	22......... 40	0.......... 0
	21......... 39	
	20......... 38	

Part 1 Credits for 55 Questions (Total Credits = 65)

No. Right	Credits
55	65
54	64
53	63
52	63
51	62
50	61
49	60
48	60
47	59
46	58
45	57
44	57
43	56
42	55
41	54
40	54
39	53
38	52
37	51
36	50
35	50
34	49
33	48
32	47
31	47
30	46
29	45
28	44
27	44
26	43
25	42
24	41
23	40
22	40
21	39
20	38
19	37
18	37
17	36
16	35
15	34
14	34
13	33
12	30
11	28
10	25
9	23
8	20
7	18
6	15
5	13
4	10
3	8
2	5
1	3
0	0

To determine the Total Score for the Practice Exam, add

_____ (1) the total Credits for Part I as determined from the Credit Tables

_____ (2) The Part II score

_____ (3) the Part III score

_____ Total Score

Physics Practice Exam 1 – June 1991

Part I Answer all 60 questions in this part. [70]

Directions (1-60): For *each* statement or question, select the word or expression that, of those given, best completes the statement or answers the question.

1 Which is a scalar quantity?
1 displacement 2 distance 3 force 4 acceleration

2 What is the approximate mass of a chicken egg?
(1) 1×10^1 kg (2) 1×10^2 kg (3) 1×10^{-1} kg (4) 1×10^{-4} kg

3 Compared to the mass of an object at the surface of the Earth, the mass of the object a distance of two Earth radii from the center of the Earth is
1 the same 3 one-half as great
2 twice as great 4 one-fourth as great

4 A runner completed the 100.-meter dash in 10.0 seconds. Her average speed was
(1) 0.100 m/s (2) 10.0 m/s (3) 100. m/s (4) 1,000 m/s

5 Which pair of graphs represent the same motion?

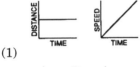

(1) (3)

(2) (4)

6 A child riding a bicycle at 15 meters per second decelerates at the rate of 3.0 meters per second^2 for 4.0 seconds. What is the child's speed at the end of the 4.0 seconds?
(1) 12 m/s (2) 27 m/s (3) 3.0 m/s (4) 7.0 m/s

7 A skier starting from rest skis straight down a slope 50. meters long in 5.0 seconds. What is the magnitude of the acceleration of the skier?
(1) 20. m/s^2 (2) 9.8 m/s^2 (3) 5.0 m/s^2 (4) 4.0 m/s^2

8 In the graph at the right, the acceleration of an object is plotted against the unbalanced force on the object. What is the object's mass?
(1) 1 kg (3) 0.5 kg
(2) 2 kg (4) 0.2 kg

9 What is the gravitational force of attraction between a planet and a 17-kilogram mass that is freely falling toward the surface of the planet at 8.8 meters per second^2?
(1) 150 N (2) 8.8 N (3) 1.9 N (4) 0.52 N

10 Two perpendicular forces act on an object as shown in the diagram at the right. What is the magnitude of the resultant force on the object?
(1) 17 N   (2) 13 N   (3) 7.0 N   (4) 5.0 N

5.0 N

12 N

11 If the Earth were twice as massive as it is now, then the gravitational force between it and the Sun would be
1   the same   3   half as great
2   twice as great   4   four times as great

12 A 0.025-kilogram bullet is fired from a rifle by an unbalanced force of 200 newtons. If the force acts on the bullet for 0.1 second, what is the maximum speed attained by the bullet?
(1) 5 m/s   (2) 20 m/s   (3) 400 m/s   (4) 800 m/s

13 What is an essential characteristic of an object in equilibrium?
1   zero velocity   3   zero potential energy
2   zero acceleration   4   zero kinetic energy

14 A 2.0-kilogram ball traveling north at 4.0 meters per second collides head on with a 1.0-kilogram ball traveling south at 8.0 meters per second. What is the magnitude of the total momentum of the two balls after collision?
(1) 0 kg·m/s   (2) 8.0 kg·m/s (3) 16 kg·m/s   (4) 32 kg·m/s

15 The magnitude of the force that a baseball bat exerts on a ball is 50. newtons. The magnitude of the force that the ball exerts on the bat is
(1) 5.0 N   (2) 10. N   (3) 50.N   (4) 250 N

16 Which quantity and unit are correctly paired?

1   velocity — m/s^2 3   energy — $\dfrac{kg \bullet m^2}{s^2}$

2   momentum — $\dfrac{kg \bullet m}{s^2}$ 4   work — kg/m

17 A jack exerts a force of 4,500 newtons to raise a car 0.25 meter. What is the approximate work done by the jack?
(1) 5.6 x 10^{-5} J   (2) 1.1 x 10^3 J   (3) 4.5 x 10^3 J   (4) 1.8 x 10^4 J

18 A 6.0 x 10^2-newton man climbing a rope at a speed of 2.0 meters per second develops power at the rate of
(1) 1.2 x 10^1 W   (3) 3.0 x 10^2 W
(2) 6.0 x 10^2 W   (4) 1.2 x 10^3 W

19 Three people of equal mass climb a mountain using paths A, B, and C shown in the diagram. Along which path(s) does a person gain the greatest amount of gravitational potential energy from start to finish?
(1) A, only
(2) B, only
(3) C, only
(4) The gain is the same along all paths.

FINISH

A

B

C

START

20 Which graph best represents the relationship between the elongation of an ideal spring and the applied force?

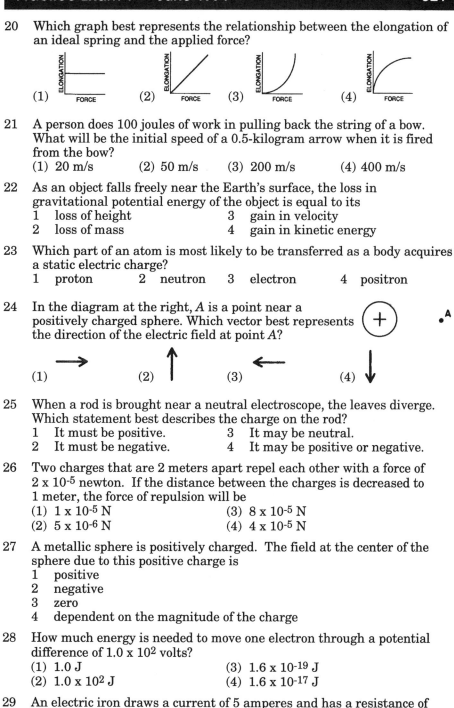

(1)   (2)   (3)   (4)

21 A person does 100 joules of work in pulling back the string of a bow. What will be the initial speed of a 0.5-kilogram arrow when it is fired from the bow?
(1) 20 m/s   (2) 50 m/s   (3) 200 m/s   (4) 400 m/s

22 As an object falls freely near the Earth's surface, the loss in gravitational potential energy of the object is equal to its
1   loss of height   3   gain in velocity
2   loss of mass   4   gain in kinetic energy

23 Which part of an atom is most likely to be transferred as a body acquires a static electric charge?
1   proton   2   neutron   3   electron   4   positron

24 In the diagram at the right, A is a point near a positively charged sphere. Which vector best represents the direction of the electric field at point A?

(1)   (2)   (3)   (4)

25 When a rod is brought near a neutral electroscope, the leaves diverge. Which statement best describes the charge on the rod?
1   It must be positive.   3   It may be neutral.
2   It must be negative.   4   It may be positive or negative.

26 Two charges that are 2 meters apart repel each other with a force of 2×10^{-5} newton. If the distance between the charges is decreased to 1 meter, the force of repulsion will be
(1) 1×10^{-5} N   (3) 8×10^{-5} N
(2) 5×10^{-6} N   (4) 4×10^{-5} N

27 A metallic sphere is positively charged. The field at the center of the sphere due to this positive charge is
1   positive
2   negative
3   zero
4   dependent on the magnitude of the charge

28 How much energy is needed to move one electron through a potential difference of 1.0×10^2 volts?
(1) 1.0 J   (3) 1.6×10^{-19} J
(2) 1.0×10^2 J   (4) 1.6×10^{-17} J

29 An electric iron draws a current of 5 amperes and has a resistance of 20 ohms. The amount of energy used by the iron in 40 seconds is
(1) 100 J   (2) 500 J   (3) 4,000 J   (4) 20,000 J

30 The diagram at the right represents a
    series circuit containing three resistors.
    What is the current through resistor R_2?
    (1) 1.0 A
    (2) 0.33 A
    (3) 3.0 A
    (4) 9.0 A

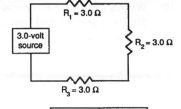

31 In the circuit diagram at the right, what is the
    potential difference across the 3.0-ohm
    resistor?
    (1) 1.0 V
    (2) 2.0 V
    (3) 3.0 V
    (4) 1.5 V

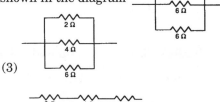

32 Which circuit segment below has the same total
    resistance as the circuit segment shown in the diagram
    at the right.

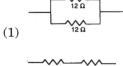

(1) 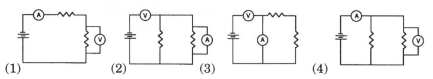   (3)

(2)   (4)

33 While operating at 120 volts, an electric toaster has a resistance of
    15 ohms. The power used by the toaster is
    (1) 8.0 W   (2) 120 W   (3) 960 W   (4) 1,800 W

?   Which circuit shown below could be used to determi   e total current
    and potential difference of a parallel circuit?

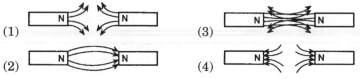

(1)   (2)   (3)   (4)

35 A wire carries a current of 6.0 amperes. How much charge passes a
    point in the wire in 120 seconds?
    (1) 6.0 C   (2) 20. C   (3) 360 C   (4) 720 C

36 Which diagram best represents the magnetic field between two magnetic
    north poles?

(1)   (3)

(2)   (4)

37 A volt is to electric potential as a tesla is to
    1   electrical energy   3   magnetic flux density
    2   electric field intensity   4   charge density

38 The diagrams at the right show cross sections of   ⊙ CURRENT OUT
    conductors with electrons flowing into or out of the page.
    In which diagram below will the magnetic flux density at   ⊗ CURRENT IN
    point *A* be greater than the magnetic flux density at
    point *B*?

    ⊙ A ⊗ B   ⊗ A ⊗ B   ⊙ A ⊙ B   A ⊙ B
      •   •   •   •   •   •   •   •
    (1)   (2)   (3)   (4)

39 The diagram at the right represents a conductor carrying   X X X X X X X X
    a current in which the electron flow is from left to right.   X X X X X X X X
    The conductor is located in a magnetic field which is   X X X X X X X X
    directed into the page. The direction of the magnetic   〈 e⁻ ⟶ 〉
    force on the conductor will be
    1   into the page   X X X X X X X X
    2   out of the page   X X X X X X X X
    3   toward the top of the page   X X X X X X X X
    4   toward the bottom of the page   MAGNETIC FIELD

40 The diagram at the right shows the direction of water waves   BARRIER
    moving along path *XY* toward a barrier. Which arrow   X
    represents the direction of the waves after they have   Y
    reflected from the barrier?

    (1) ⟵   (2) ⟍   (3) ⟋   (4) ⟍

41 What is the frequency of a wave if its period is 0.25 second?
    (1) 1.0 Hz   (2) 0.25 Hz   (3) 12 Hz   (4) 4.0 Hz

42 The diagram at the right shows a transverse water
    wave moving in the direction shown by velocity
    vector *v*. At the instant shown, a cork at point *P* on
    the water's surface is moving toward
    (1) *A*   (3) *C*
    (2) *B*   (4) *D*

43 To the nearest order of magnitude, how many times greater than the
    speed of sound is the speed of light?
    (1) 10^4 (2) 10^6 (3) 10^{10} (4) 10^{12}

44 Diagram I shows a glass tube containing undisturbed air molecules.
    Diagram II shows the same glass tube when a wave passes through it.

    Diagram I   GLASS TUBE   Diagram II   GLASS TUBE

                    ⟵ MOLECULES   ⟵ MOLECULES
                       OF AIR   OF AIR

    Which type of wave produced the disturbance shown in diagram II?
    1   longitudinal 2   torsional   3   transverse   4   elliptical

45 An opera singer's voice is able to break a thin crystal glass if the singer's
    voice and the glass have the same natural
    1   frequency   2   speed   3   amplitude   4   wavelength

46   The diagram at the right shows a ray of light, R, entering glass from air. Which path is the ray most likely to follow in the glass?
(1) A
(2) B
(3) C
(4) D

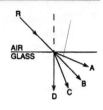

47   What occurs when light passes from water into flint glass?
1   Its speed decreases, its wavelength becomes smaller, and its frequency remains the same.
2   Its speed decreases, its wavelength becomes smaller, and its frequency increases.
3   Its speed increases, its wavelength becomes larger, and its frequency remains the same
4   Its speed increases, its wavelength becomes larger, and its frequency decreases.

48   As shown in the diagram at the right, a beam of light can pass through the length of a curved glass fiber. This phenomenon is possible due to the effect of
1   dispersion   3   polarization
2   internal reflection   4   diffraction

49   Compared to visible light, ultraviolet radiation is more harmful to human skin and eyes because ultraviolet radiation has a
1   higher frequency   3   higher speed
2   longer period   4   longer wavelength

50   Polychromatic light passing through a glass prism is separated into its component frequencies. This phenomenon is called
1   diffraction   3   reflection
2   dispersion   4   polarization

51   Which is a characteristic of light produced by a monochromatic light source?
1   It can be dispersed by a prism.
2   It is a longitudinal wave.
3   Its frequency is constantly changing.
4   It can be polarized.

52   In the diagram at the right, two speakers are connected to a sound generator. The speakers produce a sound pattern of constant frequency such that a listener will hear the sound very well at A and C, but not as well at point B. Which wave phenomenon is illustrated by this experiment?
1   interference   3   reflection
2   polarization   4   refraction

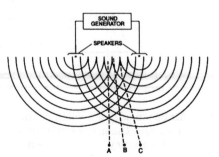

53 Blue light has a frequency of approximately 6.0×10^{14} hertz. A photon of blue light will have an energy of approximately
(1) 1.1×10^{-48} J
(3) 5.0×10^{-7} J
(2) 6.0×10^{-34} J
(4) 4.0×10^{-19} J

54 An electron in a mercury atom that is changing from the a to the g level absorbs a photon with an energy of
(1) 12.86 eV
(3) 7.90 eV
(2) 10.38 eV
(4) 2.48 eV

55 Which phenomenon can be explained by both the particle model and wave model?
1 reflection
3 diffraction
2 polarization
4 interference

56 What do alpha-particle scattering experiments indicate about an atom's structure?
1 Electrons occupy distinct energy levels.
2 Positive and negative charges are evenly distributed.
3 Negative charge fills the space around the core.
4 Positive charge is concentrated in a small, dense core.

57 A box initially at rest on a level floor is being acted upon by a variable horizontal force, as shown in the diagram at the right. Compared to the force required to start the box moving, the force required to keep it moving at constant speed is

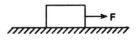

1 less
2 greater
3 the same

58 A police officer's stationary radar device indicates that the frequency of the radar wave reflected from an automobile is less than the frequency emitted by the radar device. This indicates that the automobile is
1 moving toward the police officer
2 moving away from the police officer
3 not moving

59 If the diameter of a wire were to increase, its electrical resistance would
1 decrease
2 increase
3 remain the same

60 Electromagnetic radiation of constant frequency incident on a photoemissive material causes the emission of photoelectrons. If the intensity of this radiation is increased, the rate of emission of photoelectrons will
1 decrease
2 increase
3 remain the same

Part II

This part consists of six groups, each containing ten questions. Each group
tests an optional area of the course. Choose two of these six groups. Be sure
to answer all ten questions in each group chosen. [20]

Group 1 — Motion in a Plane
If you chose this group, be sure to answer questions 61 - 70.

61 Four cannonballs, each with mass M and initial velocity V, are fired
from a cannon at different angles relative to the Earth. Neglecting air
friction, which angular direction of the cannon produces the greatest
projectile height?
(1) 90°   (2) 70°   (3) 45°   (4) 20°

62 A projectile is launched at an angle of 60.° to the horizontal at an initial
speed of 10. meters per second. What is the magnitude of the vertical
component of its initial speed?
(1) 2.5 m/s   (2) 4.3 m/s   (3) 5.0 m/s   (4) 8.7 m/s

63 The diagram at the right represents a ball
undergoing uniform circular motion as it travels
clockwise on a string. At the moment shown in the
diagram, what are the correct directions of both the
velocity and centripetal acceleration of the ball?

(1) $v\uparrow$ $\xrightarrow{a}$ (3) $v\downarrow$ $\xleftarrow{a}$

(2) $\xrightarrow{v}$ $a\uparrow$ (4) $\xrightarrow{v}$ $a\downarrow$

64 A 3.0-kilogram mass is traveling in a circle of 0.20-meter radius with a
speed of 2.0 meters per second. What is its centripetal acceleration?
(1) 10. m/s^2 (2) 20. m/s^2 (3) 60. m/s^2 (4) 6.0 m/s^2

65 A car going around a curve is acted upon by a centripetal force, F. If the
speed of the car were twice as great, the centripetal force necessary to
keep it moving in the same path would be
(1) F (2) $2F$ (3) $F_{/2}$ (4) $4F$

66 The shape of the path of the Earth about the Sun is
1   a circle with the Sun at the center
2   an ellipse with the Sun at one focus
3   an ellipse with the Moon at one focus
4   an ellipse with nothing at either focus

67 The increase in a planet's speed as it approaches the Sun is described by
Kepler's second law. What is the best explanation for this empirical
law?
1   Since kinetic energy depends on temperature, the planet must move
     faster as it nears the Sun.
2   Ions from the Sun (solar wind) speed up the planet.
3   The days are shorter in the winter, causing the planet to move faster
     as it nears the Sun.
4   The gravitational potential energy lost as the planet nears the Sun
     becomes kinetic energy.

68 An object is thrown into the air and follows the path
    shown in the diagram at the right. Which vector best
    represents the acceleration of the object at point *A*?
    [Neglect air friction.]

    (1) ╱   (2) ╱   (3) ↓   (4) ⟵

69 Above a flat horizontal
    plane, an arrow, *A*, is shot
    horizontally from a bow at
    a speed of 50 meters per
    second, as shown in the
    diagram at the right. *A*
    second arrow, *B*, is
    dropped from the same

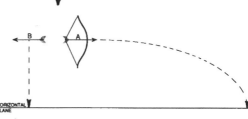

    height and at the same instant as *A* is fired. Neglecting air friction,
    compared to the amount of time *A* takes to strike the plane, the amount
    of time *B* takes to strike the plane is
    1   less   2   greater   3   the same

70 Two satellites, *A* and *B*, are traveling around the Earth in nearly
    circular orbits. The radius of satellite *A*'s orbit is greater than the
    radius of satellite *B*'s orbit. Compared to the orbital period of satellite
    A, the orbital period of satellite *B* is
    1   less   2   greater   3   the same

Group 2 — Internal Energy
If you chose this group, be sure to answer questions 71 - 80.

Base your answer to
question 71 on the
graph at the right
which represents the
temperature of
2.0 kilograms of a
material as a function
of the heat added to
the substance.

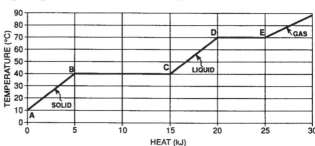

71 During which two intervals shown on the graph is the average potential
    energy of the molecules of the material increasing?
    (1) *AB* and *CD*   (3) *CD* and *DE*
    (2) *BC* and *DE*   (4) *AB* and *DE*

72 A temperature reading of absolute zero for a system would mean that
    the system's
    1   temperature is –273 K   3   kinetic energy is at a maximum
    2   temperature is 273° C   4   internal energy is at a minimum

73 Equal masses of aluminum, silver, tungsten, and zinc, initially at room
    temperature, are cooled 10 C°. Which metal gives off the most heat?
    1   aluminum   2   silver   3   tungsten   4   zinc

74 Two objects, *A* and *B*, are in contact with one another. Initially, the temperature of *A* is 300 K and the temperature of *B* is 400 K. Which diagram best represents the net flow of heat in the closed system? [Arrows represent the direction of heat flow.]

(1) [A | B]   (2) [A | B]   (3) [A | B]   (4) [A | B]

75 What is the final temperature of a 10.-kilogram sample of lead, initially at 0°C, after it has absorbed 39 kilojoules of heat energy?
(1) 1.3°C   (2) 3.9°C   (3) 30.°C   (4) 300°C

76 Which substance is a liquid at 0°C?
1   alcohol   3   ammonia
2   aluminum   4   copper

77 Rock salt is thrown on icy pavement to make roadways safer for driving in winter. This process works because the dissolved salt
1   raises the temperature of water
2   raises the freezing point of water
3   lowers the freezing point of water
4   lowers the boiling point of water

78 One kilogram of an ideal gas is heated from 27°C to 327°C. If the volume of the gas remains constant, the ratio of the pressure of the gas before heating to the pressure after heating is
(1) 1:1   (2) 1:2   (3) 1:3   (4) 1:4

79 A quantitative measure of the disorder of a system is called
1   entropy   2   fusion   3   equilibrium   4   vaporization

80 A force causes an object on a horizontal surface to overcome friction and begin to move. As this happens, the object's internal energy will
1   decrease   2   increase   3   remain the same

Group 3 — Electromagnetic Applications
If you chose this group, be sure to answer questions 81 - 90.

81 Which type of energy conversion occurs in an electric motor?
1   rotational mechanical energy to electrical energy
2   electrical energy to rotational mechanical energy
3   chemical energy to induced electrical energy
4   induced electrical energy to stored chemical energy

82 In the diagram at the right, a portion of a wire is being moved upward through a magnetic field. The direction of the induced electron current in the wire is toward point
(1) *A*
(2) *B*
(3) *C*
(4) *D*

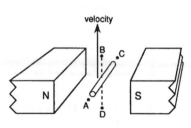

83   The calculation of the mass of an electron is based on the results of the
     Millikan oil drop experiment and on the
     1   charge on the proton
     2   charge-to-mass ratio of the electron
     3   mass of an oil drop
     4   mass-to-weight ratio of the neutron

84   Electrons are ejected from the filament of a cathode-ray tube when it
     becomes very hot. This phenomenon is an example of
     1   thermionic emission   3   back electromotive force
     2   photoelectric emission   4   torque

Base your answers to questions 85 & 86 on the diagram which represents an
electron entering the region between two oppositely charged parallel plates.

85   In which direction will the electron be deflected by the electric field?
     1   toward the bottom of the page
     2   toward the top of the page
     3   into the page
     4   out of the page

86   If the magnitude of the electric field between the plates is 4.0 x 10^3 volts
     per meter, the electric force on the electron is
     (1) 2.5 x 10^{22} N   (3) 6.4 x 10^{-16} N
     (2) 4.0 x 10^3 N   (4) 4.0 x 10^{-23} N

87   In the diagram at the right, a wire 0.50 meter long is
     moved at a speed of 2.0 meters per second
     perpendicularly through a uniform magnetic field with a
     flux density of 3.0 teslas directed into the page. What is
     the induced electromotive force?
     (1) 1.0 V   (3) 3.0 V
     (2) 1.5 V   (4) 12 V

Base your answers to questions 88 and
89 on the diagram at the right which
represents a 100% efficient device
connected to an alternating current
source of 220 volts. The primary coil has
50 turns and the secondary coil has 25
turns. When the device operates, a
0.50-ampere current flows through the
primary.

88   The device represented by this diagram is
     1   an induction coil   3   a generator
     2   a motor   4   a transformer

89   What is the potential difference across resistor R in the secondary coil?
     (1) 110 V   (2) 220 V   (3) 440 V   (4) 2,800 V

90   Compared to the internal resistance of a voltmeter, the internal
     resistance of an ammeter is
     1   smaller   2   greater   3   the same

Group 4 — Geometric Optics
If you chose this group, be sure to answer questions 91 - 100.

91 The image of an object is viewed in a plane mirror. What is the ratio of the object size to the image size?
(1) 1:1   (2) 2:1   (3) 1:2   (4) 1:4

92 In the diagram at the right, how far from the mirror is the light bulb (object) most likely located?
1   closer than the focal length of the mirror
2   at the principal focus of the mirror
3   at twice the focal length of the mirror
4   farther than twice the focal length of the mirror

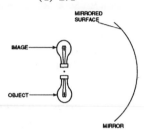

93 A light ray is incident upon a cylindrical reflecting surface as shown in the diagram at the right. The ray will most likely be reflected toward letter
(1) *A*
(2) *B*
(3) *C*
(4) *D*

94 A searchlight consists of a high-intensity light source at the focal point of a concave (converging) mirror. The light reflected from the mirror will
1   diverge uniformly   3   scatter in all directions
2   converge to a point   4   form a nearly parallel beam

95 The image produced by a convex (diverging) mirror must be
1   real and erect   3   virtual and erect
2   real and inverted   4   virtual and inverted

96 Which phenomenon of light accounts for the formation of images by a lens?
1   reflection   2   refraction   3   dispersion   4   polarization

Base your answers to questions 97 and 98 on the diagram at the right which represents a convex lens being used to form the image of an object. The distance from the center of the lens to the object is 0.060 meter. The distance from the center of the lens to the image is 0.120 meter.

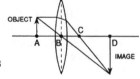

97 If the height of the object is 2.6 x 10^{-2} meter, the height of the image is
(1) 1.3 x 10^{-2} m   (3) 2.6 x 10^{-2} m
(2) 2.0 x 10^{-2} m   (4) 5.2 x 10^{-2} m

98 The focal length of the lens is
(1) 25 m   (3) 0.12 m
(2) 2.0 m   (4) 0.040 m

99 The same frequency of monochromatic light is incident from air upon four lenses having the same curvature, but made of different materials. Which lens has the shortest focal length?

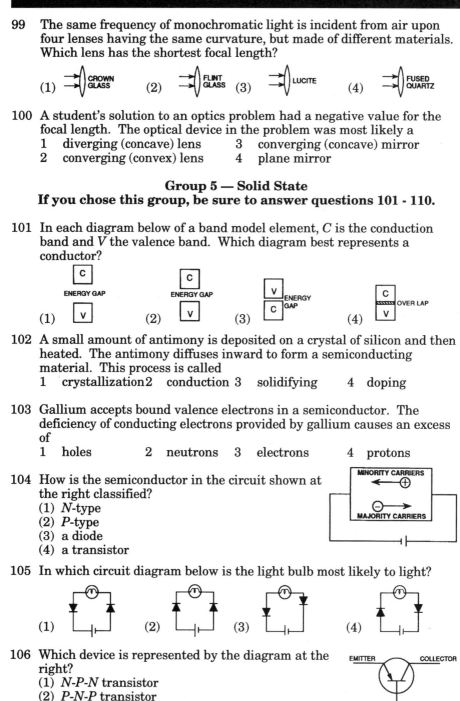

(1) CROWN GLASS   (2) FLINT GLASS   (3) LUCITE   (4) FUSED QUARTZ

100 A student's solution to an optics problem had a negative value for the focal length. The optical device in the problem was most likely a
1   diverging (concave) lens   3   converging (concave) mirror
2   converging (convex) lens   4   plane mirror

Group 5 — Solid State
If you chose this group, be sure to answer questions 101 - 110.

101 In each diagram below of a band model element, C is the conduction band and V the valence band. Which diagram best represents a conductor?

(1) C ENERGY GAP V   (2) C ENERGY GAP V   (3) V ENERGY GAP C   (4) C OVER LAP V

102 A small amount of antimony is deposited on a crystal of silicon and then heated. The antimony diffuses inward to form a semiconducting material. This process is called
1   crystallization   2   conduction   3   solidifying   4   doping

103 Gallium accepts bound valence electrons in a semiconductor. The deficiency of conducting electrons provided by gallium causes an excess of
1   holes   2   neutrons   3   electrons   4   protons

104 How is the semiconductor in the circuit shown at the right classified?
(1) N-type
(2) P-type
(3) a diode
(4) a transistor

MINORITY CARRIERS
MAJORITY CARRIERS

105 In which circuit diagram below is the light bulb most likely to light?

(1)   (2)   (3)   (4)

106 Which device is represented by the diagram at the right?
(1) N-P-N transistor
(2) P-N-P transistor
(3) zener diode
(4) solenoid

EMITTER   COLLECTOR
BASE

107 How is current affected when a semiconductor diode is reverse biased?
    1 Current conduction increases because the junction electric field
       barrier increases.
    2 Current conduction increases because the junction electric field
       barrier decreases.
    3 Current conduction decreases because the junction electric field
       barrier increases.
    4 Current conduction decreases because the junction electric field
       barrier decreases.

108 The graph at the right represents the
    relationship between the current and
    applied potential for a diode. The
    avalanche region of the graph is in
    Quadrant
    (1) I
    (2) II
    (3) III
    (4) IV

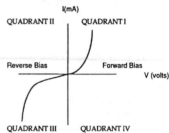

109 In an *N–P–N* transistor, the emitter-base combination is forward biased.
    In this transistor, there is a flow of
    1 electrons from the base to the emitter
    2 electrons from the emitter to the base
    3 holes from the base to the collector
    4 holes from the emitter to the base

110 As the number of free charges per unit value of a solid increases, its
    electrical conductivity
    1 decreases   2 increases   3 remains the same

Group 6 — Nuclear Energy
If you chose this group, be sure to answer questions 111 - 120.

111 What is the total number of neutrons in the nucleus of a $^{234}_{91}$ Pa atom ?

    (1) 91   (2) 143   (3) 234   (4) 325

112 Approximately how much energy would be generated if the mass in a
    nucleus of a $^{2}_{1}$H atom were completely converted to energy? [The mass
    of $^{2}_{1}$H is 2.0 atomic mass units.]

    (1) 3.2×10^{-19} J   (3) 9.3×10^{2} MeV
    (2) 1.5×10^{-10} J   (4) 1.9×10^{3} MeV

113 Which two symbols represent isotopes of the same element?

    (1) $^{8}_{2}X$ & $^{8}_{4}X$ (2) $^{6}_{2}X$ & $^{8}_{2}X$ (3) $^{3}_{2}X$ & $^{2}_{3}X$ (4) $^{2}_{2}X$ & $^{4}_{4}X$

114 The half-life of $^{234}_{90}$Th is 24 days. How much of a 128-milligram sample
    of thorium will remain after 144 days?
    (1) 5.3 mg   (2) 2 mg   (3) 21.3 mg   (4) 64 mg

115 Which equation is a step in the Uranium Disintegration Series?

(1) $^{234}_{90}$ Th → $^{234}_{91}$ Pa + $^{0}_{-1}$ e   (3) $^{234}_{90}$ Th → $^{226}_{88}$ Ra + $^{4}_{2}$ He

(2) $^{234}_{90}$ Th → $^{230}_{88}$ Ra + $^{4}_{2}$ He   (4) $^{234}_{90}$ Th → $^{226}_{89}$ Ac + $^{0}_{-1}$ e

116 Which particles will *not* increase in kinetic energy in a particle accelerator?
1   alpha particles   3   protons
2   beta particles   4   neutrons

117 Which is an example of electron capture (*K*-capture)?

(1) $^{27}_{13}$Al + $^{4}_{2}$ He → $^{30}_{15}$ P + $^{1}_{0}$ n

(2) $^{238}_{92}$U + $^{1}_{0}$ n → $^{239}_{92}$ U → $^{239}_{93}$ Np + $^{0}_{-1}$ e

(3) $^{40}_{19}$K + $^{0}_{-1}$ e → $^{40}_{18}$Ar

(4) $^{64}_{29}$ Cu → $^{64}_{28}$ Ni + $^{0}_{+1}$ e

118 Which equation represents nuclear fission?

(1) $^{226}_{88}$ Ra → $^{222}_{86}$ Rn + $^{4}_{2}$ He

(2) $^{24}_{11}$Na → $^{24}_{12}$ Mg + $^{0}_{-1}$ e

(3) $^{9}_{4}$Be + $^{4}_{2}$ He → $^{12}_{6}$ C + $^{1}_{0}$ n

(4) $^{235}_{92}$ U + $^{1}_{0}$ n → $^{141}_{56}$ Ba + $^{92}_{36}$ Kr + 3^{1}_{0} n + Q

119 The critical mass of nuclear reactor materials is defined as the mass of
1   shielding material needed to reflect neutrons back into the reactor
2   moderating material needed to control the core's temperature
3   fissionable material necessary for a chain reaction to take place
4   control rods needed to absorb excess neutrons

120 Which factor associated with radioactive waste has the biggest impact on its storage time?
1   half-life   2   quantity   3   temperature   4   density

Part III
You must answer all questions in this part. [10]

121 Base your answers to parts *a* through *c* on the information below.

A student pulls a cart across a horizontal floor by exerting a force of 50. newtons at an angle of 35° to the horizontal.

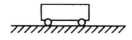

a   On the diagram, *using a protractor and a straightedge,* construct a scaled vector showing the 50.-newton force acting on the cart at the appropriate angle. The force *must* be drawn to a scale of 1.0 centimeter = 10. N. Label the 50.-newton force and the 35° angle on your diagram. *Be sure your final answer appears with the correct labels (numbers and units).* [2]

b   Construct the horizontal component of the force vector to scale on your diagram and label it H. [1]

c   What is the magnitude of the horizontal component of the force? [1]

122 Base your answers to parts a and b on the information and diagram.

The sonar of a stationary ship sends a signal with a frequency of 5.0 x 10^3 hertz down through water. The speed of the signal is 1.5 x 10^3 meters per second. The echo from the bottom is detected 4.0 seconds later.

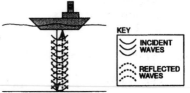

a   What is the wavelength of the sonar wave? [Show all calculations, including the equation and substitution with units.] [2]

b   What is the depth of the water under the ship? [Show all calculations, including the equation and substitution with units.] [2]

Base your answer to question 123 on the information and graph below.

A student performed an experiment measuring the maximum kinetic energy of emitted photoelectrons as the frequency of light shining on a photoemissive surface was increased. A graph of the student's data appears below.

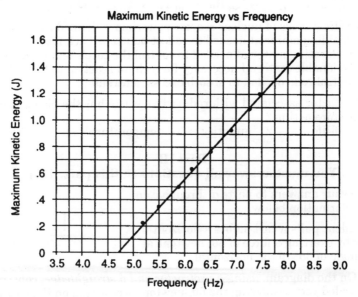

123 On the graph, draw a line representing the relationship between maximum kinetic energy and frequency when a photoemissive surface having a larger work function is used. [2]

Physics Practice Exam 2 – June 1992

Part I Answer all 55 questions in this part.   **[65]**
Directions (1-55): For *each* statement or question, select the word or expression that, of those given, best completes the statement or answers the question.

1   The velocity of a car changes from 60. meters per second north to 45 meters per second north in 5.0 seconds. The magnitude of the car's acceleration is
(1) 9.8 m/s^2 (3) 3.0 m/s^2
(2) 15 m/s^2 (4) 53 m/s^2

2   The height of a doorknob above the floor is approximately
(1) 1×10^2 m   (3) 1×10^0 m
(2) 1×10^1 m   (4) 1×10^{-2} m

3   Which two terms represent a vector quantity and the scalar quantity of the vector's magnitude, respectively?
1   acceleration and velocity   3   speed and time
2   weight and force   4   displacement and distance

4   A group of bike riders took a 4.0-hour trip. During the first 3.0 hours, they traveled a total of 50. kilometers, but during the last hour they traveled only 10. kilometers. What was the group's average speed for the entire trip?
(1) 15 km/hr   (3) 40. km/hr
(2) 30. km/hr   (4) 60. km/hr

5   The graph at the right represents the motion of a body moving along a straight line. According to the graph, which quantity related to the motion of the body is constant?
1   speed   3   acceleration
2   velocity   4   displacement

6   A student walks 1.0 kilometer due east and 1.0 kilometer due south. Then she runs 2.0 kilometers due west. The magnitude of the student's resultant displacement is closest to
(1) 0 km   (3) 3.4 km
(2) 1.4 km   (4) 4.0 km

7   A locomotive starts from rest and accelerates at 0.12 meter per second^2 to a speed of 2.4 meters per second in 20. seconds. This motion could best be described as
1   constant acceleration and constant velocity
2   increasing acceleration and constant velocity
3   constant acceleration and increasing velocity
4   increasing acceleration and increasing velocity

8   A clam dropped by a sea gull takes 3.0 seconds to hit the ground. What is the sea gull's approximate height above the ground at the time the clam was dropped?
(1) 15 m   (3) 45 m
(2) 30. m   (4) 90. m

9   Which graph best represents the relationship between mass and acceleration due to gravity for objects near the surface of the Earth? [Neglect air resistance.]

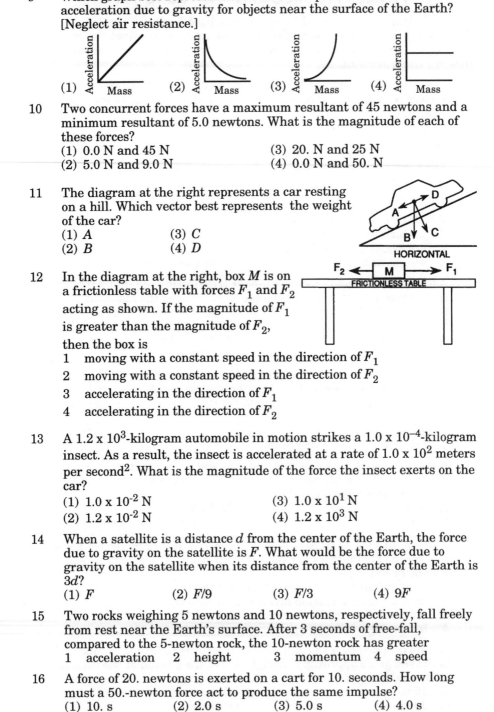

10 Two concurrent forces have a maximum resultant of 45 newtons and a minimum resultant of 5.0 newtons. What is the magnitude of each of these forces?
(1) 0.0 N and 45 N   (3) 20. N and 25 N
(2) 5.0 N and 9.0 N   (4) 0.0 N and 50. N

11 The diagram at the right represents a car resting on a hill. Which vector best represents the weight of the car?
(1) *A*   (3) *C*
(2) *B*   (4) *D*

12 In the diagram at the right, box *M* is on a frictionless table with forces F_1 and F_2 acting as shown. If the magnitude of F_1 is greater than the magnitude of F_2, then the box is
1   moving with a constant speed in the direction of F_1
2   moving with a constant speed in the direction of F_2
3   accelerating in the direction of F_1
4   accelerating in the direction of F_2

13 A 1.2×10^3-kilogram automobile in motion strikes a 1.0×10^{-4}-kilogram insect. As a result, the insect is accelerated at a rate of 1.0×10^2 meters per second^2. What is the magnitude of the force the insect exerts on the car?
(1) 1.0×10^{-2} N   (3) 1.0×10^1 N
(2) 1.2×10^{-2} N   (4) 1.2×10^3 N

14 When a satellite is a distance *d* from the center of the Earth, the force due to gravity on the satellite is *F*. What would be the force due to gravity on the satellite when its distance from the center of the Earth is 3*d*?
(1) *F*   (2) *F*/9   (3) *F*/3   (4) 9*F*

15 Two rocks weighing 5 newtons and 10 newtons, respectively, fall freely from rest near the Earth's surface. After 3 seconds of free-fall, compared to the 5-newton rock, the 10-newton rock has greater
1   acceleration   2   height   3   momentum   4   speed

16 A force of 20. newtons is exerted on a cart for 10. seconds. How long must a 50.-newton force act to produce the same impulse?
(1) 10. s   (2) 2.0 s   (3) 5.0 s   (4) 4.0 s

17   A 20.-kilogram object strikes the ground with 1,960 joules of kinetic energy after falling freely from rest. How far above the ground was the object when it was released?

(1) 10. m   (2) 14 m   (3) 98 m   (4) 200 m

18   The graph at the right shows the elongation of a spring as a function of the force. What is the value of the spring constant?

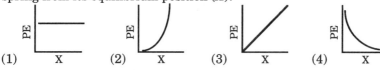

(1) 0.1 m/N
(2) 0.1 N/m
(3) 10 m/N
(4) 10 N/m

19   Which graph best represents the relationship between the potential energy stored in a spring (PE) and the change in the length of the spring from its equilibrium position (X)?

(1)   X   (2)   X   (3)   X   (4)   X

20   An object with +10 elementary charges is grounded and becomes neutral. What is the best explanation for this occurrence?

1   The object gained 10 electrons from the ground.
2   The object lost 10 electrons to the ground.
3   The object gained 10 protons from the ground.
4   The object lost 10 protons to the ground.

21   Two identical spheres carry charges of +0.6 coulomb and −0.2 coulomb, respectively. If these spheres tough, the resulting charge on the first sphere will be

(1) +0.8 C   (2) +0.2 C   (3) −0.3 C   (4) +0.4 C

22   The diagram at the right represents two charges, q_1 and q_2, separated by distance d. Which change would produce the greatest increase in the electrical force between the two charges?

1   doubling charge q_1, only
2   doubling d, only
3   doubling d and charge q_1, only
4   doubling d and charges q_1 and q_2

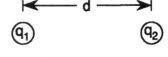

23   A helium ion with +2 elementary charges is accelerated by a potential difference of 5.0×10^3 volts. What is the kinetic energy acquired by the ion?

(1) 3.2×10^{-19} eV   (3) 5.0×10^3 eV
(2) 2.0 eV   (4) 1.0×10^4 eV

24   Which is the unit of electrical power?

1   volt/ampere   3   ampere^2/ohm
2   ampere/ohm   4   volt^2/ohm

25   Two equal positive point charges, A and B,
      are positioned as shown in the diagram. At
      which location is the electric field intensity
      due to these two charges equal to zero?
      (1) A (2) B (3) X (4) Y

26   In the circuit shown at the right, how many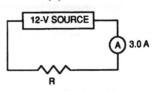
      coulombs of charge will pass through resistor
      R in 2.0 seconds?
      (1) 36 V   (3) 3.0 C
      (2) 6.0 C   (4) 4.0 C

27   Which graph best represents how the resistance (R) of a series of copper
      wires of uniform length and temperature varies with crossectional area
      (A)?

      (1)   (2)   (3)   (4)

28   A physics student is given three 12-ohm resistors with instructions to
      create the circuit that would have the lowest possible resistance. The
      correct circuit would be a
      1   series circuit with a total resistance of 36 Ω
      2   series circuit with a total resistance of 4 Ω
      3   parallel circuit with a total resistance of 36 Ω
      4   parallel circuit with a total resistance of 4 Ω

29   Ammeters A_1, A_2, and A_3 are placed in a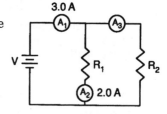
      circuit as shown in the diagram. What is the
      reading on ammeter A_3?
      (1) 1.0 A
      (2) 2.0 A
      (3) 3.0 A
      (4) 5.0 A

30   What is the approximate amount of
      electrical energy needed to operate a 1,600-watt toaster for 60. seconds?
      (1) 27 J   (2) 1,500 J   (3) 1,700 J   (4) 96,000 J

31   An electric motor uses 15 amperes of current in a 440-volt circuit to
      raise an elevator weighing 11,000 newtons. What is the average speed
      attained by the elevator?
      (1) 0.0027 m/s   (2) 0.60 m/s   (3) 27 m/s   (4) 6,000 m/s

32   Which phenomenon does *not* occur when a sound wave reaches the
      boundary between air and a steel block?
      1   reflection   2   refraction   3   polarization   4   absorption

33   The speaker in the diagram makes use of a
      current-carrying coil of wire. The N-pole of the coil
      would be closest to
      (1) A (3) C
      (2) B (4) D

34   An electron current (e⁻) moving upward through a straight conductor creates a magnetic field. Which diagram correctly represents this magnetic field?

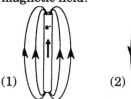 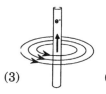

(1)   (2)   (3)   (4)

35   What is the period of a periodic wave that has a frequency of 60. hertz?
(1) 1.7×10^{-2} s   (3) 3.0×10^{-3} s
(2) 2.0×10^{4} s   (4) 3.3×10^{2} s

36   Which point on the wave diagram is in phase with point A?
(1) E
(2) B
(3) C
(4) D

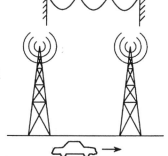

37   What is the wavelength of 30.-hertz periodic wave moving at 60. meters per second?
(1) 0.50 m   (2) 2.0 m   (3) 20. m   (4) 1,800 m

38   How many nodes are represented in the standing wave diagram at the right?
(1) 1   (3) 3
(2) 6   (4) 4

39   A car radio is tuned to the frequency being emitted from two transmitting towers. As the car moves at constant speed past the towers, as shown in the diagram, the sound from the radio repeatedly fades in and out. This phenomenon can best be explained by
1   refraction
2   interference
3   reflection
4   resonance

40   As a periodic wave travels from one medium to another, which pair of the wave's characteristics cannot change?
(1) period and frequency   (3) frequency and velocity
(2) period and amplitude   (4) amplitude and wavelength

41   When light rays from an object are incident upon an opaque, rough-textured surface, no reflected image of the object can be seen. This phenomenon occurs because of
1   regular reflection
2   diffuse reflection
3   reflected angles not being equal to incident angles
4   reflected angles not being equal to refracted angles

42   The diagram represents a light ray being reflected
     from a plane mirror. The angle between the
     incident ray and the reflected ray is 70.°. What is
     the angle of incidence for this ray?
     (1) 20.°   (3) 55°
     (2) 35°   (4) 70.°

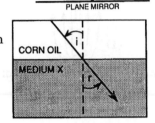

43   The diagram represents a light ray passing
     from corn oil into medium X with no change in
     velocity. In this diagram, medium X could be
     1   water
     2   diamond
     3   glycerol
     4   alcohol

44   A ray of light in air is incident on a block of Lucite at an angle of 60.°
     from the normal. The angle of refraction of this ray in the Lucite is
     closet to
     (1) 35°   (2) 45°   (3) 60.°   (4) 75°

Base your answers to questions 45 and 46 on the diagram which represents a
monochromatic light wave passing through the double slits. A pattern of
bright and dark bands is formed on the screen.

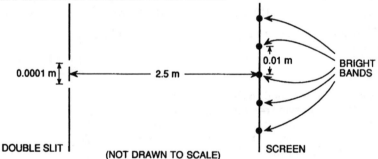

45   What is the color of the light used?
     1   violet   2   blue   3   green   4   yellow

46   If the original light wave is replaced by a wave of longer wavelength,
     the space between the bright bands on the screen will
     1   decrease   2   increase   3   remain the same

47   The graph shows the relationship between
     the frequency of radiation incident on the
     photosensitive surface and the maximum
     kinetic energy (KE_{max}) of the emitted
     photoelectrons. The point labeled A on the
     graph represents the
     1   incident photon intensity
     2   photoelectron frequency
     3   threshold frequency
     4   work function energy

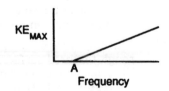

48 A metal has a work function of 1.3×10^{-18} joule. What is the threshold frequency for electromagnetic radiation incident on this metal?
    (1) 2.0×10^{15} Hz   (3) 8.6×10^{14} Hz
    (2) 2.0×10^{14} Hz   (4) 8.6×10^{-52} Hz

49 A photon emitted from an excited hydrogen atom has an energy of 3.02 electronvolts. Which electron energy-level transition would produce this photon?
    (1) $n = 1$ to $n = 6$ (3) $n = 6$ to $n = 1$
    (2) $n = 2$ to $n = 6$ (4) $n = 6$ to $n = 2$

50 The diagram represents alpha particle A approaching a gold nucleus. D is the distance between the path of the alpha particle and the path for a head-on collision. If D is decreased, the angle of deflection (θ) of the alpha particle would

    1   decrease
    2   increase
    3   remain the same

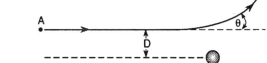

51 A lawnmower is pushed with a constant force F, as shown in the diagram. As angle θ between the lawnmower handle and the horizontal increases, the horizontal component of F
    1   decreases
    2   increases
    3   remains the same

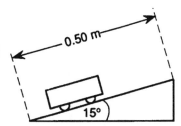

52 As the time required to do a given quantity of work decreases, the power developed
    1   decreases   2   increases   3   remains the same

53 As the speed of a bicycle moving along a level horizontal surface changes from 2 meters per second to 4 meters per second, the magnitude of the bicycle's gravitational potential energy
    1   decreases   2   increases   3   remains the same

54 As shown in the diagram, pulling a 9.8-newton cart a distance of 0.50 meter along a plane inclined at 15° requires 1.3 joules of work. If the cart were raised 0.50 meter vertically instead of being pulled along the inclined plane, the amount of work done would be
    1   less
    2   more
    3   the same

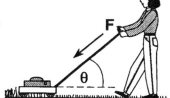

55 As the momentum of an electron increases, the electron's wavelength
    1   decreases   2   increases   3   remains the same

Part II

This part consists of six groups, each containing ten questions. Each group tests an optional area of the course. Choose two of these six groups. Be sure to answer all ten questions in each group chosen. [20]

Group 1 — Motion in a Plane
If you chose this group, be sure to answer questions 56 - 65.

Base your answers to questions 56 through 58 on the following information: An object is thrown horizontally off a cliff with an initial velocity of 5.0 meters per second. The object strikes the ground 3.0 seconds later.

56   What is the vertical speed of the object as it reaches the ground? [Neglect friction.]
      (1) 130 m/s   (2) 29 m/s   (3) 15 m/s   (4) 5.0 m/s

57   How far from the base of the cliff will the object strike the ground? [Neglect friction.]
      (1) 2.9 m   (2) 9.8 m   (3) 15 m   (4) 44 m

58   What is the horizontal speed of the object 1.0 second after it is released? [Neglect friction.]
      (1) 5.0 m/s   (2) 10. m/s   (3) 15 m/s   (4) 30. m/s

Base your answers to questions 59 and 60 on the diagram which shows a ball thrown toward the east and upward at an angle of 30° to the horizontal. Point X represents the ball's highest point.

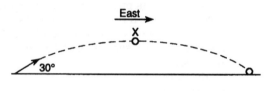

59   What is the direction of the ball's velocity at point X? [Neglect friction.]
      1   down   2   up   3   west   4   east

60   What is the direction of the ball's acceleration at point X? [Neglect friction.]
      1   down   2   up   3   west   4   east

61   A batted softball leaves the bat with an initial velocity of 44 meters per second at an angle of 37° above the horizontal. What is the magnitude of the initial vertical component of the softball's velocity?
      (1) 0 m/s   (2) 26 m/s   (3) 35 m/s   (4) 44 m/s

62   The diagram shows the path of a satellite in an elliptical orbit around the Earth. As the satellite moves from point A to point B, what changes occur in its potential and kinetic energies?

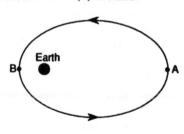

      1   Both potential energy and kinetic energy increase.
      2   Both potential energy and kinetic energy decrease.
      3   Potential energy increases and kinetic energy decreases.
      4   Potential energy decreases and kinetic energy increases.

63   A satellite is in geosynchronous orbit around the Earth. The period of
      the satellite's orbit is closest to
      (1) 6 hours   (2) 12 hours   (3) 24 hours   (4) 48 hours

Base your answers to questions 64 and 65 on the
diagram which represents a space station shaped
like a wheel and having a radius of 40. meters. The
station is rotating clockwise with a speed of 12
meters per second at its outer wall. A 50-kilogram
astronaut is standing inside the space station
against the outer wall.

64   What is the apparent weight of the astronaut?
      [Consider only the interacting forces between
      the space station and the astronaut.]
      (1) 3.6 N   (3) 180 N
      (2) 15 N   (4) 490 N

65   If the speed of rotation of the space station were doubled, the
      centripetal acceleration of the astronaut would
      1   decrease   2   increase   3   remain the same

Group 2 — Internal Energy
If you chose this group, be sure to answer questions 66 - 75.

66   What is the difference between 15°C and 6°C, expressed in Kelvin?
      (1) 282 K   (2) 273 K   (3) 15 K   (4) 9 K

67   A certain mass of lead requires 1 kilojoule of heat energy to raise its
      temperature from 20.°C to 44°C. If 1 kilojoule of heat energy is added to
      the same mass of copper at 20.°C, what will be the final temperature of
      the copper?
      (1) 28°C   (2) 32°C   (3) 44°C   (4) 92°C

68   Maximum average molecular kinetic energy in the solid phase of
      mercury is reached at a temperature of
      (1) –39°C   (2) 0°C   (3) 357°C   (4) 396°C

69   If two substances are placed in contact with each other and no net
      exchange of internal energy occurs between them, the substances must
      have the same
      1   specific heat   3   temperature
      2   melting point   4   heat of fusion

70   What is the approximate amount of heat energy needed to change 5.0
      kilograms of ice at 0°C to water at 0°C?
      (1) 21 kJ   (2) 340 kJ   (3) 1,700 kJ   (4) 11,000 kJ

71   According to the kinetic theory, pressure exerted by a gas is caused by
      the
      1   collision of gas molecules with each other
      2   collision of gas molecules with the walls of the container
      3   negligible volume of the gas molecules
      4   large forces between gas molecules

72   Which graph best represents the relationship between pressure (P) and absolute temperature (T_K) for a fixed mass of an ideal gas at constant volume?

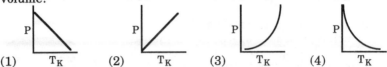

(1) T_K (2) T_K (3) T_K (4) T_K

73   When a student drops a beaker, it shatters, spreading randomly shaped pieces of glass over a large area of the floor. According to the 2nd law of thermodynamics, the measure of the disorder of this system is known as

1   vaporization   3   molecular collision
2   absolute order   4   entropy

74   Compared to the quantity of heat needed to raise the temperature of 10 grams of liquid water 5 C°, the heat needed to raise the temperature of 10 grams of ice 5 C° is

1   less   2   greater   3   the same

75   The graph below shows temperature versus time for 1 kilogram of water at constant pressure as heat is added at a constant rate. Dissolving a salt in the water will cause interval AB to occur at

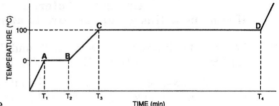

1   a lower temperature
2   a higher temperature
3   the same temperature

Group 3 — Electromagnetic Applications
If you chose this group, be sure to answer questions 76 - 85.

76   Which electrical device must have high resistance in order to function properly?

1   electric motor   3   voltmeter
2   ammeter   4   galvanometer

Base your answers to questions 77 and 78 on the diagram which represents an electron with a velocity of 2.0 x 10^6 meters per second directed into a region between two large, flat charged parallel plates.

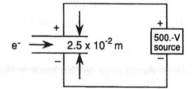

77   The magnitude of the electric field intensity between the plates is

(1) 1.0×10^2 N/C   (3) 5.0×10^{-5} N/C
(2) 2.0×10^4 N/C   (4) 2.5×10^{-11} N/C

78 The direction of the acceleration of the electron in the region between the plates is
1 into the page
2 out of the page
3 toward the bottom of the page
4 toward the top of the page

Base your answers to questions 79 and 80 on the diagram which represents a beam of electrons moving through a uniform magnetic field. The magnetic field is directed into the page.

magnetic field

X X X X
X X X X
X X▲X X
X X│X X
e⁻

79 As the beam of electrons moves through the magnetic field, the electrons will be deflected
1 into the page
2 out of the page
3 toward the left
4 toward the right

80 If the speed of an electron in the magnetic field is 6.0 x 10^6 meters per second and a force of 5.0 x 10^{-14} newton acts on the electron, what is the flux density of the magnetic field?
(1) 5.2 x 10^{-2} T
(2) 8.3 x 10^{-21} T
(3) 3.0 x 10^{-7} T
(4) 1.9 x 10^1 T

81 The thermionic emission of electrons from the metal filament of an operating light bulb is caused by
1 heat
2 magnetism
3 radio waves
4 sound waves

82 A conductor 0.10 meter long moves with a velocity of 8.0 meters per second perpendicular to a magnetic field measuring 4.0 x 10^{-2} newton per ampere-meter. What is the magnitude of the electromotive force induced in the conductor?
(1) 5.0 x 10^{-3} V
(2) 2.0 x 10^{-1} V
(3) 3.2 x 10^{-2} V
(4) 4.0 x 10^{-3} V

83 Some fluorescent ceiling lights operate at higher voltage than that supplied by household circuits. Which device is used to increased the voltage for these lights?
1 laser 2 transformer 3 motor 4 generator

84 Compared to the power developed in the primary coil of a 100% efficient transformer, the power developed in the secondary coil is
1 less 2 greater 3 the same

85 The only difference between two motors is the material of their armature cores. Motor *A* has its coil wrapped around a piece of soft iron, and motor *B* has its coil wrapped around a piece of wood. Compared to the force exerted on the armature of motor *A*, the force exerted on the armature of motor *B* is
1 less 2 greater 3 the same

Group 4 — Geometric Optics
If you chose this group, be sure to answer questions 86 - 95.

86 When the calculated image distance for an image formed using a curved mirror has a negative value, the image must be
1 real 2 virtual 3 reduced 4 enlarged

87   The diagram represents two light rays emerging from a candle flame and being reflected from a plane mirror. What does point *P* represent?

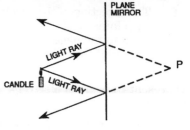

1   the virtual image point of the candle flame
2   the real image point of the candle flame
3   the focal point of the mirror
4   the center of curvature of the mirror

88   A pencil 0.10 meter long is placed 1.0 meter in front of a concave (converging) mirror whose focal length is 0.50 meter. The image of the pencil is
1   erect and 0.030 meter long   3   inverted and 0.030 meter long
2   erect and 0.10 meter long   4   inverted and 0.10 meter long

89   If the distance between an object and a concave (converging) mirror is more than twice the focal length of the mirror, the image formed will be
1   real and behind mirror   3   virtual and behind mirror
2   real and in front of mirror   4   virtual and in front of mirror

90   In the diagram, an object is located in front of a convex (diverging) mirror. *F* is the virtual focal point of the mirror, and *C* is its center of curvature. Ray *R* is parallel to the principal axis. Ray *R* will most likely be reflected along path

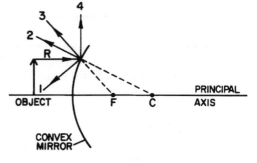

(1) 1   (3) 3
(2) 2   (4) 4

91   Which phenomenon is represented by the diagram?
1   reflection
2   refraction
3   diffraction
4   polarization

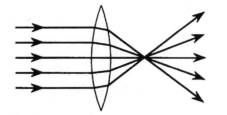

92   As a teacher showed slides by projecting them on a fixed screen, a student complained that the image was too small. The teacher enlarged the image by moving the projector away from the screen, but the image blurred. The image should then have been brought into focus by
1   moving the lens closer to the slide
2   moving the lens away from the slide
3   decreasing the amount of light in the room
4   increasing the power of the projector lamp

93 Which ray diagram is *incorrect*?

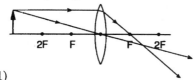

(1)

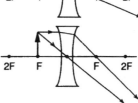

(3)

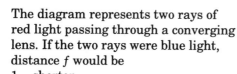

(2)

(4)

94 The diagram at the right shows a convex (converging) lens with focal length f. Where should an object be placed to produce a virtual image?
1 at f
2 at $2f$
3 between f and the lens
4 between $2f$ and f

95 The diagram represents two rays of red light passing through a converging lens. If the two rays were blue light, distance f would be
1 shorter
2 longer
3 the same

Group 5 — Solid State
If you chose this group, be sure to answer questions 96 - 105.

96 Which is an important factor in determining whether two materials can be used to form a semiconductor material?
1 One material must be able to donate electrons and the other must be able to accept electrons.
2 Both materials must have the same number of electrons in the valence shells.
3 One material must have positive electrons and the other must have negative electrons.
4 Both materials must have more neutrons than protons in their nuclei.

97 Which graph best represents the relationship between the conductivity and the temperature of semiconductors?

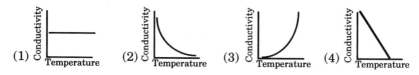

(1) (2) (3) (4)

98   In the circuit shown in the diagram, ammeter
      A reads 4 milliamperes when connected to an
      N-type semiconductor. If the connections to the
      battery are reversed, the reading of ammeter *A*
      will be

      (1) 8 mA   (3) 0 mA
      (2) 2 mA   (4) 4 mA

99   Why do current carriers have difficulty crossing a *P–N* junction?
      1   The junction area has a large positive charge.
      2   The junction area has a large negative charge.
      3   The junction acts as an electric field barrier.
      4   The resistance across the junction is extremely low.

100   In the *P–N* junction diode shown, in
      which direction do both the holes in the
      P–type material and the electrons in
      the *N*–type material move?
      1   away from the junction
      2   toward the junction
      3   toward the right
      4   toward the left

101   Which diagram shows both a forward and a reverse bias?

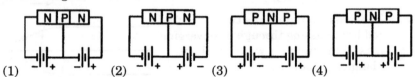

102   If the transistor shown in the diagram is to serve as an
      amplifier, most of the electrons must pass from
      (1) *B* to *E*   (3) *C* to *E*
      (2) *C* to *B*   (4) *E* to *C*

103   As the amount of doping material used in a diode increases, the
      potential difference needed to cause the avalanche
      1   decreases   2   increases   3   remains the same

104   When a semiconductor is replaced in a circuit by an insulator, the
      resistance of that section of the circuit
      1   decreases   2   increases   3   remains the same

105   Adding small amounts of an impurity such as phosphorus (5 valence
      electrons) to a semiconductor will cause the net charge of the
      semiconductor to
      1   decrease   2   increase   3   remain the same

Group 6 — Nuclear Energy
If you chose this group, be sure to answer questions 106 - 115.

106   A neutral atom of an isotope of element *X* has 44 electrons and 63
      neutrons. What is the mass number of this isotope?
      (1) 19   (2) 44   (3) 63   (4) 107

107 Compared to electrostatic forces, nuclear forces are
  1 weaker and of shorter range   3 stronger and of shorter range
  2 weaker and of longer range   4 stronger and of longer range

108 The half-life of $^{223}_{88}$Ra is 11.2 days. If mass M of this radium isotope is present initially, how much $^{223}_{88}$Ra remains at the end of 56 days?

  (1) $\frac{1}{2}$M   (2) $\frac{1}{4}$M   (3) $\frac{1}{5}$M   (4) $\frac{1}{32}$M

109 Which nuclear symbol represents an isotope of polonium (Po) that is part of the Uranium Disintegration Series?

  (1) $^{214}_{83}$Po   (2) $^{214}_{84}$Po   (3) $^{218}_{82}$Po   (4) $^{222}_{84}$Po

110 A decrease in both mass number and atomic number of a nucleus occurs due to the emission of
  1 an alpha particle   3 a neutron
  2 a beta particle   4 a positron

111 What kind of nuclear reaction is shown here? $^{1}_{1}$H + $^{3}_{1}$H − $^{4}_{2}$He + Q

  1 alpha decay   3 fusion
  2 beta decay   4 fission

112 If a proton were absorbed by $^{222}_{86}$Rn, the symbol for the resulting nucleus would be

  (1) $^{222}_{87}$Fr   (2) $^{223}_{87}$Fr   (3) $^{222}_{85}$At   (4) $^{223}_{86}$Rn

113 In nuclear reactors, the function of a moderator is to decrease the neutrons'
  1 binding energy   3 potential energy
  2 electromagnetic energy   4 kinetic energy

114 As two nuclei are moved closer together, the electrostatic force of repulsion between them
  1 decreases   2 increases   3 remains the same

115 As a star gives off energy in a thermonuclear reaction, the mass of the star
  1 decreases   2 increases   3 remains the same

Part III
You must answer all questions in this part. [15]

116 Base your answers to parts *a* through *c* on the following information: A student performs a laboratory activity in which a 15-newton force acts on a 2.0-kilogram mass. The work done over time is summarized in the table.

Data Table	
Time (s)	Work (J)
0	0
1.0	32
2.0	59
3.0	89
4.0	120

  a Using the information in the data table, construct a graph on the grid, following the directions on the next page.

(1) Develop an appropriate scale for work, and plot the points for a *work*-versus-*time* graph. [1]

(2) Draw the best-fit line. [1]

b Calculate the value of the slope of the graph constructed in part a. (Show all calculations, including equations and substitutions with units.) [2]

c Based on your graph, how much time did it take to do 75 joules of work? [1]

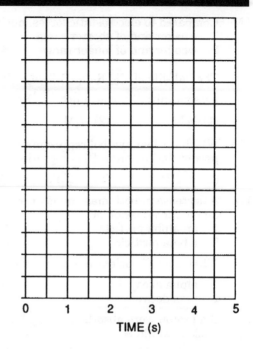

117 Base your answers to parts a through c on the following diagram and information: Two railroad carts, *A* and *B*, are on a frictionless, level track. Cart *A* has a mass of 2.0×10^3 kilograms and a velocity of 3.0 meters per second toward the right. Cart *B* has a velocity of 1.5 meters per second toward the left. The magnitude of the momentum of cart *B* is 6.0×10^3 kilogram-meters per second. When the two carts collide, they lock together.

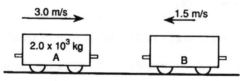

a What is the magnitude of the momentum of cart A before the collision? (Show all calculations, including equations and substitutions with units.) [2]

b On the diagram, construct a scaled vector that represents the momentum of cart A before the collision. The momentum vector **must** be drawn to a scale of 1.0 centimeter = 1,000 kilogram-meters per second. [1]

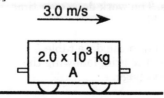

c   In one or more *complete sentences*, describe the momentum of the two carts after collision and justify your answer based on the initial momenta of both carts. [2]

118 Base your answers to parts *a* and *b* on the following information: Two resistors are connected in parallel to a 12-volt battery. One resistor, R_1, has a value of 18 ohms. The other resistor, R_2, has a value of 9 ohms. The total current in the circuit is 2 amperes. A student wishes to measure the current through R_1 and the potential difference across R_2.

a   Using the symbols at the right for a battery, an ammeter, a voltmeter, and resistors, draw and label a circuit diagram that will enable the student to make the desired measurements. [3]

SYMBOLS:

—||⊢— BATTERY

—(A)— AMMETER

—(V)— VOLTMETER

—⋀⋀⋀— RESISTOR

b   Calculate the value of the current in resistor R_1. (Show all calculations, including equations and substitutions with units.) [2]

Physics Practice Exam 3 – June 1993

Part I Answer all 55 questions in this part. [65]
Directions (1-55): For *each* statement or question, select the word or expression that, of those given, best completes the statement or answers the question.

1   A car travels a distance of 98 meters in 10. seconds. What is the average speed of the car during this 10.-second interval?
   (1) 4.9 m/s   (2) 9.8 m/s   (3) 49 m/s   (4) 98 m/s

2   Which measurement of an average classroom door is closest to 1 meter?
   1   thickness   2   width   3   height   4   surface area

3   A boat initially traveling at 10. meters per second accelerates uniformly at the rate of 5.0 meters per second² for 10. seconds. How far does the boat travel during this time?
   (1) 50. m   (2) 250 m   (3) 350 m   (4) 500 m

4   The graph at the right represents the relationship between distance and time for an object. What is the instantaneous speed of the object at $t = 5.0$ seconds?
   (1) 0 m/s
   (2) 2.0 m/s
   (3) 5.0 m/s
   (4) 4.0 m/s

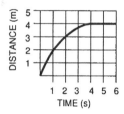

5   An object accelerates uniformly from rest to a speed of 50. meters per second in 5.0 seconds. The average speed of the object during the 5.0-second interval is
   (1) 5.0 m/s   (2) 10. m/s   (3) 25 m/s   (4) 50. m/s

6   A 5-newton ball and a 10-newton ball are released simultaneously from a point 50 meters above the surface of the Earth. Neglecting air resistance, which statement is true?
   1   The 5-N ball will have a greater acceleration than the 10-N ball.
   2   The 10-N ball will have a greater acceleration than the 5-N ball.
   3   At the end of 3 seconds of free-fall, the 10-N ball will have a greater momentum than the 5-N ball.
   4   At the end of 3 seconds of free-fall, the 5-N ball will have a greater momentum than the 10-N ball.

7   In the diagram at the right , the weight of a box on a plane inclined at 30.° is represented by the vector W. What is the magnitude of the component of the weight (W) that acts parallel to the incline?
   (1) W (3) $0.87W$
   (2) $0.50W$ (4) $1.5W$

8   The diagram at the right represents a force acting at point P. Which pair of concurrent forces would produce equilibrium when added to the force acting at point P?

(1)

(2)

(3)

(4)

9   A boat heads directly eastward across a river at 12 meters per second.
     If the current in the river is flowing at 5.0 meters per second due south,
     what is the magnitude of the boat's resultant velocity?
     (1) 7.0 m/s   (2) 8.5 m/s   (3) 13 m/s   (4) 17 m/s

10   A bird feeder with two birds has a total mass of 2.0
     kilograms and is supported by wire as shown in the
     diagram at the right. The force in the top wire is
     approximately
     (1) 10. N   (3) 20. N
     (2) 14 N   (4) 39 N

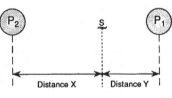

11   A 50.-kilogram woman wearing a seat belt is traveling in
     a car that is moving with a velocity of +10. meters per
     second. In an emergency, the car is brought to a stop in
     0.50 second. What force does the seat belt exert on the woman so that
     she remains in her seat?
     (1) -1.0×10^3 N   (2) -5.0×10^2 N   (3) -5.0×10^1 N   (4) -2.5×10^1 N

12   A 0.10-kilogram ball dropped vertically from a height of 1.0 meter
     above the floor bounces back to a height of 0.80 meter. The mechanical
     energy lost by the ball as it bounces is approximately
     (1) 0.080 J   (2) 0.20 J   (3) 0.30 J   (4) 0.78 J

13   A student rides a bicycle up a 30.° hill at a constant speed of 6.00
     meters per second. The combined mass of the student and bicycle is
     70.0 kilograms. What is the kinetic energy of the student-bicycle
     system during this ride?
     (1) 210. J   (2) 420. J   (3) 1,260 J   (3) 2,520 J

Base your answers to questions 14 and 15 on the information and diagram at
the right.

Spacecraft S is traveling from planet P_1
toward planet P_2. At the position
shown, the magnitude of the
gravitational force of planet P_1 on the
spacecraft is equal to the magnitude of
the gravitational force of planet P_2 on
the spacecraft.

Note that questions 14 and 15 have only three choices.

14   If distance X is greater than distance Y, then the mass of P_1 must be
     1   less than the mass of P_2 3   equal to the mass of P_2
     2   greater than the mass of P_2

15   As the spacecraft moves from the position shown toward planet P_2, the
     ration of the gravitational force of P_2 on the spacecraft to the
     gravitational force of P_1 on the spacecraft will
     1   decrease   2   increase   3   remain the same

16   The graph at the right shows the relationship between
     weight and mass for a series of objects. The slope of this
     graph represents
     1   change of position   3   momentum
     2   normal force   4   acceleration due to gravity

17 Each diagram at the right shows a different block being pushed by a force across a surface at a constant velocity.

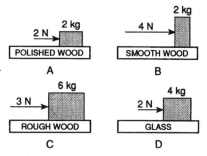

In which two diagrams is the force of friction the same?
(1) *A* and *B*
(2) *B* and *D*
(3) *A* and *D*
(4) *C* and *D*

18 A student running up a flight of stairs increases her speed at a constant rate. Which graph best represents the relationship between work and time for the student's run up the stairs?

(1)   (2)   (3)   (4)

19 A net force of 5.0 newtons moves a 2.0-kilogram object a distance of 3.0 meters in 3.0 seconds. How much work is done on the object?
(1) 1.0 J   (2) 10. J   (3) 15 J   (4) 30. J

20 Which graph best represents the relationship between the elongation of a spring whose elastic limit has not been reached and the force applied to it?

(1)   (2)   (3)   (4)

21 If a positively charged rod is brought near the knob of a positively charged electroscope, the leaves of the electroscope will
1 converge, only   3 first diverge, then converge
2 diverge, only   4 first converge, then diverge

22 The diagram at the right shows four charged metal spheres suspended by strings. The charge of each sphere is indicated. If spheres *A*, *B*, *C*, and *D* simultaneously come into contact, the net charge on the four spheres will be
(1) +1 C   (3) +3 C
(2) +2 C   (4) +4 C

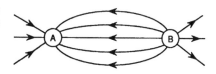
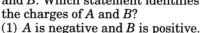

23 The diagram at the right represents the electric field lines in the vicinity of two isolated electrical charges, *A* and *B*. Which statement identifies the charges of *A* and *B*?
(1) *A* is negative and *B* is positive.
(2) *A* is positive and *B* is negative.
(3) *A* and *B* are both positive.
(4) *A* and *B* are both negative.

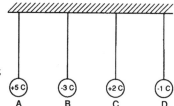

Base your answers to
questions 24 through 26
on the diagram at the
right which represents a
frictionless track. A
10-kilogram block starts
from rest at point A and
slides along the track.

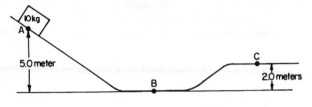

24   As the block moves from point A to point B, the total amount of gravitational potential energy changed to kinetic energy is approximately
(1) 5 J   (2) 20 J   (3) 50 J   (4) 500 J

25   What is the approximate speed of the block at point B?
(1) 1 m/s   (2) 10 m/s   (3) 50 m/s   (4) 100 m/s

26   What is the approximate potential energy of the block at point C?
(1) 20 J   (2) 200 J   (3) 300 J   (4) 500 J

27   If the potential difference between two oppositely charged parallel
metal plates is doubled, the electric field intensity at a point between
them is
1   halved   2   unchanged   3   doubled   4   quadrupled

28   Moving a point charge of 3.2×10^{-19} coulomb between points A and B in
an electric field requires 4.8×10^{-19} joule of energy. What is the
potential difference between these tow points?
(1) 0.67 V   (2) 2.0 V   (3) 3.0 V   (4) 1.5 V

29   The slope of the line on the graph at the right represents
1   resistance of a material
2   electric field intensity
3   power dissipated in a resistor
4   electrical energy

30   In the diagrams below, l represents a unit length of copper wire and A
represents a unit cross-sectional area. Which copper wire has the
smallest resistance at room temperature?

(1)   (2)    (3)   (4)

31   Two resistors are connected to a source of
voltage as shown in the diagram at the
right. At which position should an
ammeter be placed to measure the current
passing only through resistor R_1?
(1) 1   (3) 3
(2) 2   (4) 4

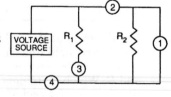

32   A toaster dissipates 1,500 watts of power in 90. seconds. The amount of
electric energy used by the toaster is approximately
(1) 1.4×10^5 J   (3) 5.2×10^8 J
(2) 1.7×10^1 J   (4) 6.0×10^{-2} J

33 In the diagram at the right, a steel paper clip is attached to a string, which is attached to a table. The clip remains suspended beneath a magnet. As the magnet is lifted, the paper clip begins to fall as a result of

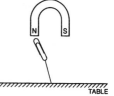

1 an increase in the potential energy of the clip
2 an increase in the gravitational field strength near the magnet
3 a decrease in the magnetic properties of the clip
4 a decrease in the magnetic field strength near the clip

34 The diagram at the right shows the magnetic field that results when a piece of iron is placed between unlike magnetic poles. At which point is the magnetic field strength greatest?

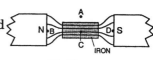

(1) *A*   (2) *B*   (3) *C*   (4) *D*

35 A wire carrying an electron current (e⁻) is placed between the poles of a magnet, as shown in the diagram at the right. Which arrow represents the direction of the magnetic force on the current?

(1) *A*   (3) *C*
(2) *B*   (4) *D*

36 The diagram at the right shows a coil of wire connected to a battery. The *N*-pole of this coil is closest to

(1) *A*   (3) *C*
(2) *B*   (4) *D*

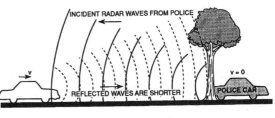

37 The diagram art the right shows radar waves being emitted from a stationary police car and reflected by a moving car back to the police car. The difference in apparent frequency between the incident and reflected waves is an example of

1 constructive interference   3 the Doppler effect
2 refraction   4 total internal reflection

38 The diagram at the right shows a transverse pulse moving to the right in a string. Which diagram best represents the motion of point *P* as the pulse passes point *P*?

(1)   (2)   (3)   (4)

39 Light is to brightness as sound is to

1 color   2 loudness   3 period   4 speed

40 The periodic wave in the diagram at the right has a frequency of 40. hertz. What is the speed of the wave?

    (1) 13 m/s   (2) 27 m/s   (3) 60. m/s   (4) 120 m/s

41 Two waves have the same frequency. Which wave characteristic must also be identical for both waves?
    1   phase   2   amplitude   3   intensity   4   period

42 A typical microwave oven produces radiation at a frequency of 1.0×10^{10} hertz. What is the wavelength of this microwave radiation?
    (1) 3.0×10^{-1} m   (2) 3.0×10^{-2} m   (3) 3.0×10^{10} m   (4) 3.0×10^{18} m

43 Two wave sources operating in phase in the same medium produce the circular wave patterns shown in the diagram at the right . The solid lines represent wave crests and the dashed lines represent wave troughs. Which point is at a position of maximum destructive interference?
    (1) *A*   (2) *B*   (3) *C*   (4) *D*

44 The distance between successive antinodes in the standing wave pattern shown at the right is equal to
    (1) 1 wavelength   (3) ½ wavelength
    (2) 2 wavelengths   (4) ⅓ wavelength

45 The diagram at the right shows a ray of light passing from medium *X* into air. What is the absolute index of refraction of medium *X*?
    (1) 0.50   (3) 1.7
    (2) 2.0   (4) 0.58

46 In the diagram at the right, a ray of monochromatic light (*A*) and a ray of polychromatic light (*B*) are both incident upon an air-glass interface. Which phenomenon could occur with ray *B*, but *not* with ray *A*?
    1   dispersion   3   polarization
    2   reflection   4   refraction

47 If the critical angle for a substance is 44°, the index of refraction of the substance is equal to
    (1) 1.0   (3) 1.4
    (2) 0.69   (4) 0.023

48 The diagram at the right shows a beam of light entering and leaving a "black box". The box most likely contains a
    1   prism
    2   converging lens
    3   double slit
    4   polarizer

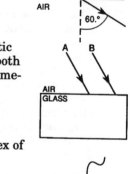

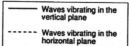

——— Waves vibrating in the vertical plane

- - - - - Waves vibrating in the horizontal plane

49   Which graph best represents the relationship between the intensity of
      light that falls on a photoemissive surface and the number of
      photoelectrons that the surface emits?

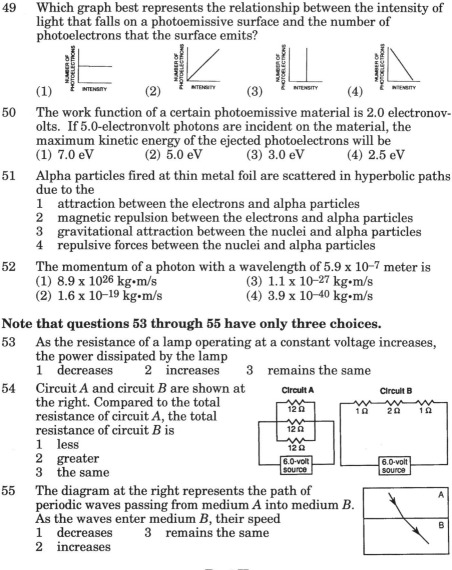

      (1)   (2)   (3)   (4)

50   The work function of a certain photoemissive material is 2.0 electronov-
      olts. If 5.0-electronvolt photons are incident on the material, the
      maximum kinetic energy of the ejected photoelectrons will be
      (1) 7.0 eV   (2) 5.0 eV   (3) 3.0 eV   (4) 2.5 eV

51   Alpha particles fired at thin metal foil are scattered in hyperbolic paths
      due to the
      1   attraction between the electrons and alpha particles
      2   magnetic repulsion between the electrons and alpha particles
      3   gravitational attraction between the nuclei and alpha particles
      4   repulsive forces between the nuclei and alpha particles

52   The momentum of a photon with a wavelength of 5.9×10^{-7} meter is
      (1) 8.9×10^{26} kg•m/s   (3) 1.1×10^{-27} kg•m/s
      (2) 1.6×10^{-19} kg•m/s   (4) 3.9×10^{-40} kg•m/s

Note that questions 53 through 55 have only three choices.

53   As the resistance of a lamp operating at a constant voltage increases,
      the power dissipated by the lamp
      1   decreases   2   increases   3   remains the same

54   Circuit *A* and circuit *B* are shown at
      the right. Compared to the total
      resistance of circuit *A*, the total
      resistance of circuit *B* is
      1   less
      2   greater
      3   the same

55   The diagram at the right represents the path of
      periodic waves passing from medium *A* into medium *B*.
      As the waves enter medium *B*, their speed
      1   decreases   3   remains the same
      2   increases

Part II
This part consists of six groups, each containing ten questions. Each group
tests an optional area of the course. Choose two of these six groups. Be sure
to answer all ten questions in each group chosen. [20]

Group 1 — Motion in a Plane
If you chose this group, be sure to answer questions 56 - 65.

56   A ball is thrown horizontally at a speed of 20. meters per second from
      the top of a cliff. How long does the ball take to fall 19.6 meters to the
      ground?
      (1) 1.0 s   (2) 2.0 s   (3) 9.8 s   (4) 4.0 s

57   A book is pushed with an initial horizontal velocity of 5.0 meters per
     second off the top of a desk. What is the initial vertical velocity of the
     book?
     (1) 0 m/s   (2) 2.5 m/s   (3) 5.0 m/s   (4) 10. m/s

58   The diagram at the right shows a
     baseball being hit with a bat. Angle
     θ represents the angle between the
     horizontal and the ball's initial
     direction of motion. Which value of θ
     would result in the ball traveling
     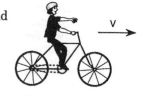
     the longest horizontal distance? [Neglect air resistance.]
     (1) 25°   (2) 45°   (3) 60°   (4) 90°

59   The diagram at the right represents a bicycle and
     rider traveling to the right at a constant speed.
     A ball is dropped from the hand of the cyclist.
     Which set of graphs best represents the
     horizontal motion of the ball relative to the
     ground? [Neglect air resistance.]

60   Pluto is sometimes closer to the Sun than Neptune is. Which statement
     is the best explanation for this phenomenon?
     1   Neptune's orbit is elliptical and Pluto's orbit is circular.
     2   Pluto's orbit is elliptical and Neptune's orbit is circular.
     3   Pluto and Neptune have circular orbits that overlap.
     4   Pluto and Neptune have elliptical orbits that overlap.

Base your answers to questions 61 through 63 on
the diagram at the right which shows a
2.0-kilogram model airplane attached to a wire.
The airplane is flying clockwise in a horizontal
circle of radius 20. meters at 30. meters per
second.

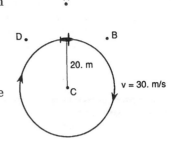

61   The centripetal force acting on the airplane
     at the position shown is directed toward
     point
     (1) *A*   (3) *C*
     (2) *B*   (4) *D*

62   What is the magnitude of the centripetal acceleration of the airplane?
     (1) 0 m/s²   (2) 1.5 m/s²   (3) 45 m/s²   (4) 90. m/s²

63   If the wire breaks when the airplane is at the position shown, the
     airplane will move toward point
     (1) *A*   (2) *B*   (3) *C*   (4) *D*

Note that questions 64 and 65 have only three choices.

64   A motorcycle travels around a flat circular track. If the speed of the motorcycle is increased, the force required to keep it in the same circular path

1   decreases   2   increases   3   remains the same

65   The diagram represents the path taken by planet P as it moves in an elliptical orbit around the sun S. The time it takes to go from point A to point B is t_1, and from point C to point D is t_2. If the two shaded areas are equal, then t_1, is

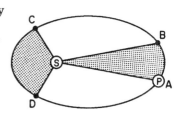

1   less than t_2
2   greater than t_2
3   the same as t_2

Group 2 — Internal Energy
If you chose this group, be sure to answer questions 66 - 75.

66   What is the difference between the melting point and boiling point of ethyl alcohol on the Kelvin scale?

(1) 38   (2) 196   (3) 352   (4) 469

67   A kilogram of each of the substances below is condensed from a gas to a liquid. Which substance releases the most energy?

1   alcohol   2   mercury   3   water   4   silver

68   Which sample of metal will gain net internal energy when placed in contact with a block of lead at 100°C?

1   platinum at 60°C   3   lead at 125°C
2   iron at 100°C   4   silver at 200°C

69   Which graph best represents the relationship between absolute temperature (T) and the product of pressure and volume ($P{\cdot}V$) for a given mass of ideal gas?

(1)   (2)   (3)   (4)

Base your answers to questions 70 through 72 on the information below.

Ten kilograms of water initially at 20°C is heated to its boiling point (100°C). Then 5.0 kilograms of water is converted into stream at 100°C.

70   What was the approximate amount of heat energy needed to raise the temperature of the water to its boiling point?

(1) 840 kJ   (2) 3,400 kJ   (3) 4,200 kJ   (4) 6,300 kJ

71   The amount of heat energy needed to convert the 5.0 kilograms of water at 100°C into steam at 100°C is approximately

(1) 1,700 kJ   (2) 2,100 kJ   (3) 5,500 kJ   (4) 11,000 kJ

Note that question 72 has only three choices.

72   If salt is added to the water, the temperature at which the water boils will

1   decrease   2   increase   3   remain the same

73   The graph at the right shows temperature versus time for 1.0 kilogram of a substance at constant pressure as heat is added at a constant rate of 100 kilojoules per minute. The substance is a solid at 20°C. How much heat was added to 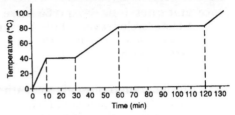 change the substance from a liquid at its melting point to a vapor at its boiling point?
      (1) 3,000 kJ   (2) 6,000 kJ   (3) 9,000 kJ   (4) 11,000 kJ

Note that questions 74 and 75 have only three choices.

74   As pressure is applied to a snowball, the melting point of the snow
      1   decreases   2   increases   3   remains the same

75   Oxygen molecules are about 16 times more massive than hydrogen molecules. An oxygen gas sample is in a closed container and a hydrogen gas sample is in a second closed container of different size. Both samples are at room temperature. Compared to the average speed of the oxygen molecules, the average speed of the hydrogen molecules will be
      1   less   2   greater   3   the same

Group 3 — Electromagnetic Applications
If you chose this group, be sure to answer questions 76 - 85.

Base your answers to questions 76 through 78 on the information and data table below.
      During a laboratory investigation of transformers, a group of students obtained the following data during four trials, using a different pair of coils in each trial.

	Primary Coil		Secondary Coil	
	V_p (volts)	I_p (amperes)	V_s (volts)	I_s (amperes)
Trial 1	3.0	12.0	16.0	2.0
Trial 2	6.0	3.0	8.0	2.2
Trial 3	9.0	4.3	54.0	0.7
Trial 4	12.0	2.5	5.0	9.0

76   What is the efficiency of the transformer in trial 1?
      (1) 75%   (2) 89%   (3) 100%   (4) 113%

77   What is the ratio of the number of turns in the primary coil to the number of turns in the secondary coil in trial 3?
      (1) 1:6   (2) 1:9   (3) 6:1   (4) 9:1

78   In which trial was an error most likely made in recording the data?
      (1) 1   (2) 2   (3) 3   (4) 4

79   A wire 0.50 meter long cuts across a magnetic field with a magnetic flux density of 20. teslas. The wire moves at a speed of 4.0 meters per second and travels in a direction perpendicular to the magnetic flux lines. What is the maximum potential difference induced between the ends of the wire?
(1) 2.5 V   (2) 10. V   (3) 40. V   (4) 160 V

80   Compared to the resistance of the circuit being measured, the internal resistance of a voltmeter is designed to be very high so that the meter will draw
1   no current from the circuit
2   little current from the circuit
3   most of the current from the circuit
4   all the current from the circuit

81   A proton and an electron traveling with the same velocity enter a uniform electric field. Compared to the acceleration of the proton, the acceleration of the electron is
1   less, and in the same direction
2   less, but in the opposite direction
3   greater, and in the same direction
4   greater, but in the opposite direction

82   The diagram at the right shows an end view of a straight conducting wire, *W*, moving with constant speed in uniform magnetic field *B*. As the conductor moves through position *P*, the electron current induced in the wire is directed
1   toward the bottom of the page
2   toward the top of the page
3   into the page
4   out of the page

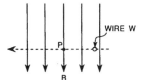

83   An electron moves at 3.0×10^7 meters per second perpendicularly to a magnetic field that has a flux density of 2.0 teslas. What is the magnitude of the force on the electron?
(1) $9.6 z 10^{-19}$ N   (3) 9.6×10^{-12} N
(2) 3.2×10^{-19} N   (4) 4.8×10^{-12} N

84   In each diagram below, an electron travels to the right between points A and B. In which diagram would the electron be deflected toward the bottom of the page?

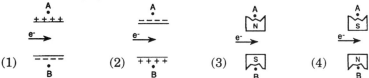

85   What is one characteristic of a light beam produced by a monochromatic laser?
1   It consists of coherent waves.
2   It can be dispersed into a complete continuous spectrum.
3   It cannot be reflected or refracted.
4   It does not exhibit any wave properties.

Group 4 — Geometric Optics
If you chose this group, be sure to answer questions 86 - 95.

86   An object is placed in front of a plane mirror as shown in
      the diagram at the right. Which diagram below best
      represents the image that is formed?

87   The diagram at the right shows light ray
      R parallel to the principal axis of a
      spherical concave (converging) mirror.
      Point F is the focal point of the mirror
      and C is the center of curvature. After
      reflecting, the light ray will pass through
      point

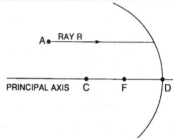

      (1) A   (3) C
      (2) F   (4) D

88   The tip of a person's nose is 12 centimeters from a concave (converging)
      spherical mirror that has a radius of curvature of 16 centimeters. What
      is the distance from the mirror to the image of the tip of the person's
      nose?
      (1) 8.0 cm   (2) 12 cm   (3) 16 cm   (4) 24 cm

89   The image of a shoplifter in a department store is viewed in a cones
      (diverging) mirror. The image is
      1   real and smaller than the shoplifter
      2   real and larger than the shoplifter
      3   virtual and smaller than the shoplifter
      4   virtual and larger than the shoplifter

90   When light rays pass through the film in a movie projector, an image of
      the film is produced on a screen. In order to produce the image on the
      screen, what type of lens does the projector use and how far from the
      lens must the film be placed?
      1   converging lens, at a distance greater than the focal length
      2   converging lens, at a distance less than the focal length
      3   diverging lens, at a distance greater than the focal length
      4   diverging lens, at a distance less than the focal length

91   Two light rays from a common point are refracted by a lens. A real
      image is formed when these two refracted rays
      1   converge to a single point
      2   diverge and appear to come from a single point
      3   travel in parallel paths
      4   totally reflect inside the lens

92   The diagram at the right represents
     a convex (converging) lens with focal
     point *F*. If an object is placed at 2F,
     the image will be
     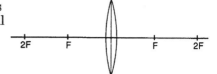
     1   virtual, erect, and smaller than
         the object
     2   real, inverted, and the same size as the object
     3   real, inverted, and larger than the object
     4   virtual, erect, and the same size as the object

93   The diagram at the right shows the
     refraction of the blue and red compo-
     nents of a white light beam. Which
     phenomenon does the diagram illus-
     trate?
     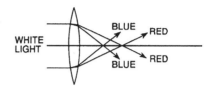
     1   total internal reflection
     2   critical angle reflection
     3   spherical aberration
     4   chromatic aberration

94   When a 2.0-meter-tall object is placed 4.0 meters in front of a lens, an
     image is formed on a screen located 0.050 meter behind the lens. What
     is the size of the image?
     (1) 0.10 m   (2) 0.025 m   (3) 2.5 m   (4) 0.40 m

Note that question 95 has only three choices.

95   As the distance between a man and a plane mirror increases, the size of
     the image of the man produced by the mirror
     1   decreases   2   increases   3   remains the same

Group 5 — Solid State
If you chose this group, be sure to answer questions 96 - 105.

96   A particular solid has a small energy gap between its valence and
     conduction bands. This solid is most likely classified as
     1   a good conductor   3   a type of glass
     2   a semiconductor   4   an insulator

97   The diagram at the right shows an
     electron moving through a
     semiconductor. Toward which letter
     will the hole move?

     (1) *A*   (3) *C*
     (2) *B*   (4) *D*

98   Diagram *A* represents the wave form of an
     electron current entering a semiconductor
     device, and diagram *B* represents the wave
     form as the current leaves the device.
     What is this device?
     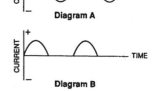
     1   resistor   3   cathode
     2   anode   4   diode

99   Compared to an insulator, a conductor of electric current has
      1   more free electrons per unit volume
      2   fewer free electrons per unit volume
      3   more free atoms per unit volume
      4   fewer free atoms per unit volume

Base your answers to questions 100 through 102 on
the diagram at the right which represents a diode.

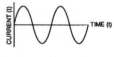

100   The P-N junction in the diagram is biased
      (1) reverse   (3) B to C
      (2) forward   (4) A to D

101   In the diagram, B represents the
      (1) N-type silicon   (3) cathode
      (2) P-type silicon   (4) diode

Note that question 102 has only three choices.

102   If the positive and negative wires of the circuit in the diagram were
      reversed, the current would
      1   decrease   2   increase   3   remain the same

103   The graph at the right represents the alternating
      current signal input to a transistor amplifier.
      Which graph below best represents the amplified
      output signal from this transistor?

      (1)   (2)   (3)   (4)

104   Which device contains a large number of transistors on a single block of
      silicon?
      (1) junction diode   (3) integrated circuit
      (2) conductor   (4) N-type semiconductor

**Note that question 105 has only three
choices.**

105   The diagram below represents an
      operating N-P-N transistor circuit.
      Ammeter A_c reads the collector current
      and ammeter A_b reads the base current.
      Compared to the reading of ammeter A_c,
      the reading of ammeter A_b is
      1   less
      2   greater
      3   the same

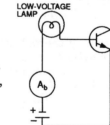

Group 6 — Nuclear Energy
If you chose this group, be sure to answer questions 106 - 115.

106   An element has an atomic number of 63 and a mass number of 155.
      How many protons are in the nucleus of the element?
      (1) 63   (2) 92   (3) 155   (4) 218

107   Which particle would generate the greatest amount of energy if its
      entire mass were converted into energy?
      1   electron   3   alpha particle
      2   proton   4   neutron

108   Which particles can be accelerated by a linear accelerator?
      1   protons and gamma rays   3   electrons and protons
      2   neutrons and electrons   4   neutrons and alpha particles

109   The equation below represents an unstable radioactive nucleus that is
      transmuted into another isotope (*X*) by the emission of a beta particle.

      $$_{90}^{234}\text{TH} \rightarrow X + _{-1}^{0}\text{e}$$

      Which new isotope is formed?
      (1) $_{91}^{234}\text{Pa}$ (2) $_{91}^{234}\text{Th}$ (3) $_{90}^{235}\text{Pa}$ (4) $_{90}^{235}\text{Th}$

110   In 4.0 years, 40.0 kilograms of element *A* decays to 5.0 kilograms. The
      half-life of element *A* is
      (1) 1.3 years   (2) 2.0 years   (3) 0.7 year   (4) 4.0 years

111   The subatomic particles that make up both protons and neutrons are
      known as
      1   electrons   2   nuclides   3   positrons   4   quarks

112   Which equation is an example of positron emission?
      (1) $_{88}^{226}\text{Ra} \rightarrow _{86}^{222}\text{Rn} + _{2}^{4}\text{He}$ (3) $_{29}^{64}\text{Cu} \rightarrow _{28}^{64}\text{Ni} + _{+1}^{0}\text{e}$
      (2) $_{82}^{210}\text{Pb} \rightarrow _{83}^{210}\text{Bi} + _{-1}^{0}\text{e}$ (4) $_{7}^{14}\text{N} + _{2}^{4}\text{He} \rightarrow _{8}^{17}\text{O} + _{1}^{1}\text{H}$

113   Which process occurs during nuclear fission?
      1   Light nuclei are forced together to form a heavier nucleus.
      2   A heavy nucleus splits into lighter nuclei.
      3   An atom is converted to a different isotope of the same element.
      4   Transmutation is produced by the emission of alpha particles.

114   In order to increase the likelihood that a neutron emitted from a
      nucleus will be captured by another nucleus, the neutron should be
      1   accelerated through a potential difference
      2   heated to a higher temperature
      3   slowed down to decrease its kinetic energy
      4   absorbed by a control rod

115   The energy emitted by the Sun originates from the process of
      1   fission   2   fusion   3   alpha decay   4   beta decay

Part III
You must answer all questions in this part. [15]

116   Base your answers to parts *a* through *c* on the information below.
      A newspaper carrier on her delivery route travels 200. meters due north
      and then turns and walks 300. meters due east.
      a   *D*raw a vector diagram following the directions below.
      (1) Using a ruler and protractor and starting at point *P*, construct the
          sequence of two displacement vectors for the newspaper carrier's
          route. Use a scale of 1.0 centimeter = 50. meters. Label the vectors.
          [3]

(2) Construct and label the vector that represents the carrier's resultant displacement from point P. [1]

b   What is the magnitude of the carrier's resultant displacement? [1]

c   What is the angle (in degrees) between north and the carrier's resultant displacement? [1]

117 The diagram at the right shows a spring compressed by a force of 6.0 newtons from its rest position to its compressed position. Calculate the spring constant for this spring. [Show all calculations, including equations and substitutions with units.] [2]

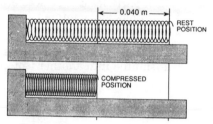

118 Base your answers to parts a through c on the diagram and information below.

Monochromatic light is incident on a two-slit apparatus. The distance between the slits is 1.0 x 10-3 meter, and the distance from the two-slit apparatus to a screen displaying the interference pattern is 4.0 meters. The distance between the central maximum and the first-order maximum is 2.4 x 10-3meter.

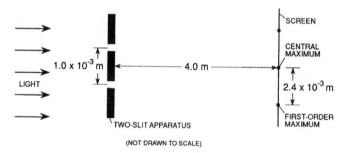

(NOT DRAWN TO SCALE)

a   What is the wavelength of the monochromatic light? [Show all calculations, including equations and substitutions with units] [2]

b   What is the color of the monochromatic light? [1]

c   List two ways the variables could be changed that would cause the distance between the central maximum and the first-order maximum to increase. [2]

119 Infrared electromagnetic radiation incident on a material produces no photoelectrons. When red light of equal intensity is shone on the same material, photoelectrons are emitted from the surface.

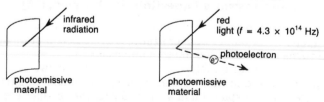

Using one or more complete sentences, explain why the visible red light causes photoelectric emission, but the infrared radiation does not. [2]